WORLD HEALTH ORGANIZATION

INTERNATIONAL AGENCY FOR RESEARCH ON CANCER

IARC MONOGRAPHS
ON THE
EVALUATION OF THE CARCINOGENIC RISK OF CHEMICALS TO HUMANS

Polynuclear Aromatic Compounds,
Part 4,
Bitumens, Coal-tars and Derived Products,
Shale-oils and Soots

VOLUME 35

This publication represents the views and expert opinions
of an IARC Working Group on the
Evaluation of the Carcinogenic Risk of Chemicals to Humans
which met in Lyon,

21-28 February 1984

January 1985

IARC MONOGRAPHS

In 1969, the International Agency for Research on Cancer (IARC) initiated a programme on the evaluation of the carcinogenic risk of chemicals to humans involving the production of critically evaluated monographs on individual chemicals. In 1980, the programme was expanded to include the evaluation of the carcinogenic risk associated with employment in specific occupations.

The objective of the programme is to elaborate and publish in the form of monographs critical reviews of data on carcinogenicity for chemicals and complex mixtures to which humans are known to be exposed, and on specific occupational exposures, to evaluate these data in terms of human risk with the help of international working groups of experts in chemical carcinogenesis and related fields, and to indicate where additional research efforts are needed.

This projects is supported by PHS Grant No. 1 UO1 CA33193-02 awarded by the US National Cancer Institute, Department of Health and Human Services.

© International Agency for Research on Cancer 1985

ISBN 92 832 1235 5 (soft-cover edition)

ISBN 92 832 1535 4 (hard-cover edition)

PRINTED IN FRANCE

Distributed for the International Agency for Research on Cancer by the Secretariat of the World Health Organization

CONTENTS

NOTE TO THE READER 5

LIST OF PARTICIPANTS 7

PREAMBLE
    Background 11
    Objective and Scope 11
    Selection of Chemicals and Complex Exposures for Monographs 12
    Working Procedures 12
    Data for Evaluations 13
    The Working Group 13
    General Principles Applied by the Working Group in Evaluating the
        Carcinogenic Risk of Chemicals or Complex Mixtures 13
    Explanatory Notes on the Contents of Monographs on Chemicals and Complex
        Mixtures 20

GENERAL REMARKS ON THE SUBSTANCES CONSIDERED
    Background 31
    Contents of this Volume 31
    Historical Perspectives 32
    Special Issues 33

THE MONOGRAPHS
    Bitumens 39
    Coal-tars and Derived Products 83
    Shale-oils 161
    Soots 219
    APPENDIX : Some case reports of skin cancer in workers exposed to soots,
    coal-tars, pitch, creosote, bitumens and shale-oils 243

SUPPLEMENTARY CORRIGENDA TO VOLUMES 1-34 249

CUMULATIVE INDEX TO THE MONOGRAPH SERIES 251

NOTE TO THE READER

The term 'carcinogenic risk' in the *IARC Monographs* series is taken to mean the probability that exposure to the chemical will lead to cancer in humans.

Inclusion of a chemical in the monographs does not imply that it is a carcinogen, only that the published data have been examined. Equally, the fact that a chemical has not yet been evaluated in a monograph does not mean that it is not carcinogenic.

Anyone who is aware of published data that may alter the evaluation of the carcinogenic risk of a chemical to humans is encouraged to make this information available to the Unit of Carcinogen Identification and Evaluation, International Agency for Research on Cancer, 150 cours Albert Thomas, 69372 Lyon Cedex 08, France, in order that the chemical may be considered for re-evaluation by a future Working Group.

Although every effort is made to prepare the monographs as accurately as possible, mistakes may occur. Readers are requested to communicate any errors to the Unit of Carcinogen Identification and Evaluation, so that corrections can be reported in future volumes.

IARC WORKING GROUP ON THE EVALUATION OF THE CARCINOGENIC RISK OF CHEMICALS TO HUMANS: POLYNUCLEAR AROMATIC COMPOUNDS, PART 4, BITUMENS, COAL-TARS AND DERIVED PRODUCTS, SHALE-OILS AND SOOTS

Lyon, 21-28 February, 1984

Members[1]

R. Bass, Bundesgesundheitsamt, Institut für Arzneimittel, Berlin 33, Federal Republic of Germany

E. Bingham, Vice-President and University Dean, Professor of Environmental Health, University of Cincinnati, Cincinnati, OH 45267, USA (*Chairman*)

P. Bogovski, Director, Institute of Experimental and Clinical Medicine, 42 Hiiu Street, Tallinn, Estonia 200016, USSR (*Vice-Chairman*)

I. Chouroulinkov, Institut de Recherches Scientifiques sur le Cancer, CNRS, 7 rue Guy Moquet, BP 8, 94800 Villejuif, France

J. Dement, National Institute of Environmental Health Sciences, PO Box 12233, Research Triangle Park, NC 27511, USA

E. Dybing, Department of Toxicology, National Institute of Public Health, Postuttak, Oslo 1, Norway

E.A. Emmett, Director, Division of Occupational Medicine, Johns Hopkins School of Hygiene and Public Health, 615 N Wolfe, Baltimore, MD 21205, USA

---

[1]Unable to attend: I. Öpik, Estonian SSR Academy of Sciences, 6 Kohtu Street, Tallinn, Estonia 200016, USSR; R. Lauwerys, Faculté de Médecine, Unité de Toxicologie industrielle et médicale, Clos Chapelle aux Champs, 30, 1200 Brussels, Belgium; J. Melius, Chief, Hazard Evaluations and Technical Assistance Branch, National Institute for Occupational Health and Safety, 4676 Columbia Parkway, Cincinnati, OH 45226, USA; S. Nesnow, Carcinogenesis and Metabolism Branch (MD-68), Health Effects Research Laboratory, US Environmental Protection Agency, Research Triangle Park, NC 27711, USA

S. de Flora, Institute of Hygiene, University of Genoa, via Pastore 1, 16132 Genoa, Italy

L. Garfinkel, Vice-President for Epidemiology and Statistics, Director of Cancer Prevention, American Cancer Society, Inc., 4 West 35th Street, New York, NY 10001, USA

S. Goldfarb, Medical School, Department of Pathology, University of Wisconsin, 470 North Charter Street, Madison, WI 53706, USA

R.A. Griesemer, Director, Biology Division, Oak Ridge National Laboratory, PO Box Y, Oak Ridge, TN 37831, USA

K.M. Hargis, Associate Group Leader, Industrial Hygiene Group, Los Alamos National Laboratory, Los Alamos, NM 87545, USA

C. Hogstedt, National Board of Occupational Safety and Health, Occupational Health Department, 17184 Solna, Sweden

P.J. Kalliokoski, Environmental and Occupational Hygiene, University of Kuopio, 70101 Kuopio 10, Finland

T. Kuroki, Department of Cancer Cell Research, Institute of Medical Sciences, University of Tokyo, 4-6-1 Shirokanedai, Minato-ku, Tokyo 10𝒞, Japan

J.C. McDonald, McGill University, School of Occupational Health, Charles Meredith House, 1130 Pine Avenue West, Montreal H3A 1A3, Quebec, Canada

M. Sorsa, Institute of Occupational Health, Haartmaninkatu 1, 00290 Helsinki 29, Finland

R. Waxweiler, MSK 404, Los Alamos National Laboratory, Los Alamos, NM 87545, USA

Representative of the National Cancer Institute

A. Blair, Environmental Epidemiology Branch, National Cancer Institute, Landow Building, Room 4C-16, Bethesda, MD 20205, USA

Representative of SRI International

K.E. McCaleb, Director, Chemical-Environmental Department, SRI International, 333 Ravenswood Avenue, Menlo Park, CA 94025, USA

Representatives of the American Petroleum Institute

N.K. Weaver, American Petroleum Institute, 1220 L Street Northwest, Washington DC 20005, USA

K.W. Whitehead, Union Science and Technology Division, Union Oil Company of California, 376 S Valencia, Brea, CA 92621, USA

PARTICIPANTS

Representative of the British Tar Industry Association

B. Thomas, National Smokeless Fuels Limited, PO Box 16, Wingerworth, Chesterfield, Derbyshire S42 6JT, UK

Representative of the Commission of the European Communities

G. Aresini, Commission of the European Communities, Health and Safety Directorate, Bâtiment Jean Monnet, Plateau du Kirchberg, BP 1907, Luxembourg, Grand Duchy of Luxembourg

Representative of CONCAWE

B.J. Simpson, Shell International Petroleum Company, Toxicology Division, MDTL/321, Shell Centre, London SE1 7NA, UK

Representative of the Federal Office for Workers' Safety (Federal Republic of Germany)

E. Lehmann, Bundesanstalt für Arbeitsschutz, Vogelpothsweg 50-52, 4600 Dortmund 1 (Dorstfeld), Federal Republic of Germany

Representative of the Institute of Petroleum (UK)

E.H. Chipperfield, Senior Consultant, BP Oil International Limited, Product Management, Britannic House, Moor Lane, London EC2Y 9BU, UK

OBSERVERS TO THE PREPARATORY MEETING OF 16-18 FEBRUARY 1984

Representative of the British Tar Industry Association

M. Holden, Development Manager, Thomas Ness Limited, Eastwood Hall, Eastwood, Nottingham NG16 3EB, UK

Representative of the Union des Chambres syndicales de l'Industrie du Pétrole

A. Deyon, Direction technique, UCSIP, 16 avenue Kléber, 75116 Paris, France

Secretariat

M. Friesen, Division of Environmental Carcinogenesis
L. Haroun, Division of Environmental Carcinogenesis (*Co-Secretary*)
E. Heseltine, Editorial and Publications Services
J. Kaldor, Division of Epidemiology and Biostatistics
A. Likhachev, Division of Environmental Carcinogenesis
D. Mietton, Division of Environmental Carcinogenesis
R. Montesano, Division of Environmental Carcinogenesis
I. O'Neill, Division of Environmental Carcinogenesis
C. Partensky, Division of Environmental Carcinogenesis
I. Peterschmitt, Division of Environmental Carcinogenesis, Geneva, Switzerland
S. Poole, Birmingham, UK
R. Saracci, Division of Epidemiology and Biostatistics
L. Simonato, Division of Epidemiology and Biostatistics
L. Tomatis, Director
H. Vainio, Division of Environmental Carcinogenesis (*Head of the Programme*)
J. Wahrendorf, Division of Epidemiology and Biostatistics
J. Wilbourn, Division of Environmental Carcinogenesis (*Co-Secretary*)
H. Yamasaki, Division of Environmental Carcinogenesis

Secretarial assistance

J. Cazeaux
S. Cotterell
M.-J. Ghess
M. Lézère
S. Reynaud

IARC MONOGRAPHS PROGRAMME ON THE EVALUATION OF THE CARCINOGENIC RISK OF CHEMICALS TO HUMANS[1]

PREAMBLE

1. BACKGROUND

In 1969, the International Agency for Research on Cancer (IARC) initiated a programme to evaluate the carcinogenic risk of chemicals to humans and to produce monographs on individual chemicals. Following the recommendations of an ad-hoc Working Group, which met in Lyon in 1979 to prepare criteria to select chemicals for *IARC Monographs*(1), the *Monographs* programme was expanded to include consideration of exposures to complex mixtures which occur, for example, in many occupations.

The criteria established in 1971 to evaluate carcinogenic risk to humans were adopted by all the working groups whose deliberations resulted in the first 16 volumes of the *IARC Monographs* series. This preamble reflects subsequent re-evaluation of those criteria by working groups which met in 1977(2), 1978(3), 1982(4) and 1983(5).

2. OBJECTIVE AND SCOPE

The objective of the programme is to elaborate and publish in the form of monographs critical reviews of data on carcinogenicity for chemicals, groups of chemicals and industrial processes to which humans are known to be exposed, to evaluate the data in terms of human risk with the help of international working groups of experts, and to indicate where additional research efforts are needed. These evaluations are intended to assist national and international authorities in formulating decisions concerning preventive measures. No recommendation is given concerning legislation, since this depends on risk-benefit evaluations, which seem best made by individual governments and/or other international agencies.

---

[1]This project is supported by PHS Grant No. 1 U01 CA33193-02 awarded by the US National Cancer Institute, Department of Health and Human Services.

The *IARC Monographs* are recognized as an authoritative source of information on the carcinogenicity of environmental and other chemicals. A users' survey, made in 1984, indicated that the monographs are consulted by various agencies in 45 countries. As of October 1984, 34 volumes of the *Monographs* had been published or were in press. Four supplements have been published: two summaries of evaluations of chemicals associated with human cancer, an evaluation of screening assays for carcinogens, and a cross index of synonyms and trade names of chemicals evaluated in the series(6).

3. SELECTION OF CHEMICALS AND COMPLEX EXPOSURES FOR MONOGRAPHS

The chemicals (natural and synthetic, including those which occur as mixtures and in manufacturing processes) and complex exposures are selected for evaluation on the basis of two main criteria: (a) there is evidence of human exposure, and (b) there is some experimental evidence of carcinogenicity and/or there is some evidence or suspicion of a risk to humans. In certain instances, chemical analogues are also considered. The scientific literature is surveyed for published data relevant to the *Monographs* programme; and the IARC *Survey of Chemicals Being Tested for Carcinogenicity*(7) often indicates those chemicals that may be scheduled for future meetings.

As new data on chemicals for which monographs have already been prepared become available, re-evaluations are made at subsequent meetings, and revised monographs are published.

4. WORKING PROCEDURES

Approximately one year in advance of a meeting of a working group, a list of the substances or complex exposures to be considered is prepared by IARC staff in consultation with other experts. Subsequently, all relevant biological data are collected by IARC; recognized sources of information on chemical carcinogenesis and systems such as CANCERLINE, MEDLINE and TOXLINE are used in conjunction with US Public Health Service Publication No. 149(8). The major collection of data and the preparation of first drafts for the sections on chemical and physical properties, on production and use, on occurrence, and on analysis are carried out by SRI International, Menlo Park, CA, USA, under a separate contract with the US National Cancer Institute. Most of the data so obtained refer to the USA and Japan; IARC supplements this information with that from other sources in Europe. Representatives from industrial associations may assist in the preparation of sections describing industrial processes. Bibliographical sources for data on mutagenicity and teratogenicity are the Environmental Mutagen Information Center and the Environmental Teratology Information Center, both located at the Oak Ridge National Laboratory, TN, USA.

Six months before the meeting, reprints of articles containing relevant biological data are sent to an expert(s), or are used by IARC staff, to prepare first drafts of monographs. These drafts are then compiled by IARC staff and sent, prior to the meeting, to all participants of the Working Group for their comments.

The Working Group then meets in Lyon for seven to eight days to discuss and finalize the texts of the monographs and to formulate the evaluations. After the meeting, the master copy of each monograph is verified by consulting the original literature, then edited by a professional editor and prepared for reproduction. The aim is to publish monographs within nine months of the Working Group meeting. Each volume of monographs is printed in 4000 copies for distribution to governments, regulatory agencies and interested scientists. The monographs are also available *via* the WHO Distribution and Sales Service.

5. DATA FOR EVALUATIONS

With regard to biological data, only reports that have been published or accepted for publication are reviewed by the working groups, although a few exceptions have been made: in certain instances, reports from government agencies that have undergone peer review and are widely available are considered. The monographs do not cite all of the literature on a particular chemical or complex exposure: only those data considered by the Working Group to be relevant to the evaluation of carcinogenic risk to humans are included.

Anyone who is aware of data that have been published or are in press which are relevant to the evaluations of the carcinogenic risk to humans of chemicals or complex exposures for which monographs have appeared is asked to make them available to the Unit of Carcinogen Identification and Evaluation, Division of Environmental Carcinogenesis, International Agency for Research on Cancer, Lyon, France.

6. THE WORKING GROUP

The tasks of the Working Group are five-fold: (a) to ascertain that all data have been collected; (b) to select the data relevant for evaluation; (c) to ensure that the summaries of the data enable the reader to follow the reasoning of the Working Group; (d) to judge the significance of the results of experimental and epidemiological studies; and (e) to make an evaluation of the carcinogenicity of the chemical or complex exposure.

Working Group participants who contributed to the consideration and evaluation of chemicals or complex exposures within a particular volume are listed, with their addresses, at the beginning of each publication. Each member serves as an individual scientist and not as a representative of any organization or government. In addition, observers are often invited from national and international agencies and industrial associations.

7. GENERAL PRINCIPLES APPLIED BY THE WORKING GROUP IN EVALUATING THE CARCINOGENIC RISK OF CHEMICALS OR COMPLEX MIXTURES

The widely accepted meaning of the term 'chemical carcinogenesis', and that used in these monographs, is the induction by chemicals (or complex mixtures of chemicals) of neoplasms that are not usually observed, the earlier induction of neoplasms that are commonly observed, and/or the induction of more neoplasms than are usually found - although fundamentally different mechanisms may be involved in these three situations. Etymologically, the term 'carcinogenesis' means the induction of cancer, that is, of malignant neoplasms; however, the commonly accepted meaning is the induction of various types of neoplasms or of a combination of malignant and benign tumours. In the monographs, the words 'tumour' and 'neoplasm' are used interchangeably. (In the scientific literature, the terms 'tumorigen', 'oncogen' and 'blastomogen' have all been used synonymously with 'carcinogen', although occasionally 'tumorigen' has been used specifically to denote a substance that induces benign tumours.)

(a) Experimental Evidence

(i) *Evidence for carcinogenicity in experimental animals*

The Working Group considers various aspects of the experimental evidence reported in the literature and formulates an evaluation of that evidence.

Qualitative aspects: Both the interpretation and evaluation of a particular study as well as the overall assessment of the carcinogenic activity of a chemical (or complex mixture) involve several considerations of qualitative importance, including: (a) the experimental parameters under which the chemical was tested, including route of administration and exposure, species, strain, sex, age, etc.; (b) the consistency with which the chemical has been shown to be carcinogenic, e.g., in how many species and at which target organ(s); (c) the spectrum of neoplastic response, from benign neoplasm to multiple malignant tumours; (d) the stage of tumour formation in which a chemical may be involved: some chemicals act as complete carcinogens and have initiating and promoting activity, while others may have promoting activity only; and (e) the possible role of modifying factors.

There are problems not only of differential survival but of differential toxicity, which may be manifested by unequal growth and weight gain in treated and control animals. These complexities are also considered in the interpretation of data.

Many chemicals induce both benign and malignant tumours. Among chemicals that have been studied extensively, there are few instances in which the neoplasms induced are only benign. Benign tumours may represent a stage in the evolution of a malignant neoplasm or they may be 'end-points' that do not readily undergo transition to malignancy. If a substance is found to induce only benign tumours in experimental animals, it should nevertheless be suspected of being a carcinogen, and it requires further investigation.

Hormonal carcinogenesis: Hormonal carcinogenesis presents certain distinctive features: the chemicals involved occur both endogenously and exogenously; in many instances, long exposure is required; and tumours occur in the target tissue in association with a stimulation of non-neoplastic growth, although in some cases hormones promote the proliferation of tumour cells in a target organ. For hormones that occur in excessive amounts, for hormone-mimetic agents and for agents that cause hyperactivity or imbalance in the endocrine system, evaluative methods comparable with those used to identify chemical carcinogens may be required; particular emphasis must be laid on quantitative aspects and duration of exposure. Some chemical carcinogens have significant side-effects on the endocrine system, which may also result in hormonal carcinogenesis. Synthetic hormones and anti-hormones can be expected to possess other pharmacological and toxicological actions in addition to those on the endocrine system, and in this respect they must be treated like any other chemical with regard to intrinsic carcinogenic potential.

Complex mixtures: There is an increasing amount of data from long-term carcinogenicity studies on complex mixtures and on crude materials obtained by sampling in an occupational environment. The representativity of such samples must be considered carefully.

Quantitative aspects: Dose-response studies are important in the evaluation of carcinogenesis: the confidence with which a carcinogenic effect can be established is strengthened by the observation of an increasing incidence of neoplasms with increasing exposure.

The assessment of carcinogenicity in animals is frequently complicated by recognized differences among the test animals (species, strain, sex, age) and route and schedule of administration; often, the target organs at which a cancer occurs and its histological type may vary with these parameters. Nevertheless, indices of carcinogenic potency in particular experimental systems (for instance, the dose-rate required under continuous exposure to halve the probability of the animals remaining tumourless[9]) have been formulated in the hope that, at least among categories of fairly similar agents, such indices may be of some predictive value in other species, including humans.

Chemical carcinogens share many common biological properties, which include metabolism to reactive (electrophilic (10,11)) intermediates capable of interacting with DNA. However, they may differ widely in the dose required to produce a given level of tumour induction. The reason for this variation in dose-response is not understood, but it may be due to differences in metabolic activation and detoxification processes, to different DNA-repair capacities among various organs and species or to the operation of qualitatively distinct mechanisms.

Statistical analysis of animal studies: It is possible that an animal may die prematurely from unrelated causes, so that tumours that would have arisen had the animal lived longer may not be observed; this possibility must be allowed for. Various analytical techniques have been developed which use the assumption of independence of competing risks to allow for the effects of intercurrent mortality on the final numbers of tumour-bearing animals in particular treatment groups.

For externally visible tumours and for neoplasms that cause death, methods such as Kaplan-Meier (i.e., 'life-table', 'product-limit' or 'actuarial') estimates(9), with associated significance tests(12,13), have been recommended. For internal neoplasms that are discovered 'incidentally'(12) at autopsy but that did not cause the death of the host, different estimates(14) and significance tests(12,13) may be necessary for the unbiased study of the numbers of tumour-bearing animals.

The design and statistical analysis of long-term carcinogenicity experiments were reviewed in Supplement 2 to the *Monographs* series(15). That review outlined the way in which the context of observation of a given tumour (fatal or incidental) could be included in an analysis yielding a single combined result. This method requires information on time to death for each animal and is therefore comparable to only a limited extent with analyses which include global proportions of tumour-bearing animals.

Evaluation of carcinogenicity studies in experimental animals: The evidence of carcinogenicity in experimental animals is assessed by the Working Group and judged to fall into one of four groups, defined as follows:

(1) *Sufficient evidence* of carcinogenicity is provided when there is an increased incidence of malignant tumours: (a) in multiple species or strains; or (b) in multiple experiments (preferably with different routes of administration or using different dose levels); or (c) to an unusual degree with regard to incidence, site or type of tumour, or age at onset. Additional evidence may be provided by data on dose-response effects.

(2) *Limited evidence* of carcinogenicity is available when the data suggest a carcinogenic effect but are limited because: (a) the studies involve a single species, strain or experiment; or (b) the experiments are restricted by inadequate dosage levels, inadequate duration of exposure to the agent, inadequate period of follow-up, poor survival, too few animals, or inadequate reporting; or (c) the neoplasms produced often occur spontaneously and, in the past, have been difficult to classify as malignant by histological criteria alone (e.g., lung adenomas and adenocarcinomas and liver tumours in certain strains of mice).

(3) *Inadequate evidence* is available when, because of major qualitative or quantitative limitations, the studies cannot be interpreted as showing either the presence or absence of a carcinogenic effect.

(4) *No evidence* applies when several adequate studies are available which show that, within the limits of the tests used, the chemical or complex mixture is not carcinogenic.

It should be noted that the categories *sufficient evidence* and *limited evidence* refer only to the strength of the experimental evidence that these chemicals or complex mixtures are carcinogenic and not to the extent of their carcinogenic activity nor to the mechanism involved. The classification of any chemical may change as new information becomes available.

(ii) *Evidence for activity in short-term tests*[1]

Many short-term tests bearing on postulated mechanisms of carcinogenesis or on the properties of known carcinogens have been developed in recent years. The induction of cancer is thought to proceed by a series of steps, some of which have been distinguished experimentally(16-20). The first step - initiation - is thought to involve damage to DNA, resulting in heritable alterations in or rearrangements of genetic information. Most short-term tests in common use today are designed to evaluate the genetic activity of a substance. Data from these assays are useful for identifying potential carcinogenic hazards, in identifying active metabolites of known carcinogens in human or animal body fluids, and in helping to elucidate mechanisms of carcinogenesis. Short-term tests to detect agents with tumour-promoting activity are, at this time, insufficiently developed.

Because of the large number of short-term tests, it is difficult to establish rigid criteria for adequacy that would be applicable to all studies. General considerations relevant to all tests, however, include (a) that the test system be valid with respect to known animal carcinogens and noncarcinogens; (b) that the experimental parameters under which the chemical (or complex mixture) is tested include a sufficiently wide dose range and duration of exposure to the agent and an appropriate metabolic system; (c) that appropriate controls be used; and (d) that the purity of the compound or, in the case of complex mixtures, that the source and representativity of the sample being tested be specified. Confidence in positive results is increased if a dose-response relationship is demonstrated and if this effect has been reported in two or more independent studies.

Most established short-term tests employ as end-points well-defined genetic markers in prokaryotes and lower eukaryotes and in mammalian cell lines. The tests can be grouped according to the end-point detected:

Tests of *DNA damage*. These include tests for covalent binding to DNA, induction of DNA breakage or repair, induction of prophage in bacteria and differential survival of DNA repair-proficient/-deficient strains of bacteria.

Tests of *mutation* (measurement of heritable alterations in phenotype and/or genotype). These include tests for detection of the loss or alteration of a gene product, and change of function through forward or reverse mutation, recombination and gene conversion; they may involve the nuclear genome, the mitochondrial genome and resident viral or plasmid genomes.

Tests of *chromosomal effects*. These include tests for detection of changes in chromosome number (aneuploidy), structural chromosomal aberrations, sister chromatid exchanges, micronuclei and dominant-lethal events. This classification does not imply that some chromosomal effects are not mutational events.

---

[1]Based on the recommendations of a working group which met in 1983(5).

PREAMBLE

Tests for *cell transformation*, which monitor the production of preneoplastic or neoplastic cells in culture, are also of importance because they attempt to simulate essential steps in cellular carcinogenesis. These assays are not grouped with those listed above since the mechanisms by which chemicals induce cell transformation may not necessarily be the result of genetic change.

The selection of specific tests and end-points for consideration remains flexible and should reflect the most advanced state of knowledge in this field.

The data from short-term tests are summarized by the Working Group and the test results tabulated according to the end-points detected and the biological complexities of the test systems. The format of the table used is shown below. In these tables, a '+' indicates that the compound was judged by the Working Group to be significantly positive in one or more assays for the specific end-point and level of biological complexity; '-' indicates that it was

Overall assessment of data from short-term tests

	Genetic activity			Cell transformation
	DNA damage	Mutation	Chromosomal effects	
Prokaryotes				
Fungi/ Green plants				
Insects				
Mammalian cells (*in vitro*)				
Mammals (*in vivo*)				
Humans (*in vivo*)				

judged to be negative in one or more assays; and '?' indicates that there were contradictory results from different laboratories or in different biological systems, or that the result was judged to be equivocal. These judgements reflect the assessment by the Working Group of the quality of the data (including such factors as the purity of the test compound, problems of metabolic activation and appropriateness of the test system) and the relative significance of the component tests.

An overall assessment of the evidence for *genetic activity* is then made on the basis of the entries in the table, and the evidence is judged to fall into one of four categories, defined as follows:

(*i*) *Sufficient evidence* is provided by at least three positive entries, one of which must involve mammalian cells *in vitro* or *in vivo* and which must include at least two of three end-points - DNA damage, mutation and chromosomal effects.

(*ii*) *Limited evidence* is provided by at least two positive entries.

(*iii*) *Inadequate evidence* is available when there is only one positive entry or when there are too few data to permit an evaluation of an absence of genetic activity or when there are unexplained, inconsistent findings in different test systems.

(*iv*) *No evidence* applies when there are only negative entries; these must include entries for at least two end-points and two levels of biological complexity, one of which must involve mammalian cells *in vitro* or *in vivo*.

It is emphasized that the above definitions are operational, and that the assignment of a chemical or complex mixture into one of these categories is thus arbitrary.

In general, emphasis is placed on positive results; however, in view of the limitations of current knowledge about mechanisms of carcinogenesis, certain cautions should be respected: (i) At present, short-term tests should not be used by themselves to conclude whether or not an agent is carcinogenic, nor can they predict reliably the relative potencies of compounds as carcinogens in intact animals. (ii) Since the currently available tests do not detect all classes of agents that are active in the carcinogenic process (e.g., hormones), one must be cautious in utilizing these tests as the sole criterion for setting priorities in carcinogenesis research and in selecting compounds for animal bioassays. (iii) Negative results from short-term tests cannot be considered as evidence to rule out carcinogenicity, nor does lack of demonstrable genetic activity attribute an epigenetic or any other property to a substance(5).

(b) *Evaluation of Carcinogenicity in Humans*

Evidence of carcinogenicity can be derived from case reports, descriptive epidemiological studies and analytical epidemiological studies.

An analytical study that shows a positive association between an exposure and a cancer may be interpreted as implying causality to a greater or lesser extent, on the basis of the following criteria: (a) There is no identifiable positive bias. (By 'positive bias' is meant the operation of factors in study design or execution that lead erroneously to a more strongly positive association between an exposure and disease than in fact exists. Examples of positive bias include, in case-control studies, better documentation of the exposure for cases than for controls, and, in cohort studies, the use of better means of detecting cancer in exposed individuals than in individuals not exposed.) (b) The possibility of positive confounding has been considered. (By 'positive confounding' is meant a situation in which the relationship between an exposure and a disease is rendered more strongly positive than it truly is as a result of an association between that exposure and another exposure which either causes or prevents the disease. An example of positive confounding is the association between coffee consumption and lung cancer, which results from their joint association with cigarette smoking.) (c) The association is unlikely to be due to chance alone. (d) The association is strong. (e) There is a dose-response relationship.

In some instances, a single epidemiological study may be strongly indicative of a cause-effect relationship; however, the most convincing evidence of causality comes when several independent studies done under different circumstances result in 'positive' findings.

Analytical epidemiological studies that show no association between an exposure and a cancer ('negative' studies) should be interpreted according to criteria analogous to those listed above: (a) there is no identifiable negative bias; (b) the possibility of negative confounding has been considered; and (c) the possible effects of misclassification of exposure or outcome have been weighed. In addition, it must be recognized that the probability that a given study can detect a certain effect is limited by its size. This can be perceived from the confidence limits around the estimate of association or relative risk. In a study regarded as 'negative', the upper confidence limit may indicate a relative risk substantially greater than unity; in that case, the study excludes only relative risks that are above the upper limit. This usually means that a 'negative' study must be large to be convincing. Confidence in a 'negative' result is increased when several independent studies carried out under different circumstances are in agreement. Finally, a 'negative' study may be considered to be relevant only to dose levels within or below the range of those observed in the study and is pertinent only if sufficient time has elapsed since first human exposure to the agent. Experience with human cancers of known etiology suggests that the period from first exposure to a chemical carcinogen to development of clinically observed cancer is usually measured in decades and may be in excess of 30 years.

The evidence for carcinogenicity from studies in humans is assessed by the Working Group and judged to fall into one of four groups, defined as follows:

1. *Sufficient evidence* of carcinogenicity indicates that there is a causal relationship between the exposure and human cancer.

2. *Limited evidence* of carcinogenicity indicates that a causal interpretation is credible, but that alternative explanations, such as chance, bias or confounding, could not adequately be excluded.

3. *Inadequate evidence*, which applies to both positive and negative evidence, indicates that one of two conditions prevailed: (a) there are few pertinent data; or (b) the available studies, while showing evidence of association, do not exclude chance, bias or confounding.

4. *No evidence* applies when several adequate studies are available which do not show evidence of carcinogenicity.

(c) *Relevance of Experimental Data to the Evaluation of Carcinogenic Risk to Humans*

Information compiled from the first 29 volumes of the *IARC Monographs*(4,21,22) shows that, of the chemicals or groups of chemicals now generally accepted to cause or probably to cause cancer in humans, all (with the possible exception of arsenic) of those that have been tested appropriately produce cancer in at least one animal species. For several of the chemicals (e.g., aflatoxins, 4-aminobiphenyl, diethylstilboestrol, melphalan, mustard gas and vinyl chloride), evidence of carcinogenicity in experimental animals preceded evidence obtained from epidemiological studies or case reports.

For many of the chemicals (or complex mixtures) evaluated in the *IARC Monographs* for which there is *sufficient evidence* of carcinogenicity in animals, data relating to carcinogenicity for humans are either insufficient or nonexistent. **In the absence of adequate data on humans, it is reasonable, for practical purposes, to regard chemicals for which there is sufficient evidence of carcinogenicity in animals as if they presented a carcinogenic risk to humans.** The use of the expressions 'for practical purposes' and 'as if they presented a carcinogenic risk' indicates that, at the present time, a correlation between carcinogenicity in animals and possible human risk cannot be made on a purely scientific basis, but only pragmatically. Such a pragmatic correlation may be useful to regulatory agencies in making decisions related to the primary prevention of cancer.

In the present state of knowledge, it would be difficult to define a predictable relationship between the dose (mg/kg bw per day) of a particular chemical required to produce cancer in test animals and the dose that would produce a similar incidence of cancer in humans. Some data, however, suggest that such a relationship may exist(23,24), at least for certain classes of carcinogenic chemicals, although no acceptable method is currently available for quantifying the possible errors that may be involved in such an extrapolation procedure.

8. EXPLANATORY NOTES ON THE CONTENTS OF MONOGRAPHS ON CHEMICALS AND COMPLEX MIXTURES

The sections 1 and 2, as outlined below, are those used in monographs on individual chemicals. When relevant, similar information is included in monographs on complex mixtures; additional information is provided as considered necessary.

(a) Chemical and Physical Data (Section 1)

The Chemical Abstracts Services Registry Number, the latest Chemical Abstracts Primary Name (9th Collective Index)(25) and the IUPAC Systematic Name(26) are recorded in section 1. Other synonyms and trade names are given, but no comprehensive list is provided. Some of the trade names are those of mixtures in which the compound being evaluated is only one of the ingredients.

The structural and molecular formulae, molecular weight and chemical and physical properties are given. The properties listed refer to the pure substance, unless otherwise specified, and include, in particular, data that might be relevant to carcinogenicity (e.g., lipid solubility) and those that concern identification.

A separate description of the composition of technical products includes available information on impurities and formulated products.

(b) Production, Use, Occurrence and Analysis (Section 2)

The purpose of section 2 is to provide indications of the extent of past and present human exposure to the chemical.

PREAMBLE

(i) *Synthesis*

Since cancer is a delayed toxic effect, the dates of first synthesis and of first commercial production of the chemical are provided. This information allows a reasonable estimate to be made of the date before which no human exposure could have occurred. In addition, methods of synthesis used in past and present commercial production are described.

(ii) *Production*

Since Europe, Japan and the USA are reasonably representative industrialized areas of the world, most data on production, foreign trade and uses are obtained from those countries. It should not, however, be inferred that those areas or nations are the sole or even the major sources or users of any individual chemical.

Production and foreign trade data are obtained from both governmental and trade publications by chemical economists in the three geographical areas. In some cases, separate production data on organic chemicals manufactured in the USA are not available because their publication could disclose confidential information. In such cases, an indication of the minimum quantity produced can be inferred from the number of companies reporting commercial production. Each company is required to report on individual chemicals if the sales value or the weight of the annual production exceeds a specified minimum level. These levels vary for chemicals classified for different uses, e.g., medicinals and plastics; in fact, the minimal annual sales value is between $1000 and $50 000, and the minimal annual weight of production is between 450 kg and 22 700 kg. Data on production in some European countries are obtained by means of general questionnaires sent to companies thought to produce the compounds being evaluated. Information from the completed questionnaires is compiled, by country, and the resulting estimates of production are included in the individual monographs.

(iii) *Use*

Information on uses is meant to serve as a guide only and is not complete. It is usually obtained from published data but is often complemented by direct contact with manufacturers of the chemical. In the case of drugs, mention of their therapeutic uses does not necessarily represent current practice nor does it imply judgement as to their clinical efficacy.

Statements concerning regulations and standards (e.g., pesticide registrations, maximum levels permitted in foods, occupational standards and allowable limits) in specific countries are mentioned as examples only. They may not reflect the most recent situation, since such legislation is in a constant state of change; nor should it be taken to imply that other countries do not have similar regulations.

(iv) *Occurrence*

Information on the occurrence of a chemical in the environment is obtained from published data, including that derived from the monitoring and surveillance of levels of the chemical in occupational environments, air, water, soil, foods and tissues of animals and humans. When no published data are available to the Working Group, unpublished reports, deemed appropriate, may be considered. When available, data on the generation, persistence and bioaccumulation of a chemical are also included.

(v) *Analysis*

The purpose of the section on analysis is to give the reader an indication, rather than a complete review, of methods cited in the literature. No attempt is made to evaluate critically or to recommend any of the methods.

(c) *Biological Data Relevant to the Evaluation of Carcinogenic Risk to Humans (Section 3)*

In general, the data recorded in section 3 are summarized as given by the author; however, comments made by the Working Group on certain shortcomings of reporting, of statistical analysis or of experimental design are given in square brackets. The nature and extent of impurities/contaminants in the chemicals being tested are given when available.

(i) *Carcinogenicity studies in animals*

The monographs are not intended to cover all reported studies. Some studies are purposely omitted (a) because they are inadequate, as judged from previously described criteria(27-30) (e.g., too short a duration, too few animals, poor survival); (b) because they only confirm findings that have already been fully described; or (c) because they are judged irrelevant for the purpose of the evaluation. In certain cases, however, such studies are mentioned briefly, particularly when the information is considered to be a useful supplement to other reports or when it is the only data available. Their inclusion does not, however, imply acceptance of the adequacy of their experimental design or of the analysis and interpretation of their results.

Mention is made of all routes of administration by which the test material has been adequately tested and of all species in which relevant tests have been done(30). In most cases, animal strains are given. Quantitative data are given to indicate the order of magnitude of the effective carcinogenic doses. In general, the doses and schedules are indicated as they appear in the original; sometimes units have been converted for easier comparison. Experiments in which the compound was administered in conjunction with known carcinogens and experiments on factors that modify the carcinogenic effect are also reported. Experiments on the carcinogenicity of known metabolites and derivatives are also included.

(ii) *Other relevant biological data*

LD_{50} data are given when available, and other data on toxicity are included when considered relevant.

Data on effects on reproduction, on teratogenicity and embryo- and fetotoxicity and on placental transfer, from studies in experimental animals and from observations in humans, are included when considered relevant.

Information is given on absorption, distribution and excretion. Data on metabolism are usually restricted to studies that show the metabolic fate of the chemical in experimental animals and humans, and comparisons of data from animals and humans are made when possible.

Data from short-term tests are also included. In addition to the tests for genetic activity and cell transformation described previously (see pages 16-17), data from studies of related effects, but for which the relevance to the carcinogenic process is less well established, may also be mentioned.

The criteria used for considering short-term tests and for evaluating their results have been described (see pages 16-18). In general, the authors' results are given as reported. An assessment of the data by the Working Group which differs from that of the authors, and comments concerning aspects of the study that might affect its interpretation are given in square brackets. Reports of studies in which few or no experimental details are given, or in which the data on which a reported positive or negative result is based are not available for examination, are cited, but are identified as 'abstract' or 'details not given' and are not considered in the summary tables or in making the overall assessment of genetic activity.

For several recent reviews on short-term tests, see IARC(30), Montesano et al.(31), de Serres and Ashby(32), Sugimura et al.(33), Bartsch et al.(34) and Hollstein et al.(35).

(iii) *Case reports and epidemiological studies of carcinogenicity in humans*

Observations in humans are summarized in this section. These include case reports, descriptive epidemiological studies (which correlate cancer incidence in space or time to an exposure) and analytical epidemiological studies of the case-control or cohort type. In principle, a comprehensive coverage is made of observations in humans; however, reports are excluded when judged to be clearly not pertinent. This applies in particular to case reports, in which either the clinico-pathological description of the tumours or the exposure history, or both, are poorly described; and to published routine statistics, for example, of cancer mortality by occupational category, when the categories are so broadly defined as to contribute virtually no specific information on the possible relation between cancer occurrence and a given exposure. Results of studies are assessed on the basis of the data and analyses that are presented in the published papers. Some additional analyses of the published data may be performed by the Working Group to gain better insight into the relation between cancer occurrence and the exposure under consideration. The Working Group may use these analyses in its assessment of the evidence or may actually include them in the text to summarize a study; in such cases, the results of the supplementary analyses are given in square brackets. Any comments by the Working Group are also reported in square brackets; however, these are kept to a minimum, being restricted to those instances in which it is felt that an important aspect of a study, directly impinging on its interpretation, should be brought to the attention of the reader.

(d) Summary of Data Reported and Evaluation (Section 4)

Section 4 summarizes the relevant data from animals and humans and gives the critical views of the Working Group on those data.

(i) *Exposures*

Human exposure to the chemical or complex mixture is summarized on the basis of data on production, use and occurrence.

(ii) *Experimental data*

Data relevant to the evaluation of the carcinogenicity of the test material in animals are summarized in this section. The animal species mentioned are those in which the carcinogenicity of the substance was clearly demonstrated. Tumour sites are also indicated. If the substance has produced tumours after prenatal exposure or in single-dose experiments, this is indicated. Dose-response data are given when available.

Significant findings on effects on reproduction and prenatal toxicity, and results from short-term tests for genetic activity and cell transformation assays are summarized, and the latter are presented in tables. An overall assessment is made of the degree of evidence for genetic activity in short-term tests.

(iii) *Human data*

Case reports and epidemiological studies that are considered to be pertinent to an assessment of human carcinogenicity are described. Other biological data that are considered to be relevant are also mentioned.

(iv) *Evaluation*

This section comprises evaluations by the Working Group of the degrees of evidence for carcinogenicity of the exposure to experimental animals and to humans. An overall evaluation is then made of the carcinogenic risk of the chemical or complex mixture to humans. This section should be read in conjunction with page 15-16 and 19 of this Preamble for definitions of degrees of evidence.

When no data are available from epidemiological studies but there is *sufficient evidence* that the exposure is carcinogenic to animals, a footnote is included, reading: 'In the absence of adequate data on humans, it is reasonable, for practical purposes, to regard chemicals for which there is *sufficient evidence* of carcinogenicity in animals as if they presented a carcinogenic risk to humans.'

References

1. IARC (1979) Criteria to select chemicals for *IARC Monographs*. *IARC intern. tech. Rep. No. 79/003*

2. IARC (1977) IARC Monograph Programme on the Evaluation of the Carcinogenic Risk of Chemicals to Humans. Preamble. *IARC intern. tech. Rep. No. 77/002*

3. IARC (1978) Chemicals with *sufficient evidence* of carcinogenicity in experimental animals - IARC Monographs volumes 1-17. *IARC intern. tech. Rep. No. 78/003*

4. IARC (1982) *IARC Monographs on the Evaluation of the Carcinogenic Risk of Chemicals to Humans*, Supplement 4, *Chemicals, Industrial Processes and Industries Associated with Cancer in Humans* (IARC Monographs, Volumes 1 to 29)

5. IARC (1983) Approaches to classifying chemical carcinogens according to mechanism of action. *IARC intern. tech. Rep. No. 83/001*

6. IARC (1972-1983) *IARC Monographs on the Evaluation of the Carcinogenic Risk of Chemicals to Humans*, Volumes 1-35, Lyon, France

    Volume 1 (1972) Some Inorganic Substances, Chlorinated Hydrocarbons, Aromatic Amines, *N*-Nitroso Compounds and Natural Products (19 monographs), 184 pages

Volume 2 (1973) Some Inorganic and Organometallic Compounds (7 monographs), 181 pages

Volume 3 (1973) Certain Polycyclic Aromatic Hydrocarbons and Heterocyclic Compounds (17 monographs), 271 pages

Volume 4 (1974) Some Aromatic Amines, Hydrazine and Related Substances, N-Nitroso Compounds and Miscellaneous Alkylating Agents (28 monographs), 286 pages

Volume 5 (1974) Some Organochlorine Pesticides (12 monographs), 241 pages

Volume 6 (1974) Sex Hormones (15 monographs), 243 pages

Volume 7 (1974) Some Anti-thyroid and Related Substances, Nitrofurans and Industrial Chemicals (23 monographs), 326 pages

Volume 8 (1975) Some Aromatic Azo Compounds (32 monographs), 357 pages

Volume 9 (1975) Some Aziridines, N-, S- and O-Mustards and Selenium (24 monographs), 268 pages

Volume 10 (1976) Some Naturally Occurring Substances (22 monographs), 353 pages

Volume 11 (1976) Cadmium, Nickel, Some Epoxides, Miscellaneous Industrial Chemicals and General Considerations on Volatile Anaesthetics (24 monographs), 306 pages

Volume 12 (1976) Some Carbamates, Thiocarbamates and Carbazides (24 monographs), 282 pages

Volume 13 (1977) Some Miscellaneous Pharmaceutical Substances (17 monographs), 255 pages

Volume 14 (1977) Asbestos (1 monograph), 106 pages

Volume 15 (1977) Some Fumigants, the Herbicides, 2,4-D and 2,4,5-T, Chlorinated Dibenzodioxins and Miscellaneous Industrial Chemicals (18 monographs), 354 pages

Volume 16 (1978) Some Aromatic Amines and Related Nitro Compounds - Hair Dyes, Colouring Agents, and Miscellaneous Industrial Chemicals (32 monographs), 400 pages

Volume 17 (1978) Some N-Nitroso Compounds (17 monographs), 365 pages

Volume 18 (1978) Polychlorinated Biphenyls and Polybrominated Biphenyls (2 monographs), 140 pages

Volume 19 (1979) Some Monomers, Plastics and Synthetic Elastomers, and Acrolein (17 monographs), 513 pages

Volume 20 (1979) Some Halogenated Hydrocarbons (25 monographs), 609 pages

Volume 21 (1979) Sex Hormones (II) (22 monographs), 583 pages

Volume 22 (1980) Some Non-Nutritive Sweetening Agents (2 monographs), 208 pages

Volume 23 (1980) Some Metals and Metallic Compounds (4 monographs), 438 pages

Volume 24 (1980) Some Pharmaceutical Drugs (16 monographs), 337 pages

Volume 25 (1981) Wood, Leather and Some Associated Industries (7 monographs), 412 pages

Volume 26 (1981) Some Antineoplastic and Immunosuppressive Agents (18 monographs), 411 pages

Volume 27 (1981) Some Aromatic Amines, Anthraquinones and Nitroso Compounds, and Inorganic Fluorides Used in Drinking-Water and Dental Preparations (18 monographs), 344 pages

Volume 28 (1982) The Rubber Manufacturing Industry (1 monograph), 486 pages

Volume 29 (1982) Some Industrial Chemicals (18 monographs), 416 pages

Volume 30 (1982) Miscellaneous Pesticides (18 monographs), 424 pages

Volume 31 (1983) Some Food Additives, Feed Additives and Naturally Occurring Substances (21 monographs), 314 pages

Volume 32 (1983) Polynuclear Aromatic Compounds, Part 1, Chemical, Environmental and Experimental Data (42 monographs), 477 pages

Volume 33 (1984) Polynuclear Aromatic Compounds, Part 2, Carbon Blacks, Mineral Oils and Some Nitroarenes (8 monographs), 245 pages

Volume 34 (1984) Polynuclear Aromatic Compounds, Part 3, Industrial Exposures in Aluminium Production, Coal Gasification, Coke Production and Iron and Steel Founding (4 monographs), 219 pages

Volumes 35 (1984) Polynuclear Aromatic Compounds, Part 4, Bitumens, Coal-Tars and Derived Products, Shale-Oils and Soots (4 monographs), 271 pages

Supplement No. 1 (1979) Chemicals and Industrial Processes Associated with Cancer in Humans (IARC Monographs, Volumes 1 to 20), 71 pages

Supplement No. 2 (1980) Long-term and Short-term Screening Assays for Carcinogens: A Critical Appraisal, 426 pages

Supplement No. 3 (1982) Cross Index of Synonyms and Trade Names in Volumes 1 to 26, 199 pages

Supplement No. 4 (1982) Chemicals, Industrial Processes and Industries Associated with Cancer in Humans (IARC Monographs, Volumes 1 to 29), 292 pages

7. IARC (1973-1983) *Information Bulletin on the Survey of Chemicals Being Tested for Carcinogenicity*, Numbers 1-10, Lyon, France

   Number 1 (1973) 52 pages
   Number 2 (1973) 77 pages
   Number 3 (1974) 67 pages
   Number 4 (1974) 97 pages
   Number 5 (1975) 88 pages
   Number 6 (1976) 360 pages
   Number 7 (1978) 460 pages
   Number 8 (1979) 604 pages
   Number 9 (1981) 294 pages
   Number 10 (1983) 326 pages

8. PHS 149 (1951-1983) Public Health Service Publication No. 149, *Survey of Compounds which have been Tested for Carcinogenic Activity*, Washington DC, US Government Printing Office

   1951 Hartwell, J.L., 2nd ed., Literature up to 1947 on 1329 compounds, 583 pages
   1957 Shubik, P. & Hartwell, J.L., Supplement 1, Literature for the years 1948-1953 on 981 compounds, 388 pages
   1969 Shubik, P. & Hartwell, J.L., edited by Peters, J.A., Supplement 2, Literature for the years 1954-1960 on 1048 compounds, 655 pages
   1971 National Cancer Institute, Literature for the years 1968-1969 on 882 compounds, 653 pages
   1973 National Cancer Institute, Literature for the years 1961-1967 on 1632 compounds, 2343 pages
   1974 National Cancer Institute, Literature for the years 1970-1971 on 750 compounds, 1667 pages
   1976 National Cancer Institute, Literature for the years 1972-1973 on 966 compounds, 1638 pages
   1980 National Cancer Institute, Literature for the year 1978 on 664 compounds, 1331 pages
   1983 National Cancer Institute, Literature for years 1974-1975 on 575 compounds, 1043 pages

9. Pike, M.C. & Roe, F.J.C. (1963) An actuarial method of analysis of an experiment in two-stage carcinogenesis. *Br. J. Cancer*, *17*, 605-610

10. Miller, E.C. (1978) Some current perspectives on chemical carcinogenesis in humans and experimental animals: Presidential address. *Cancer Res.*, *38*, 1479-1496

11. Miller, E.C. & Miller, J.A. (1981) Searches for ultimate chemical carcinogens and their reactions with cellular macromolecules. *Cancer*, *47*, 2327-2345

12. Peto, R. (1974) Guidelines on the analysis of tumour rates and death rates in experimental animals. *Br. J. Cancer*, *29*, 101-105

13. Peto, R. (1975) Letter to the editor. *Br. J. Cancer*, *31*, 697-699

14. Hoel, D.G. & Walburg, H.E., Jr (1972) Statistical analysis of survival experiments. *J. natl Cancer Inst.*, *49*, 361 372

15. Peto, R., Pike, M.C., Day, N.E., Gray, R.G., Lee, P.N., Parish, S., Peto, J., Richards, S. & Wahrendorf, J. (1980) *Guidelines for simple sensitive significance tests for carcinogenic effects in long-term animal experiments.* In: *IARC Monographs on the Evaluation of the Carcinogenic Risk of Chemicals to Humans*, Supplement 2, *Long-term and Short-term Screening Assays for Carcinogens: A Critical Appraisal*, Lyon, pp. 311-426

16. Berenblum, I. (1975) *Sequential aspects of chemical carcinogenesis: Skin.* In: Becker, F.F., ed., *Cancer. A Comprehensive Treatise*, Vol. 1, New York, Plenum Press, pp. 323-344

17. Foulds, L. (1969) *Neoplastic Development*, Vol. 2, London, Academic Press

18. Farber, E. & Cameron, R. (1980) The sequential analysis of cancer development. *Adv. Cancer Res., 31*, 125-126

19. Weinstein, I.B. (1981) The scientific basis for carcinogen detection and primary cancer prevention. *Cancer, 47*, 1133-1141

20. Slaga, T.J., Sivak, A. & Boutwell, R.K., eds (1978) *Mechanisms of Tumor Promotion and Cocarcinogenesis*, Vol. 2, New York, Raven Press

21. IARC Working Group (1980) An evaluation of chemicals and industrial processes associated with cancer in humans based on human and animal data: IARC Monographs Volumes 1 to 20. *Cancer Res., 40*, 1-12

22. IARC (1979) *IARC Monographs on the Evaluation of the Carcinogenic Risk of Chemicals to Humans*, Supplement 1, *Chemicals and Industrial Processes Associated with Cancer in Humans*, Lyon

23. Rall, D.P. (1977) *Species differences in carcinogenesis testing.* In: Hiatt, H.H., Watson, J.D. & Winsten, J.A., eds, *Origins of Human Cancer*, Book C, Cold Spring Harbor, NY, Cold Spring Harbor Laboratory, pp. 1383-1390

24. National Academy of Sciences (NAS) (1975) *Contemporary Pest Control Practices and Prospects: The Report of the Executive Committee*, Washington DC

25. Chemical Abstracts Services (1978) *Chemical Abstracts Ninth Collective Index (9CI), 1972-1976*, Vols 76-85, Columbus, OH

26. International Union of Pure & Applied Chemistry (1965) *Nomenclature of Organic Chemistry*, Section C, London, Butterworths

27. WHO (1958) Second Report of the Joint FAO/WHO Expert Committee on Food Additives. Procedures for the testing of intentional food additives to establish their safety and use. *WHO tech. Rep. Ser., No. 144*

28. WHO (1967) Scientific Group. Procedures for investigating intentional and unintentional food additives. *WHO tech. Rep. Ser., No. 348*

29. Sontag, J.M., Page, N.P. & Saffiotti, U. (1976) Guidelines for carcinogen bioassay in small rodents. *Natl Cancer Inst. Carcinog. tech. Rep. Ser., No.1*

30. IARC (1980) *IARC Monographs on the Evaluation of the Carcinogenic Risk of Chemicals to Humans*, Supplement 2, *Long-term and Short-term Screening Assays for Carcinogens: A Critical Appraisal*, Lyon

31. Montesano, R., Bartsch, H. & Tomatis, L., eds (1980) *Molecular and Cellular Aspects of Carcinogen Screening Tests (IARC Scientific Publications No. 27)*, Lyon

32. de Serres, F.J. & Ashby, J., eds (1981) *Evaluation of Short-Term Tests for Carcinogens. Report of the International Collaborative Program*, Amsterdam, Elsevier/North-Holland Biomedical Press

33. Sugimura, T., Sato, S., Nagao, M., Yahagi, T., Matsushima, T., Seino, Y., Takeuchi, M. & Kawachi, T. (1976) *Overlapping of carcinogens and mutagens.* In: Magee, P.N., Takayama, S., Sugimura, T. & Matsushima, T., eds, *Fundamentals in Cancer Prevention*, Tokyo/Baltimore, University of Tokyo/University Park Press, pp. 191-215

34. Bartsch, H., Tomatis, L. & Malaveille, C. (1982) *Qualitative and quantitative comparison between mutagenic and carcinogenic activities of chemicals.* In: Heddle, J.A., ed., *Mutagenicity: New Horizons in Genetic Toxicology*, New York, Academic Press, pp. 35-72

35. Hollstein, M., McCann, J., Angelosanto, F.A. & Nichols, W.W. (1979) Short-term tests for carcinogens and mutagens. *Mutat. Res.*, 65, 133-226

GENERAL REMARKS ON THE SUBSTANCES CONSIDERED

Background

This thirty-fifth volume of *IARC Monographs* comprises considerations of some complex mixtures that contain polynuclear aromatic compounds (PACs) - bitumens, certain crude coal-tars, including pharmaceutical tars, some products derived from coal-tars (e.g., creosotes and coal-tar pitches), shale-oils, and chimney soots from domestic and institutional sources. It is the fourth in a series dealing with polynuclear aromatic compounds: volume 32 of the *Monographs* covered chemical, environmental and experimental data on 48 individual PACs which occur in complex mixtures resulting from the combustion or pyrolysis of fossil fuels or products derived from them. In volume 33, experimental and epidemiological data on carbon blacks, mineral oils and some nitroarenes were evaluated. In volume 34, four industries were considered in which exposure to PACs may occur - aluminium production, coal gasification, coke production, and iron and steel founding.

Some of the early literature on creosotes, coal-tars, coal-tar pitches and shale-oils was summarized in volume 3 of the *IARC Monographs* (IARC, 1972). Since that time, new data have become available, and these are considered in the present volume.

Contents of this volume

The term 'bitumens', as used in this volume, refers to the products derived from residues resulting from vacuum distillation of selected petroleum crude oils. They are widely used for paving, roofing and industrial applications, with world sales of more than 60 million tonnes per annum. Bitumens must be distinguished from coal-tars, which are products of the destructive distillation of coals, and also from coal-tar pitches, which are residues from the distillation of coal-tars. Such coal-tar products are sometimes used in applications for which bitumens may also be employed. In North America, the terms 'asphalts' and 'petroleum asphalts' are used to describe bitumens; elsewhere, the term 'asphalts' is used to describe mixtures of bitumen and mineral matter.

Petroleum pitches are not covered in this volume as they are different from bitumens in terms of manufacture, properties and applications. Petroleum pitches are dark-coloured bituminous substances obtained as non-volatile residues from destructive distillation processes carried out during the refining of petroleum. Petroleum pitches are used primarily as binders for carbon electrodes, which are used in one process for the production of aluminium from alumina. The main occupational exposure to petroleum pitches thus occurs during aluminium production, which was evaluated in Volume 34 of the *IARC Monographs* (IARC, 1984).

Two previous monographs - on coal gasification and on coke production - dealt with the industrial processes involved in the production of gas for heating and lighting and the production of coke, mainly for use in steel blast furnaces; and biological data on samples of crude coal-tars taken directly from those industries were summarized and evaluated in those monographs. Exposures to coal-tar pitch volatiles also occur in aluminium production and in iron and steel founding, and these were discussed in the monographs on those industries (IARC, 1984).

The monographs in this volume describe the further refinement of crude coal-tars and the production of certain products for specific end-uses. A monograph on coal-tars and derived products includes crude coal-tars, pharmaceutical tars, coal-tar pitches and creosotes, although there may be other end-products from tar distilleries in addition to those considered here. Human exposure to creosotes, coal-tars and coal-tar pitches can occur in the distilleries as well as during use of these products. In certain occupations, however, workers may be exposed to both coal-derived products (creosotes and coal-tar pitches) and petroleum-derived products (bitumens), as in the case of roofers and road pavers.

Shale-oil is produced by retorting oil shale and is currently utilized only as a refinery or chemical-plant feedstock. Certain shale-oils, due to the unique nature of the oil shale from which they are derived, contain sufficient concentrations of chemical species to justify chemical recovery directly from the shale-oils. In most cases, however, the dominant end-use of shale-oil is as refined liquid fuel products similar in physical characteristics to those produced from petroleum crude oil; the biological effects of these shale-oil products have not yet been investigated fully.

All incomplete combustion of organic material produces soots as an unintentional by-product. The scope of the monograph in this volume on soots is linked to exposure patterns and to the available epidemiological data - these restrictions limit the scope to chimney soots from domestic and institutional sources. Soots from other sources, e.g., diesel and other engines, fly ash and other environmental soots, will be considered in future *IARC Monographs*.

Historical perspectives

This volume has particular interest in that the history of occupational cancer began with Percival Pott's report in 1775 (Pott, 1775) of scrotal cancer in chimney-sweeps. It took more than 100 years before, in 1912 (published in 1915), Yamagiwa and Ichikawa could produce cancer in experimental animals by repeated applications of coal-tar to the ears of rabbits. The introduction of mice as an experimental model for carcinogenicity studies by Tsutsui (1918) was another important step, along with his patient observation of susceptible animals during a sufficiently long time. Subsequently, many skin-painting experiments with various coal-tars were carried out in Europe, the first by Fibiger and Bang (1920).

Deelman (1922) was the first to report that the carcinogenicity of high-temperature (horizontal retort) coal-tars was different from that of tars from low-temperature (vertical) retorts; and these results were confirmed by Kennaway (1925), by Twort and Twort (1929) and by other researchers. Bloch and Dreifuss (1921) reported that the most carcinogenic fraction of coal-tars was that boiling at more than 300°C and devoid of low-boiling hydrocarbons and phenolic and basic components. Further studies by Kennaway and his group (Cook *et al.*, 1932) in the early 1930s led to the isolation of the first pure carcinogen, benzo[a]pyrene.

A number of the earlier studies (Tsutsi, 1918; Lipschütz, 1924; Twort & Ing, 1928; Teutschlaender, 1930; and others), including the first experimental study on shale-oil carcinogenicity (Leitch, 1922), discussed in detail the histological diagnosis of the malignant skin tumours induced. The criteria they established have been used for half a century and can still be considered as valid guidelines, even today.

The reader may be struck by the extent to which the evaluations of human carcinogenicity have depended on uncontrolled medical case reports, mostly from the early part of this century. After the outstandingly able pioneering work of Henry (1937) and the Kennaway's (Kennaway & Kennaway, 1936) in the UK and the unique contribution made by the occupational mortality analyses of the Registrar General for England and Wales (Office of Population Censuses and Surveys, 1978), later epidemiological investigations in this field have been few.

Special issues

A number of precautions should be observed in the use of experimental data on complex mixtures. Along with the obvious bias that accompanies the choice of non-representative samples, the method of storage and sample treatment may influence the results of an assay. In addition, the biological activity of a crude material collected by environmental sampling may be difficult to detect because of the overall toxicity of the mixture or the very low concentrations of the active components, or both. Therefore, the material is often concentrated and/or fractionated, so that it can be tested more accurately. Possible disadvantages include the loss or modification of specific components or the loss of possible synergistic effects.

Although the *IARC Monographs* deal with complex mixtures containing polynuclear aromatic compounds (PACs), there is a wide variety of PACs in such mixtures that have not been or cannot yet be measured. Consequently, although benzo[a]pyrene and other individual PACs considered in volume 32 of the *IARC Monographs* have been used as an indication of total PACs present in complex mixtures, reportedly low levels of certain PACs in some complex mixtures cannot be construed as meaning that the mixtures have a correspondingly low carcinogenic potential. Similarly, other substances, such as silica and arsenic, are known to occur in some of those complex mixtures; they may be of importance, but nevertheless they are usually not measured, and thus possible effects of these substances cannot be singled out, nor can modification of the effects of PACs be readily discerned.

For the substances considered in this volume of the *Monographs*, the animal carcinogenicity tests for which reports were available were predominantly skin-painting experiments in mice. In evaluating these experiments, the Working Group encountered a number of experimental variables, which indicate that caution should be exercised when comparing them. For example, different strains of mice, which may vary in sensitivity, were used. Various concentrations of fractionated materials were tested, and the amounts applied to the skin were sometimes not reported. Some solvents were employed that may strongly influence the outcome of skin-application experiments. Moreover, the test substances varied considerably in toxicity: sometimes, systemic toxicity limited the duration of the experiments and caused differential survival of dosed and control groups of animals; in other cases, local toxicity resulted in ulceration, which, when severe, might have reduced the incidence of cancers. In general, there was inadequate reporting of the possible effects on internal organs of application to the skin of the test substances.

Major obstacles to epidemiological evaluation have been the frequency of mixed exposures and the relative rarity of individuals and occupational groups confidently identifiable as having worked with only one of the materials under review. As a result, the detailed chemical and physical descriptions and classifications of the mixtures cannot be paralleled in the reviews of biological and health effects.

Quantitative estimates of human exposure to specific materials are usually lacking; occasionally, exposure-response analyses have been estimated in terms of duration of employment.

In view of the complexity of bitumens, coal-tars and derived products, and the processes used to make them, the Working Group found it useful to adopt a scheme for classifying these materials. The published studies on biological effects of bitumens and of coal-tars and derived products were classified by the Working Group to aid in understanding the data. These classifications are given in the monograph on bitumens, pp. 40-41, and in the monograph on coal-tars and derived products, p. 85. It is felt that they may be the most useful approach for understanding the hazards of some present-day materials, that they could provide a basis for description of the materials studied in the future and, in particular, that they might produce information that might be useful to design engineers in minimizing future hazards.

There are difficulties inherent in making such post-hoc judgements. These include the following. (i) The literature that was reviewed characterized these materials with varying degrees of completeness, particularly as far as their origin and composition are concerned. (ii) Accurate classification of these materials requires knowledge of the nature of the industrial processes by which they were produced, including those in different geographical areas and at different times, going back in some cases to the latter part of the last century. Detailed information of this kind was seldom available to the Working Group. (iii) The assumption is made that these products resemble present-day industrial products. Although this is presumed, it may not be true for all products. Despite these limitations, the Working Group felt that a classification would be useful, provided that its limitations are understood.

References

Bloch, B. & Dreifuss, W. (1921) On the experimental production of carcinomas with lymph node and lung metastases by coal tar constituents (Ger.). *Schw. med. Wochenschr.*, *51*, 1033-1037

Cook, J.W., Hewett, C. & Hieger, I. (1932) Coal tar constituents and cancer. *Nature, 130*, 926

Deelman, H.T. (1922) Experimental production of malignant tumours after action of coal-tar on mice (Ger.). *Z. Krebsforsch.*, *18*, 261-284

Fibiger, J. & Bang, F. (1920) Experimental production of coal-tar cancer in white mice (Fr.). *C.R. Soc. Biol.*, *83*, 1157-1160

Henry, S.A. (1937) The study of fatal cases of cancer of the scrotum from 1911 to 1935 in relation to occupation, with special reference to chimney sweeping and cotton mule spinning. *Am. J. Cancer, 31*, 28-57

IARC (1972) *IARC Monographs on the Evaluation of Carcinogenic Risk of Chemicals to Man*, Vol. 3, *Certain Polycyclic Aromatic Hydrocarbons and Heterocyclic Compounds*, Lyon

IARC (1984) *IARC Monographs on the Evaluation of the Carcinogenic Risk of Chemicals to Humans*, Vol. 34, *Aluminium Production, Coal Gasification, Coke Production, and Iron and Steel Founding*, Lyon

Kennaway, E.L. (1925) Experiments on cancer-producing substances. *Br. med. J.*, *69*, 3366-3371

Kennaway, N.M. & Kennaway, E.L. (1936) A study of the incidence of cancer of the lung and larynx. *J. Hyg.*, *36*, 236-267

Leitch, A. (1922) Paraffin cancer and its experimental production. *Br. med. J.*, *ii*, 1104-1109

Lipschütz, B. (1924) Studies on the formation of experimental coal-tar carcinomas in the mouse (Ger.). *Z. Krebsforsch.*, *21*, 50-97

Office of Population Censuses and Surveys (1978) *Occupational Mortality, 1970-1972, England and Wales, Decennial Supplement*, London, Her Majesty's Stationery Office

Pott, P. (1775) *Chirurgical Observations Relative to the Cataract, the Polypus of the Nose, the Cancer of the Scrotum, the Different Kinds of Ruptures and the Mortification of the Toes and Feet*, London, T.J. Carnegy, pp. 63-68

Teutschlaender, D. (1930) New studies on the mechanism of action of coal-tar and pitch in the formation of occupational skin tumours (Ger.). *Z. Krebsforsch.*, *30*, 573-580

Tsutsui, H. (1918) On artificially formed cancroids in the mouse (Ger.). *Gann*, *12*, 17-21

Twort, C.C. & Ing, H.R. (1928) Studies on carcinogenic agents (Ger.). *Z. Krebsforsch.*, *27*, 308-351

Twort, C.C. & Twort, J.M. (1929) The relative potency of carcinogenic tars and oils. *J. Hyg.*, *29*, 373-379

Yamagiwa, K. & Ichikawa, K. (1915) Experimental study on the pathogenesis of epithelial tumours (Ger.). *Mitt. Med. Fak. Kaiserl. Univ. Tokyo*, *15*, 295-344

THE MONOGRAPHS

BITUMENS

1. Chemical and Physical Data

This monograph concerns bitumens produced by petroleum refining. Although bitumens occur naturally as natural asphalts, rock asphalts and lake asphalts, the term 'bitumens' in this monograph refers only to the product recovered from petroleum refining. Bitumens derived from oil shale are discussed in the monograph on shale-oils (p. 161). This monograph does not include 'tar sands', which occur naturally in various parts of the world (Athabasca, West Canada; Nigeria), and are natural deposits of sands and bitumens.

The terms 'bitumen' and 'asphaltic bitumen' are used in Europe and are synonymous with the term 'asphalt' used in North America. Outside North America and in this monograph, the term *asphalt* is used to describe mixtures of bitumen with mineral materials.

A glossary of terms is given on p. 77.

1.1 Synonyms and trade names

Bitumens

Chem. Abstr. Services Reg. No.: 8052-42-4

Chem. Abstr. Name: Asphalt

Synonyms: Asphaltic bitumen; asphaltum; petroleum asphalt

Oxidized bitumens

Chem. Abstr. Services Reg. No.: 64742-93-4

Chem. Abstr. Name: Asphalt, oxidized

Synonym: Bitumen, oxidized

Trade Names: Bitumens are not generally referred to by trade names. Only limited volumes of products incorporating special additives have been given trade names by suppliers.

Chemical Abstract Services Registry Numbers other than the two given above are not available for the products considered in this monograph.

1.2 Description

Types and grades of bitumens

Bitumens are viscous liquids or solids consisting essentially of hydrocarbons and their derivatives, which are soluble in carbon disulphide; they are substantially non-volatile at ambient temperatures and soften gradually when heated. They are black or black-brown in colour and possess waterproofing and adhesive properties. Bitumens are obtained by refinery processes from petroleum and are also found as natural deposits or as components of naturally-occurring asphalts, in which they are associated with mineral matter.

Bitumens should not be confused with coal-tar products such as coal-tars or coal-tar pitches. The latter are manufactured by the high-temperature carbonization of bituminous coals and differ from bitumens substantially in composition and physical characteristics (see the monograph on coal-tars and derived products in this volume). A concise review of the differences between bitumens and coal-tar products has been given by Puzinauskas and Corbett (1978).

Similarly, bitumens should not be confused with petroleum pitches, which are often highly aromatic residues, produced by thermal cracking, coking or oxidation from selected petroleum fractions. (It should be noted that the term 'petroleum pitch' is used to describe different materials in different areas.) Petroleum pitches are principally used as binders in the manufacture of metallurgical electrodes.

Bitumens are classified in terms of specification tests related to their intended applications, for example, penetration, softening-point and viscosity. The *penetration test* measures, in tenths of a millimetre, the indentation of a specially prepared and controlled sample of bitumen at 25°C by a steel needle of specified dimensions, under a load of 100 g (British Standards Institution, 1974). In the *softening-point test*, the temperature of a sample of bitumen in the form of a disc is raised at 5°C per minute while being subjected to loading by a small steel ball. As the temperature rises, the bitumen softens, and the particular temperature at which the bitumen is deformed by a distance of 2.54 cm is recorded as the softening-point in °C (British Standards Institution, 1983a). The *viscosity* of bitumens is measured in a number of different ways, e.g., vacuum capillary viscometers, if used for definition by grade. For products of relatively low viscosity, however, simple orifice-type viscometers are generally used (British Standards Institution, 1983b).

The most important types of bitumens are described below. For the purposes of this monograph they have been categorized into eight classes, which represent the major types used in industry.

Class 1: *Penetration bitumens* are classified by their penetration value. They are usually produced from the residue from atmospheric distillation of petroleum crude oil by applying further distillation under vacuum, partial oxidation (air rectification), solvent precipitation, or a combination of these processes (see section 2.1(*a*)). In Australia and the USA, bitumens that are approximately equivalent to those described here are called *asphalt cements* or *viscosity-graded asphalts*, and are specified on the basis of viscosity measurements at 60°C.

Class 2: *Oxidized bitumens* are classified by their softening-points and penetration values. They are produced by passing air through hot, soft bitumen under controlled temperature conditions. This process alters the characteristics of the bitumen to give reduced temperature susceptibility and greater resistance to different types of imposed stress.

In the USA, bitumens produced using air blowing are known as *air-blown asphalts* or *roofing asphalts* and are similar to oxidized bitumens.

Class 3: *Cutback bitumens* are produced by mixing penetration bitumens or oxidized bitumens with suitable volatile diluents from petroleum crudes such as white spirit, kerosene or gas oil, to reduce their viscosity and render them more fluid for ease of handling. When the diluent evaporates, the initial properties of bitumen are recovered. In the USA, cutback bitumens are sometimes referred to as *road oils*.

Class 4: *Hard bitumens* are normally classified by their softening-point. They are manufactured similarly to penetration bitumens, but have lower penetration values and higher softening-points, i.e., they are more brittle.

Class 5: *Bitumen emulsions* are fine dispersions of droplets of bitumen (from classes 1, 3 or 6) in water. They are manufactured using high-speed shearing devices, such as colloid mills. The bitumen content can range from 30-70% by weight. They can be anionic, cationic or non-ionic. In the USA, they are referred to as *emulsified asphalts*.

Class 6: *Blended or fluxed bitumens* may be produced by blending bitumens (primarily penetration bitumens) with solvent extracts (aromatic by-products from the refining of base oils), thermally cracked residues or certain heavy petroleum distillates with final boiling-points above 350°C. Coal-tar products are also sometimes used as fluxes (see the monograph on coal-tars and derived products in this volume). There is only limited evaporation of the flux.

Class 7: *Modified bitumens* contain appreciable quantities (typically 3-15% by weight) of special additives, such as polymers, elastomers, sulphur and other products used to modify their properties; they are used for specialized applications.

Class 8: *Thermal bitumens* were produced by extended distillation, at high temperature, of a petroleum residue. Some cracking occurred during this process. Currently, they are not manufactured in Europe or in the USA.

1.3 Chemical composition and physical properties of bitumens

The chemical composition of bitumens depends both on the original crude oil and on the processes used during refining. Bitumens can generally be described as complex mixtures containing a large number of different chemical compounds of relatively high molecular weight: typically, 82-85% combined carbon, 12-15% hydrogen, 2-8% sulphur, 0-3% nitrogen and 0-2% oxygen. Bitumens contain predominantly cyclic hydrocarbons (aromatic and/or naphthenic) and a lesser quantity of saturated components, which, because of slow and lengthy processing at moderate temperatures, are mainly of very low chemical reactivity.

Broad chemical composition

The molecules present in bitumens are combinations of well-established structural petroleum units: alkanes, cycloalkanes, aromatics and heteromolecules containing sulphur, oxygen, nitrogen (Broome, 1973) and heavy metals.

The chemical characterization of bitumens is based on their separation into four broad classes of compounds - asphaltenes, resins, cyclics and saturates - using solvent precipitation and adsorption chromatography (Fig. 1).

Fig. 1. Chemical separation of bitumens[a]

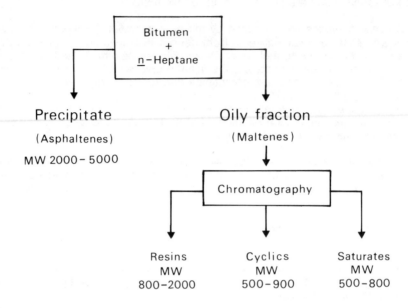

[a]Adapted from Bright et al. (1982)
[b]MW, molecular weight, number average

The most successful separation, in terms of understanding physical properties, is one in which asphaltenes are separated as insolubles in normal heptane. The components soluble in normal heptane (maltenes) are adsorbed on activated alumina/silica gel in a chromatographic column. Elution with solvents of progressively higher polarity desorbs and separates three fractions, which are designated as saturates, cyclics (naphthene-aromatics) and resins (polar aromatics).

Asphaltenes are black amorphous solids containing, in addition to carbon and hydrogen, some nitrogen, sulphur and oxygen. Trace elements such as nickel and vanadium are also present. Asphaltenes are generally considered as highly polar aromatic materials of molecular weights of 2000-5000 (number average), and constitute 5-25% of the weight of bitumens.

Saturates comprise predominantly the straight- and branched-chain aliphatic hydrocarbons present in bitumens, together with alkyl naphthenes and some alkyl aromatics. The average molecular weight range is approximately similar to that of the cyclics, and the components include both waxy and non-waxy saturates. This fraction forms 5-20% of the weight of bitumens.

Cyclics (naphthene aromatics) comprise the compounds of lowest molecular weight in bitumens and represent the major proportion of the dispersion medium for the peptized asphaltenes. They constitute 45-60% by weight of the total bitumen and are dark viscous liquids. They are compounds with aromatic and naphthenic aromatic nuclei with side chain constituents and have molecular weights of 500-900 (number average).

Resins (polar aromatics) are dark-coloured, solid or semi-solid, very adhesive fractions of relatively high molecular weight present in the maltenes. They are dispersing agents or peptizers for the asphaltenes, and the proportion of resins to asphaltenes governs to a degree the sol- or gel-type character of bitumens. Resins separated from bitumens are found to have molecular weights of 800-2000 (number average) but there is a wide molecular distribution. This component constitutes 15-25% of the weight of bitumens.

With penetration bitumens, the asphaltene content increases as penetration decreases; however, oxidized bitumens have higher asphaltene contents than those of penetration grades. During air-blowing, cyclics are converted to resins, which are in turn converted to asphaltenes, whereas vacuum distillation selectively decreases the saturate content, leading to an increased concentration of the other components.

Bitumens (penetration, oxidized and hard types) are subject to hardening from oxidation and polymerization reactions. Although minor loss of volatile components contributes to hardening, the formation of additional asphaltenes by oxidation is reported to be the main cause (Evans, 1978).

Bitumens can be regarded as colloidal systems (Witherspoon, 1962) consisting of asphaltene micelles dispersed in an oily medium of lower molecular weight (maltenes). The micelles are considered to be asphaltenes with an adsorbed sheath of aromatic resins of high molecular weight as a stabilizing solvating layer. Away from the centre of the micelle there is a gradual transition to less aromatic resins, and such layers extend outwards into the less aromatic, oily dispersion medium.

Attention has already been drawn to the presence of minor amounts of sulphur, oxygen and nitrogen in the composition of typical bitumens. The sulphur content may be 2-8% by weight in common products and can consist of many different sulphur compounds. Studies of rates of oxidation reactions (Erdman & Ramsey, 1961) have shown that the heteroatoms, sulphur, oxygen and nitrogen, occur largely in stable configurations, probably rings. X-ray diffraction (Yen *et al.*, 1961) points to the presence of naphthenic groups as well as aromatic and paraffinic groups. Other work led to the conclusion that oxygen and nitrogen as well as vanadium (80-4300 µg/g) and/or nickel (<100 µg/g) (Puzinauskas & Corbett, 1975) are combined in some form in asphaltene structure; it has been proposed that the metals are chemically associated with the 'asphaltic material' (asphaltenes), the sites being holes in the aromatic clusters, which are edged with the sulphur, oxygen and nitrogen atoms (Yen *et al.*, 1962).

Inorganic salts, such as sodium chloride, may also be present in small amounts in bitumens (the ash content is less than 0.5% by weight). There is substantially no contamination, however, with bitumens, since specifications demand greater than 99.0% solubility in solvents such as trichloroethylene.

Polynuclear aromatic hydrocarbons

Polynuclear aromatic hydrocarbons (PAHs) exist in crude oils (Bingham *et al.*, 1979) but are generally present in more limited amounts in bitumens (Lawther, 1971; Brandt & De Groot, 1985). This is because the principal refinery process used for the manufacture of bitumens, namely vacuum distillation, removes the majority of compounds of lower molecular weight with lower boiling-points, including PAHs with 3-7 fused rings, and because the maximum temperatures involved in the production of vacuum residue range from 350-450°C and are not high enough to initiate significant PAH formation. The levels of PAHs analysed in various bitumens are shown in Tables 1 and 2. Other types of bitumen (e.g., thermal bitumens [class 8]) may contain higher levels of PAHs (up to 272 µg/kg), which are formed during cracking operations (Yanysheva et al., 1963). PAHs may also be re-introduced by the flux used in blended or fluxed bitumens [class 6]. Data on PAH content are available only on bitumens in classes 1, 2 and 8.

Table 1. Content of polynuclear aromatic hydrocarbons (PAHS) (mg/kg) in eight different bitumen samples [class 1][a]

PAH	Formula	Bitumen[b]							
		A	B	C	D	E	F	G	H
Anthracene	$C_{14}H_{10}$	ND	ND	ND	ND	ND	ND	ND	ND
Phenanthrene	$C_{14}H_{10}$	2.3	0.4	3.5	1.3	0.6	35*	1.1	2.3*
Pyrene	$C_{16}H_{10}$	0.6	1.8	4.0	8.3	0.9	38	0.3	0.08
Fluoranthene	$C_{16}H_{10}$	+	+	2.0	+	+	5	ND	ND
Benzofluorenes	$C_{17}H_{12}$	+	+	+	+	+	+	+	ND
Benz[a]anthracene	$C_{18}H_{12}$	0.15	2.1	1.1	0.7	0.9	35	0.2	ND
Triphenylene	$C_{18}H_{12}$	0.25	6.1	3.1	3.4	3.8	7.6	1.0	0.3
Chrysene	$C_{18}H_{12}$	0.2	8.9	2.3	3.9	3.2	34	0.7	0.04
Benzo[a]pyrene	$C_{20}H_{12}$	0.5	1.7	1.3	2.5	1.6	27	0.1	ND
Benzo[e]pyrene	$C_{20}H_{12}$	3.8	13	2.9	3.2	6.5	52	1.6	0.03
Benzo[k]fluoranthene	$C_{20}H_{12}$	+	ND	+	+	+	ND	ND	ND
Perylene	$C_{20}H_{12}$	ND	39	2.2	6.1	2.9	3.0	0.1	ND
Anthanthrene	$C_{22}H_{12}$	ND	Tr	Tr	Tr	+	1.8	ND	ND
Benzo[ghi]perylene	$C_{22}H_{12}$	2.1	4.6	1.0	1.7	2.7	15	0.6	Tr
Indeno[1,2,3-cd]pyrene	$C_{22}H_{12}$	Tr	ND	Tr	Tr	Tr	1.0	ND	ND
Picene	$C_{22}H_{14}$	+	+	+	+	+	1.0	+	ND
Coronene	$C_{24}H_{12}$	1.9	0.8	0.5	0.2	0.9	2.8	0.9	ND

[a]Modified from Wallcave et al. (1971)
[b]ND, not detected; *, estimate includes alkyl derivatives; +, not estimated but present in small amount; Tr, trace

Table 2. Mean analytical results (µg/g) for 14 individual polynuclear aromatic hydrocarbons (PAHs) in some penetration-grade [class 1] and oxidized-grade [class 2] bitumens[a]

PAH analysed	Penetration grades[b]				Oxidized grades[b]		
	80/100	80/100	50/60	80/100	85/40	110/30	95/25
Phenanthrene	7.3	5.0	1.7	5.0	0.32	1.7	2.4
Anthracene	0.32	0.27	0.015	0.17	0.01	0.03	0.07
Fluoranthene	0.72	0.46	0.41	0.39	0.15	0.4	0.46
Pyrene	1.5	1.0	0.26	1.1	0.17	0.3	0.29
Chrysene	1.5	3.3	0.47	3.9	0.90	1.0	0.80
Benz[a]anthracene	1.1	0.89	0.14	0.63	0.33	0.3	0.23
Perylene	3.3	0.69	0.044	0.25	0.14	0.08	0.20
Benzo[k]fluoranthene	0.19	ND	0.024	ND	0.051	0.10	0.04
Benzo[a]pyrene	1.8	0.92	0.22	1.1	0.49	0.35	0.48
Benzo[ghi]perylene	4.2	2.3	1.67	2.7	1.3	1.2	2.0
Anthanthrene	0.11	0.04	0.006	0.02	0.01	ND	0.03
Dibenzo[a,l]pyrene	ND	ND	ND	ND	ND	ND	ND
Dibenzo[a,i]pyrene	0.50	ND	0.05	0.60	ND	0.3	0.10
Coronene	ND	ND	0.40	ND	ND	ND	ND

[a]Data from Brandt & De Groot (1985). These bitumens were obtained from a range of crude oils originating from the Middle East, Venezuela and Mexico.
[b]ND, not detected

Yanysheva et al. (1963) determined the benzo[a]pyrene content of three samples of a bitumen (the straight distillation residue of Ukranian crude petroleum) to be 0-0.6 µg/g. The concentrations of benzo[a]pyrene and/or benzo[e]pyrene in a roofing bitumen [class 2] were <6 µg/g (Hervin & Emmett, 1976).

1.4 Technical products and specifications

Bitumens are available in a variety of grades, as described in Section 1.2. Specifications for bitumen products are tailored to the needs of the consuming industries and are based on a series of physical tests reflecting visco-elasticity and durability. Required physical properties are related to performance rather than to chemical composition.

Grades of penetration bitumens [class 1] are defined by the upper and lower limits of penetration values. A nominal 200-PEN grade has a range of 170-230 in the British Standards Specification (British Standards Institution, 1982a,b). In France, the 180-220 grade has limits of 180-220; while in the Federal Republic of Germany, the B 200 grade may vary from 160-210 in penetration value. Penetration grades in common use vary from 15-450 PEN. Specific ranges of softening-point are required for particular penetration bitumens to ensure that the penetration index (PI), a measure of change in penetration with temperature (Pfeiffer & van Doormaal, 1936), given in a typical specification varies, for example, only from -1.0 to +1.0.

Oxidized bitumens [class 2] are defined by the mean values of the ranges allowable for penetration and softening-point. For instance, a common grade such as 85/25 has a mean value of 85 for the permissible softening-point range of 80-90°C and a mean value for the penetration range of 20-30. Common grades in this class of bitumens are 85/25, 85/40, 100/40, 105/35, 105/13 and 115/15. Oxidized bitumens are somewhat rubbery in nature and exhibit a low temperature dependence, having PIs of from +2.0 to +8.0.

Grades of cutback bitumens [class 3] are designated by a value in seconds required for a given quantity of the product to flow through a standard orifice at a fixed temperature. Typical products used in road applications are made by cutting back 100-PEN bitumens with 8-14% kerosene. They are designated by the midpoints of the viscosity limits adopted. European specifications for typical products are given by the British Standards Institution (1982a), and those for the USA by the Asphalt Institute (1983). Cutback bitumens are viscous to highly fluid materials at ambient temperatures.

Hard bitumens [class 4] have penetration values of less than 15 and are generally designated by the prefix H (HVB in the Federal Republic of Germany) and by the softening-point range, e.g., H80/90. While the nomenclature is derived from the softening-point range, each grade also has a defined penetration range, giving these materials a PI of 0 to +2.0. They are hard and brittle in nature. They are not used in the USA. Examples of specifications for oxidized and hard bitumens are given by the British Standards Institution (1982b).

Example of typical specifications for penetration [class 1], oxidized [class 2], cutback [class 3] and hard [class 4] bitumens are given in Table 3.

Table 3. Typical specifications for bitumens[a]

A. Bitumens for road purposes

Penetration bitumens [class 1]

Property		Test method	Grade									
			15 PEN	25 PEN	35 PEN	40 PEN HD	50 PEN	70 PEN	100 PEN	200 PEN	300 PEN	450 PEN
Penetration at 25°C		BS 4691	15 ± 5	25 ± 5	35 ± 7	40 ± 10	50 ± 10	70 ± 10	100 ± 10	200 ± 20	300 ± 30	450 ± 65
Softening-point °C	(min)	BS 4692	63	57	52	58	47	44	41	33	30	25
	(max)		76	69	64	68	58	54	51	42	39	34
Loss on heating for 5 h at 163°C		BS 2000: Part 45										
Loss by mass %	(max)		0.1	0.2	0.2	0.2	0.2	0.2	0.5	0.5	1.0	1.0
Drop in penetration %	(max)		20	20	20	20	20	20	20	20	25	25
Solubility in trichloroethylene % by mass	(min)	BS 4690	99.5	99.5	99.5	99.5	99.5	99.5	99.5	99.5	99.5	99.5

Table 3 (contd)

		50 secs	100 secs	200 secs
Cutback bitumens [class 3]				
Viscosity (STV[b]) at 40°C, 100-mm cup	BS 2000: Part 72	50 ± 10	100 ± 20	200 ± 40
Distillation	BS 2000: Part 27			
to 225°C (% by volume max)		1	1	1
to 360°C (% by volume)		8 to 14	6 to 12	4 to 10
Penetration at 25°C of residue from distillation to 360°C	BS 4691	100 to 350	100 to 350	100 to 350
Solubility in trichloroethylene (% by mass (min))	BS 4690	99.5	99.5	99.5

B. Bitumens for industrial uses

			75/30	85/25	85/40	95/25	105/35	115/15
Oxidized-grade bitumens [class 2]								
Softening-point, °C	(min)	BS 4692	70	80	80	90	100	110
	(max)		80	90	90	100	110	120
Penetration at 25°C		BS 4691	30 ± 5	25 ± 5	40 ± 5	25 ± 5	35 ± 5	15 ± 5
Loss on heating for 5 h at 163°C, by mass %	(max)	BS 2000: Part 45	0.2	0.2	0.5	0.2	0.5	0.2
Solubility in trichloroethylene, % by mass	(min)	BS 4690	99.5	99.5	99.5	99.5	99.5	99.5

			Grade	
			H 80/90	H 100/120
Hard-grade bitumens [class 4]				
Softening-point, °C	(min)	BS 4692	80	100
	(max)		90	120
Penetration at 25°C	(min)	BS 4691	6	2
	(max)		12	10
Loss on heating for 5 h at 163°C, % by mass	(max)	BS 2000: Part 45	0.05	0.05
Solubility in trichloroethylene, % by mass	(min)	BS 4690	99.5	99.5

[a]From British Standards Institution (1982a,b)
[b]STV, standard tar viscometer

2. Production, Use, Occurrence and Analysis

2.1 Production and use

(a) *Production*

Bitumen manufacture

Today bitumens are derived from crude oils, using manufacturing processes that avoid thermal degradation. They are produced from crude oils that give substantial amounts of heavy residue, typically from 10-50%, although crude oils giving a greater yield of residue are sometimes used.

The processes used in bitumen production are summarized below and illustrated in Figure 2. The processes are denoted in the text by letters shown in the figure. A fuller account of these processes is given in a recent article (Chipperfield, 1984).

Distillation (D): The first stage in oil refining is atmospheric distillation. The atmospheric residue of very heavy crudes is sometimes used for bitumen production and is generally distilled further to yield various products. The use of vacuum distillation prevents thermal degradation of distillates and residue. Steam is sometimes injected into the residue to aid distillation in a process known as steam stripping, and bitumens produced in this way are referred to as steam-refined bitumens.

Vacuum residues from particular crude oils meet specification requirements for penetration bitumens [class 1] or, in the USA, 'viscosity-graded asphalts'. Bitumens so produced are sometimes referred to as 'straight-reduced bitumens' or 'straight-run bitumens'.

Air blowing (B): With other vacuum residues and also to yield harder penetration bitumens and to obtain desirable properties for paving or industrial uses, air blowing is used. The blowing process dehydrogenates the residue, resulting in oxidation and condensation polymerization. The content of asphaltenes is considerably increased, while the content of cyclics is decreased.

Limited air blowing is required in some cases to produce penetration bitumens [class 1] or 'viscosity-graded asphalts' from vacuum residues.

Oxidized bitumens [class 2] are produced by extended air blowing of vacuum residues, propane-precipitated asphalt (PPA) (described below) or mixtures of vacuum residues and atmospheric residues or waxy distillates. Catalysts such as ferric chloride (0-1%) and phosphorus pentoxide (0-4%) are used in a few refineries to speed the reaction or to modify the properties of the resultant bitumens ('catalytic air-blown bitumens' [class 2]).

Solvent precipitation (P): Some crudes contain components of high boiling-point which are difficult to recover even when high vacuum is used. Such materials are therefore separated from the vacuum residue using solvent precipitation (usually with propane). The product precipitated is widely called 'propane-precipitated asphalt' (PPA) [class 1], although in a strict sense this is a bitumen as defined in this monograph. In the USA, PPA is also referred to as 'solvent-refined asphalt' or 'propane asphalt'. Solvent-precipitated asphalts have a higher content of asphaltenes than the vacuum residues from which they are produced, but a lower content of saturates than would be obtained by distillation of the vacuum residue.

Fig. 2. Main processing methods in the production of bitumens

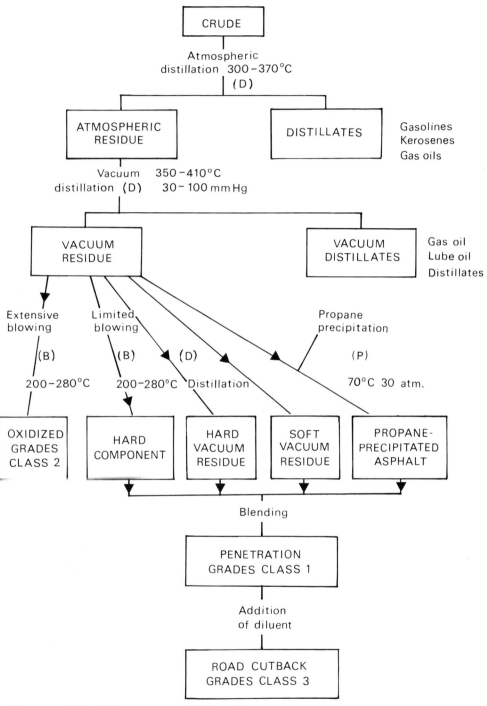

VR, vacuum residue; PPA, propane-precipitated asphalt

Typical refinery production: In Figure 2 it can be seen that these processes can be combined to produce penetration and oxidized bitumens, using blending where needed. Flux oils may be used in limited amounts to soften bitumens.

To produce hard bitumens [class 4], vacuum distillation may be employed, in some cases combined with limited air blowing. The starting material is vacuum residue [class 1] and/or propane-precipitated asphalts [class 1]. Some very hard bitumens are produced by thermal cracking followed by deep vacuum distillation for special applications (e.g., briquetting binders).

Cutback bitumens [class 3] are made by blending bitumens [classes 1 and 2] with controlled amounts of petroleum distillates (white spirit, kerosene or gas oil). This considerably reduces the viscosity of a bitumen, making the product easier to handle. Afterwards, the diluent evaporates and most of the initial properties of the bitumen are recovered. In some cases, coal-tar distillates are used [class 6].

Bitumen emulsions [class 5] are made mainly from penetration bitumens [class 1]. They may also be made using cutback bitumens [class 3] or fluxed bitumens [class 6]. They are manufactured using high-speed shearing devices, such as colloid mills. The bitumen content can range from 30-70% by weight. They can be anionic, cationic or non-ionic.

The major grades of bitumen are supplied hot in bulk. They can also be supplied in steel drums, paper sacks or fibre drums, depending on the type and grade.

Levels of production

The widespread availability of bitumens resulting from oil refining is a comparatively modern development. Bitumens have been produced in the USA by vacuum distillation of crude petroleum since 1902 (when 18 000 tonnes were produced). By 1907, the quantity made from this source equalled the amount recovered from natural bitumen sources (Evans, 1978). By 1938, annual consumption had grown to 5 million tonnes (Chipperfield, 1984). US production reached its highest level in 1978 (29 million tonnes) but has dropped steadily in recent years (20 million tonnes in 1982).

In a number of European countries, substantial quantities of bitumen were being produced by the 1920s. It is now used largely where it is manufactured.

(b) Use

It is estimated that the current annual world use of bitumens is over 60 million tonnes. The major applications are in paving for roads and airfields, hydraulic uses (such as dams, water reservoirs and sea-defence works), roofing, flooring and protection of metals against corrosion. More than 80% of bitumens is used in the many different forms of road construction and maintenance.

Recommended temperatures for handling bitumens range from 90-220°C for penetration bitumens [class 1], 150-230°C for oxidized bitumens [class 2], 65-180°C for cutback bitumens [class 3] and 160-230°C for hard bitumens [class 4] (Institute of Petroleum, 1979).

The estimated quantity of bitumens used in some countries between 1960 and 1982 is given in Table 4.

BITUMENS

Table 4. Annual use of bitumens in some countries (million tonnes)

Country	1960	1976	1980	1982
Austria	0.1	0.6	0.6	0.5
Benelux (Belgium and The Netherlands)	0.3	1.3	1.1	0.8
Canada	1.5	2.8	3.4	
Finland	-	-	-	0.4
France	1.2	3.1	2.8	2.4
Germany, Federal Republic of	1.4	3.9	3.4	3.0
Italy	0.6	-	1.9	1.9
Japan	-	-	4.7	4.4
New Zealand	0.1	0.1	0.1	0.1
Scandinavia (Denmark, Norway and Sweden)	0.4	1.3	1.0	1.3
South Africa	0.1	0.3	0.3	0.3
UK	1.1	1.9	1.8	2.0
USA	18.9	25.5	27.3	23.2

The principal uses of bitumens are shown in Table 5, and brief descriptions of some of the most important applications are given below. More detailed information is given by Chipperfield (1984).

Table 5. Principal uses of bitumens

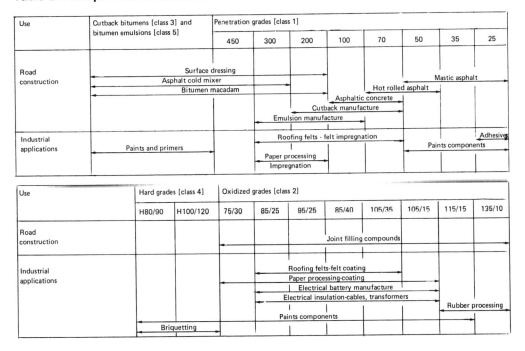

(i) *Road applications*

Various techniques are employed in using bitumens for the construction and maintenance of roads, either as a component in mixes with mineral aggregates, or as thin films to seal surfaces.

Asphalt mixes: Asphalt mixes are manufactured by heating and drying mixtures of graded crushed stone, sand and filler (the mineral aggregate) and mixing with penetration bitumen (typically 3-8% by weight) [class 1]. At the construction site, the asphalt mix is fed through a mechanical laying machine, which spreads and compacts the mix.

Asphalt mixes include asphaltic concrete, bitumen macadams and hot rolled asphalts.

Special techniques can be adopted to mix aggregate or sands with cutback bitumens [class 3] or emulsions [class 5]. These may be carried out with only minor heating or at ambient temperature and are therefore referred to as 'asphalt cold mixes'.

Surface dressing and surface treatments: These processes can be used to seal minor roads or to maintain road surfaces that have suffered abrasion and loss of skid resistance. Penetration bitumens [class 1], cutbacks [class 3] or emulsions [class 5] are sprayed onto the surface being treated to give a uniform film to which chippings are applied, followed by light rolling.

(ii) *Industrial applications*

A full range of industrial applications has been described by the Asphalt Institute (1982).

Waterproofing and roofing: By far, the largest quantity of bitumen used industrially is in this sector, especially in roofing. Roofing felt manufacture is based on the use of a hot penetration bitumen (typically 200 PEN [class 1]) to impregnate, during immersion, a dry felt made from waste paper or rags. A surface coating is then applied to both sides of the saturated felt using an oxidized bitumen [class 2] (e.g., 85/40 or 105/35), which sometimes contains added filler. A recent trend has been to use modified bitumens ([class 7], e.g., penetration bitumens with additions of styrene-butadiene rubbers) in place of oxidized bitumens. Many roofing felts are now made of glass or polyester, for which no saturant is required.

Impregnated felts are also used for damp-proof courses in masonry. Mastic asphalts, which are mixtures of finely divided limestone or other fine aggregate with a high content (12-17% by weight), of a hard penetration bitumen ([class 1] 15-30 PEN), are used in flooring and roofing.

Electrical insulation: The electrical properties of bitumens [primarily class 2] enable them to be used in wrappings and jointing compounds for heavy duty cables.

Sound insulation: Bitumens [classes 2 and 7] find wide use for sound insulation purposes, e.g., in car bodies and floor mats and in floor mountings for factory machinery.

Pipe coatings: To protect pipelines for oil, water, etc., coatings of bitumen enamel are applied, after a primer, onto the cleaned metal surface. The enamel is an oxidized bitumen ([class 2] e.g., 115/15) with addition of up to 30% of an inert filler, such as slate dust; the primer is a cutback [class 3] of the oxidized bitumen with a volatile solvent (white spirit).

Briquettes: Considerable quantities of bitumens are used as binders to form coal fines into briquettes. The bitumens are used hot, or cold (ground) or as emulsions. Bitumens are also used as pelletizing binders in the recovery of 'mill scale' (oxidized film) and 'red mud' (waste iron ore) steel wastes, to give a cooling charge additive for steel converter feeds. The most suitable bitumens are hard and of low penetration index, e.g., 85/2 or 90/1 (softening-point/penetration at 25°C). Other grades, such as 15 PEN or H80/90, may also be satisfactory.

(iii) *Use pattern*

Recent bitumen use patterns of various types of bitumens in western Europe, Japan and the USA are shown in Table 6 (see Table 4 for the estimated quantity of bitumens used in some countries).

Table 6. Bitumen use pattern (%) by grade[a]

Grade	Western Europe	Japan	USA
Penetration bitumens [class 1]	74	86	70
Oxidized bitumens [class 2]	12	6	13
Cutback bitumens [class 3]	8	} 7	10
Bitumen emulsions [class 5]	4		7
Hard bitumens / Blended grades / Modified bitumens [classes 4, 6 and 7]	} 2	} 1	0

[a]Based in part on Evans (1978) and Bright *et al.* (1982)

The data indicate that about 80% of bitumens and bitumen derivatives used in western Europe are employed in road applications, the remainder being used in many different industrial applications. Although there is considerable variation in choice of grades for road applications, the average breakdown has been estimated to be 85% or more for penetration grade bitumens (straight asphalts) [class 1], 10% for cutbacks [class 3] and 5% for emulsions [class 5]. For industrial uses, about 60% are oxidized grades [class 2] and about 30% are penetration grades [class 1]; only minor percentages of hard grades [class 4] and bitumen derivatives have been reported to be used (Bright *et al.*, 1982).

(c) *Legislation*

Several countries have occupational exposure limits for bitumen fumes. Some standards are listed in Table 7.

Table 7. National occupational exposure limits for bitumen fumes[a]

Country	Year	Concentration (mg/m^3)	Interpretation[b]	Status
Australia	1978	5	TWA	Guideline
Belgium	1978	5	TWA	Regulation
Italy	1978	5	TWA	Guideline
The Netherlands	1978	5	TWA	Guideline
Switzerland	1978	5	TWA	Regulation
USA				
ACGIH	1983	5	TWA	Guideline
		10	STEL	
NIOSH		5	Ceiling (15 min)	Guideline

[a]From American Conference of Governmental Industrial Hygienists (ACGIH) (1983); International Labour Office (1980); National Institute for Occupational Safety and Health (NIOSH) (1977)

[b]TWA, time-weighted average; STEL, short-term exposure limit

With effect from 22 February 1983, the US Occupational Safety and Health Administration (OSHA) modified its interpretation of the term 'coal-tar pitch volatiles', as used in a regulation established in 1971, which designated the permissible exposure limit to such coal-tar pitch volatiles as an eight-hour time-weighted average (TWA) not to exceed 0.2 mg/m^3. To clarify the scope of the coverage of this 1971 regulation, the 1983 modification specifically stated that bitumens and oxidized bitumens are not covered under the standard. It also indicated that OSHA was pursuing appropriate rule-making activities with regard to bitumen fumes (US Occupational Safety and Health Administration, 1983). In the Federal Republic of Germany, coal-tars, coal-tar pitches and primary coal-tar distillate fractions blended with bitumens are subject to regulations to limit occupational exposure (Federal Office for Workers' Safety, 1982) (see monograph on coal-tars and derived products).

Vacuum-distilled and steam-stripped petroleum bitumens [class 1] that meet certain specifications relating to softening-point, penetration and weight loss on heating have been approved by the US Food and Drug Administration (1980) as components of paper and paperboard in contact with aqueous and fatty foods. Such use is limited to 5% maximum of the finished dry paper and paperboard fibres in internal sizing of paper and paperboard for use in contact with raw fruits and vegetables and certain dry foods.

2.2 Occurrence

(a) Natural occurrence

Bitumens occur in natural asphalts, rock asphalts and lake asphalts.

(b) Occupational exposure

On the basis of the 1974 National Occupational Hazard Survey in the USA, the National Institute for Occupational Safety and Health (1980a, 1981) estimated that 2 000 000 and 33 000 US workers were exposed to bitumens and bitumen fumes, respectively. These estimates are based on extrapolation of the results of a survey of 174 exposures to bitumens and five exposures to bitumen fumes in selected plants within selected non-agricultural industries. The largest numbers of exposures to bitumens were observed in the following industries: highway and street construction, roofing and sheet-metal work and blast furnaces and steel mills. The largest numbers of exposures to bitumen fumes were observed in roofing and sheet-metal work.

(i) Refinery operations

Exposures measured during normal refinery operations are given in Table 8.

The techniques of distillation, blowing and blending are all conducted in closed systems. The presence of hydrogen sulphide in the vapour space in hot bitumen storage tanks and tankers used for distribution may represent a toxic hazard. Bulk loading into road tankers is conducted using special loading arms with covers for manholes to reduce emissions.

Workers operating bitumen plants may be general refinery operators or workers with duties in the vicinity of bitumen-processing units, bitumen-storage tanks, loading gantries and packaging systems.

Table 8. Exposures measured during refinery operations

Source of exposure	Job category or description	Class of bitumen handled[a]	Material and sample type[b]	Results	Reference
Vacuum distillation	General refinery workers outdoors	1	*Total PACs (23 individual or groups of PAHs and azo heterocyclics)*: area sample (4 measurements)	Mean, 9.5 µg/m^3	National Institute for Occupational Safety and Health (1980b)
Vacuum distillation (four refineries)	General refinery workers	1, 2, 3	*Total PACs (23 individual or groups of PAHs and azo heterocyclics)*: area samples (11 measurements), TWA (7-8 h)	1.8-19.0 µg/m^3	Futagaki (1981)
Deasphalting (two refineries)	General refinery workers	1	*Total PACs (23 individual or groups of PAHs and azo heterocyclics)*: personal samples (5 measurements), TWA (7-8 h) area samples (4 measurements), TWA (7-8 h)	2.5-49.8 µg/m^3 1.4-41.2 µg/m^3	Futagaki (1981)
Air blowing (three refineries)	General refinery workers	1, 2	*Total PACS (23 individual or groups of PAHs and azo heterocyclics)*: area samples (5 measurements), TWA (7-8 h)	1.6-30.6 µg/m^3	Futagaki (1981)
Refinery bitumen installation	Bitumen loading outdoors (170-210°C)	Probably includes 1, 2, 4	*Total particulates*: TWA (8 h) *Benzene solubles*: personal samples (4 measurements), TWA (8 h) *Total PACs (11 PAHs)*: TWA (8 h)	0.7-2.9 mg/m^3 <0.1-1.0 mg/m^3 3.8-95 ng/m^3	Brandt & Molyneux (1985)
Refinery bitumen unit	Bitumen loading outdoors (170-210°C)	1, 2, 3, 4	*Total particulates*: personal samples (2 measurements), TWA (8 h)	0.1-0.2 mg/m^3	Claydon et al. (1984)
Hot-mix plant (not on refinery site, but conducted for comparison)	Bitumen loading into open truck outdoors	1	*Total particulates*: area samples	1.4 mg/m^3	Claydon et al. (1984)
Indoor packaging area	Filling packages, bitumens at 220°C	3	*Total particulates*: personal samples (2 measurements), TWA (8 h)	0.20-0.32 mg/m^3	Brandt & Molyneux (1985)

[a] When direct information was not provided, the class of bitumen was derived from information describing the activity and the temperature of handling.
[b] PAC, polynuclear aromatic compound; PAH, polynuclear aromatic hydrocarbon; TWA, time-weighted average

Table 9. Exposures measured during road-paving operations

Source of exposure	Job category or description	Class of bitumen handled[a]	Material and sample type[b]	Results	Reference
Preparation of hot asphalt paving mix	Not specified	1	*Total particulates*	0.2-5.40 mg/m^3	Puzinauskas & Corbett (1975)
Asphalt plants (2)	Man in control cabin and operator on hot-mix storage bins	1	*Total PACs*: area samples (8 measurements)	Mean, 0.13 µg/m^3	Byrd & Mikkelsen (1979)
Road paving operations, asphalt mix, 130-170°C	Finishing machine operator	1	*Total particulates*: personal samples, TWA (8 h)	0-1.7 mg/m^3 (mean, 0.7 mg/m^3)	Puzinauskas (1980)
	Rakerman		*Total particulates*: personal samples (14 measurements), TWA (1-4 h)	0.15-5.6 mg/m^3 (mean, 1.26 mg/m^3)	
	Screedman		*Total particulates*: personal samples (7 measurements), TWA (1-4 h)	0.25-3.46 mg/m^3 (mean, 0.93 mg/m^3)	
			Total particulates: personal samples (7 measurements), TWA (1-3 h)	0.33-1.47 mg/m^3 (mean, 0.83 mg/m^3)	
Road paving, asphalt mix	Several categories (screedman, rakerman, foreman, roller driver)	1	*Total particulates*: personal samples (215 measurements), TWA (8 h)	Mean, 0.3-0.7 mg/m^3	Byrd & Mikkelsen (1979)
	Paver operator		*Total particulates*: personal samples (72 measurements), TWA (8 h)	Mean, 1.1 mg/m^3; max., 6.4 mg/m^3	
Road paving, asphalt mix	Machine driver, level wheel operator	1	*Total PACs (11 PAHs)*: personal samples (4 measurements)	4.32-12.99 µg/m^3	Malaiyandi *et al.* (1982)
Hot bitumen cutback road surface, 135°C	Surface dressing	3	*Total particulates*: personal samples (9 measurements), TWA (8 h)	Mean, 0.8 mg/m^3; max., 1.3 mg/m^3	Byrd & Mikkelsen (1979)
Hot bitumen cutback, 120-158°C	Surface dressing and macadam paving; chipper driver and thrower, paver driver, raker and barrowman	3	*Total particulates*: personal samples (14 measurements), TWA (8 h)	0.2-15.1 mg/m^3	Brandt & Molyneux (1985)
			Benzene solubles: personal samples (13 measurements), TWA (8 h)	0.1-0.5 mg/m^3	

BITUMENS

Operation	No.	Measurement	Value	Reference
Mastic asphalt road paving	1	*Total PACs*: personal samples (8 measurements)	0.122-0.185 mg/m^3	Gross *et al.* (1979)
Asphalt mixing plants	1	*Benzene solubles*	0.2-5.4 mg/m^3	National Institute for Occupational Safety and Health (1977)
		Benzo[a]pyrene	0.011-0.19 µg/m^3	
Road paving operators	1	*Total particulates*: personal samples (17 measurements), TWA (5-8 h)	<0.1-1.2 mg/m^3	Virtamo *et al.* (1979)
Asphalt cement paving, 145-195°C		*Cyclohexane solubles*: personal sample, TWA (6 h)	0.3 mg/m^3	
		Benzo[a]pyrene: area sample	<0.05 µg/m^3	
Coarse asphalt base	1	*Total particulates*: personal samples (10 measurements), TWA (5-6 h)	0.4-1.1 mg/m^3	
		Cyclohexane solubles: personal sample, TWA (5 h)	0.6 mg/m^3	
		Benzo[a]pyrene: area sample	<0.05 µg/m^3	
Mastic asphalt	1	*Total particulates*: personal samples (16 measurements), TWA (2-4 h)	0.2-4.2 mg/m^3	
		Cyclohexane solubles: personal samples (11 measurements), TWA (2-4 h)	<0.1-2.7 mg/m^3	
		Benzo[a]pyrene: area sample	<0.05 µg/m^3	
Heating method	1	*Total particulates*: personal samples (18 measurements), TWA (4-7 h)	0.1-3.0 mg/m^3	
		Cyclohexane solubles: personal samples (6 measurements), TWA (4-7 h)	<0.1-1.3 mg/m^3	
		Benzo[a]pyrene: area sample	<0.05 µg/m^3	
Surface dressing	3	*Total particulates*: personal samples (4 measurements), TWA (5 h)	0.8-2.5 mg/m^3	

Table 9 (contd)

Source of exposure	Job category or description	Class of bitumen handled[a]	Material and sample type[b]	Results	Reference
	Oil gravel	6	Cyclohexane solubles: personal samples (4 measurements), TWA (4-7 h)	<0.1-0.4 mg/m^3	
			Benzo[a]pyrene: area sample	<0.05 µg/m^3	
			Total particulates: personal samples (8 measurements), TWA (4-7 h)	0.1-1.4 mg/m^3	
			Cyclohexane solubles: personal samples (2 measurements), TWA (5-6 h)	<0.1-0.5 mg/m^3	
			Benzo[a]pyrene: area samples	<0.05 µg/m^3	

[a]When direct information was not provided, the class of bitumen was derived from information describing the activity and the temperature of handling.
[b]PAC, polynuclear aromatic compound; TWA, time-weighted average; PAH, polynuclear aromatic hydrocarbon

BITUMENS

(ii) *Asphalt mixing and road-paving operations*

Exposures measured during asphalt mixing and road paving are given in Table 9.

Hot-mix asphalt plants comprise a rotary drum dryer to dry and heat mineral aggregates, a pug-mill which combines the heated stone material with bitumens (3-8% by weight) at a mixing temperature of 130-180°C, and a skip-hoist feeding the hot asphalt mix to storage bins which allow trucks to be loaded directly for conveying the mix to road construction sites.

In modern asphalt-mixing plants, operators are enclosed in remote control cabins away from the mixer, hot asphalt-storage and vehicle-loading points.

Typical operations to lay and compact asphalt hot mixes as overlays on roads are as follows: A mix is delivered to a finishing machine, where it is discharged into a hopper. Conveyors in the machine then spread the asphalt mix to a fixed width, and it is screeded (levelled) and compacted to the thickness required. Further compaction is conducted using steel or rubber-tyred rollers. The finishing-machine operator is seated about two metres above road level; the rakerman corrects imperfections and completes edges on the newly-laid asphalt layer immediately behind the finisher, while a man at the back of the machine adjusts the screed (levelling device) to give the required road profile.

The temperature of the asphalt mix is up to 170°C on discharge, but the mix cools by up to 20°C while being fed through the finisher and forming a new asphalt course. Variations in fume levels occur as mix is discharged from insulated trucks into the finishing-machine hopper. Further cooling occurs with roller compaction when water is applied to prevent sticking of the asphalt to roller tyres.

(iii) *Roofing operations*

Exposures measured during roofing operations are given in Table 10.

Oxidized bitumens [class 2] may be fed directly to the roof, while other grades are melted in a roofing kettle and heated up to 350°C. In actual operations, the kettle is often uncovered. The temperatures in the container vessel used to transfer the hot bitumen vary from 240-310°C. Surface spreading of hot bitumen (usually 15-25°C lower than the kettle temperature) is usually done manually.

(iv) *Flooring operations*

Exposures measured during flooring operations are given in Table 11.

Operations include duties like heating and melting (temperatures up to 250°C) of mastic-asphalt blocks in kettles and transferring the material indoors, where it is then layered onto a floor using hand floating (a levelling technique similar to that used for smoothing cement). Activities in waterproofing include brushing of hot oxidized bitumen [class 2] as a felt interply adhesive (temperatures 160-190°C) or treatment of outside foundations with other bitumens (cutback bitumens [class 3], penetration bitumens [class 1] and oxidized bitumens [class 2]).

Table 10. Exposures measured during roofing operations

Source of exposure	Job category or description	Class of bitumen handled[a]	Material and sample type[b]	Results[c]	Reference
Bitumen heating and mopping	Kettle hood	2	Benzene solubles: area samples (4 measurements), TWA (1-3 h) Benz[a]anthracene Chrysene	170-420 mg/m^3 32-47 µg/m^3 20-68 µg/m^3	Brown & Fajen (1977a)
	Kettleman, moppers and foreman	2	Benzene solubles: personal samples (5 measurements), TWA (5 h) Benzo[a]pyrene Benz[a]anthracene, fluoranthene, pyrene, chrysene	0.58-1.3 mg/m^3 ND-0.04 µg/m^3 ND	
Bitumen heating and mopping	Kettlemen, moppers and foreman	2	Benzene solubles: personal samples (3 measurements), TWA (5 h) Pyrene Chrysene Benz[a]anthracene Benzo[a]pyrene, fluoranthene	0.43-2.25 mg/m^3 ND-1.4 µg/m^3 ND-1.0 µg/m^3 ND-0.4 µg/m^3 ND	Brown & Fajen (1977b)
Bitumen heating and mopping	Kettlemen and moppers	2	Cyclohexane solubles: personal samples (3 measurements), TWA (1-3 h) Fluoranthene Pyrene Benz[a]anthracene Chrysene Benzo[a]pyrene	0.16-0.28 mg/m^3 ND-0.85 µg/m^3 0.73-1.1 µg/m^3 ND-0.7 µg/m^3 ND-0.63 µg/m^3 ND-0.63 µg/m^3	Tharr (1982)
Spreading hot bitumen	Moppers	2	Total particulates: personal samples (8 measurements), TWA (1-5 h) Cyclohexane solubles: personal samples (8 measurements) Benzo[a]pyrene: area samples (5 measurements)	0.2-3.4 mg/m^3 0.2-2.9 mg/m^3 <0.05 µg/m^3	Priha et al. (1980)
Kettle emissions and bitumen spreading	Kettlemen and felt layers	2	Total particulates: (9 measurements), TWA (8 h) Benzene solubles: personal samples (9 measurements), TWA (8 h)	0.5-6.4 mg/m^3 0.2-5.4 mg/m^3	Brandt & Molyneux (1985)

BITUMENS

		No.	Compound measured	Concentration	Reference
Bitumen roofing	Moppers and kettlemen	2	*Total PACs (11 PAHs)*: personal samples (3 measurements), TWA (8 h)	24-364 ng/m³	Malaiyandi et al. (1982)
Bitumen roofing	Paper rollers, moppers, kettlemen	2	*Total PACs (11 PAHs)*: personal samples (6 measurements).	14.5-112.5 µg/m³	Reed (1983)
			Benzene solubles: personal samples (5 measurements), TWA (2-5 h)	0.9-1.2 mg/m³	
			Total PACs (7 PAHs): personal samples (5 measurements), TWA (2-5 h)	2.8-23.2 µg/m³	
Bitumen roofing	Felt-laying machine operators, moppers, tank operators, hot asphalt carrying	2	*Benzene solubles*: personal samples (6 measurements), TWA (2-5 h)	<0.11-2.72 mg/m³	
			Benzo[a]pyrene: personal samples (6 measurements)	ND-0.11 µg/m³	
			Benz[a]anthracene: personal samples (6 measurements)	ND-0.21 µg/m³	
			Fluoranthene, pyrene, chrysene: personal samples (6 measurements)	ND	

aWhen direct information was not provided, the class of bitumen was derived from information describing the activity and the temperature of handling.
bTWA, time-weighted average; PAC, polynuclear aromatic compound; PAH, polynuclear aromatic hydrocarbon
cND, not detected

Table 11. Exposures measured during flooring and waterproofing operations indoors

Source of exposure	Job category or description	Class of bitumen handled[a]	Material and sample type[b]	Results[c]	Reference
Mastic asphalt	Asphalt flooring indoors up to 250°C; spreaders, tippers, float finishers	1	Carbon tetrachloride extracts of total particulates by IR: personal samples (28 measurements)	1.7-260 mg/m^3	Claydon et al. (1984)
	Bucket filling and carrying	1	personal samples (6 measurements)	0.5-5.5 mg/m^3	
Mastic asphalt (265-280°C)	Bucket carrier, pourer (indoor workers)	1	*Total particulates*: personal samples (7 measurements), TWA (8 h)	3.4-17.6 mg/m^3	Brandt & Molyneux (1985)
			Benzene solubles: personal samples (7 measurements), TWA (8 h)	1.8-13.6 mg/m^3	
			Total PACs (11 PAHs): personal samples (2 measurements), TWA (8 h)	1709-1897 ng/m^3	
	Outside (kettleman)	1	*Total particulates*: personal samples (2 measurements), TWA (8 h)	2.9-7.7 mg/m^3	
			Benzene solubles: personal samples (2 measurements), TWA (8 h)	2.0-5.0 mg/m^3	
			Total PACs (11 PAHs): personal sample, TWA (8 h)	529 ng/m^3	
	Trowellers	1	*Total particulates*: personal samples (3 measurements), TWA (8 h)	10.7-18.2 mg/m^3	
			Benzene solubles: personal samples (3 measurements), TWA (8 h)	7.3-13.1 mg/m^3	
			Total PACs (11 PAHs): personal sample, TWA (8 h)	2508 ng/m^3	
Waterproofing of basements, kitchens, bathrooms and corridors	Brushing as a felt interply adhesive	2, 4	*Total particulates*: personal samples (9 measurements), TWA (1-4 h)	1.1-32 mg/m^3 (mean, 10.4 mg/m^3)	Ahonen et al. (1977)

Activity	Class of bitumen[a]	Compound measured	Concentration	Reference
Waterproofing of kitchens, bathrooms and underground spaces	Spreading hot bitumen, felt seaming	2, 4	Cyclohexane solubles: personal samples (5 measurements), TWA (1-4 h)	0.8-28 mg/m³ (mean, 7.7 mg/m³)	Priha et al. (1980)
			Total particulates	1.9-42.3 mg/m³ (mean, 17.7 mg/m³)	
			Cyclohexane solubles: personal samples (15 measurements), TWA (1-3 h)	0.3-38.9 mg/m³	
Waterproofing of basements	Spreading hot bitumen, felt seaming	2, 4	Total particulates	1.6-12.7 mg/m³ (mean, 6.1 mg/m³)	Priha et al. (1980)
			Cyclohexane solubles: personal samples (6 measurements), TWA (1-4 h)	1.6-10.8 mg/m³ (mean, 5.6 mg/m³)	

[a]When direct information was not provided, the class of bitumen was derived from information describing the activity and the temperature of handling.
[b]IR, Infra-red spectroscopy; TWA, time-weighted average; PAC, polynuclear aromatic compound; PAH, polynuclear aromatic hydrocarbon

(c) Air

Emissions from bitumens occur in the vicinity of hot-mix asphalt plants and road-laying operations. In industrial uses, they occur in and near factories. Bitumen-producing refineries are also a source of emissions into the air. It has been estimated that in the USA a total of 18 500 metric tonnes of volatile organic compounds are released annually into the air (Beggs, 1981).

In the production of roofing felts, 0.62 tonnes of bitumen are used to produce one tonne of felt. Emissions of particulates were found to be 1.35 mg/g for controlled and 3.15 mg/g for uncontrolled conditions (US Environmental Protection Agency, 1978).

In a bitumen-blowing operation, emissions of particulates ranged from 0.29-3.65 mg/g for well-controlled and uncontrolled operations (US Environmental Protection Agency, 1978).

Emissions of polynuclear aromatic compounds from batch asphalt mixers are low and mainly attributed to combustion gases. Only anthracene was identified in a sample collected directly from a mixing chamber (von Lehmden et al., 1965). High particulate emissions (up to 69.3 g/m^3) were measured in US dryer-drum hot asphalt plants without air pollution control (Kinsey, 1976).

2.3 Analysis

Methods for the analysis of bitumen fumes in air are very similar to those used for coal-tars. The National Institute for Occupational Safety and Health (1983) presented limited data that indicate that these may not be reliable. In some studies, the blank filter used in the analysis indicated higher levels of cyclohexane-extractables than all the samples taken (Todd & Timbie, 1979, 1981, 1983). In certain samples, the cyclohexane-soluble fraction exceeded the total-weight concentration (Okawa & Apol, 1977). Consequently, the results obtained by the method described in the monograph on coal-tars, p. 119, may not be an occurate indication of occupational exposure.

Claydon et al. (1984) recently recommended an improved gravimetric method for the determination of total particulates and of the benzene-soluble fraction of bitumen fumes in air.

Kawahara et al. (1974) described a qualitative method for the identification of spilled bitumen in surface-waters, based on a combination of infrared spectrophotometry and computer-assisted statistical analysis.

3. Biological Data Relevant to the Evaluation of Carcinogenic Risk to Humans

3.1 Carcinogenicity studies in animals

(a) Skin application

Mouse: A pooled sample of six steam- and air-blown (oxidized) petroleum asphalts [bitumens classes 1 and 2] liquefied with benzene [dose unspecified] was applied twice weekly onto the interscapular skin of 68 (32 males and 36 females) C57 black mice. A group of 63 mice (31 males and 32 females) were similarly treated with benzene alone. Epidermoid carcinomas appeared on the skin of 12 treated animals, the first tumour appearing during the 54th week after the beginning of the experiment. No skin tumour was found in control animals (Simmers et al., 1959).

In another experiment, heated steam-refined petroleum asphalt (a pool of three samples) [class 1] was applied to the skin of 50 (25 males and 25 females) C57 black mice three times weekly. Because only 14 males and 18 females survived an epidemic of pneumonitis after seven weeks and only one male and five females were alive after one year, eight males and five females were added to the group. The number of paintings ranged from 16-240. Topical squamous-cell carcinomas were found in three of 21 autopsied mice. A further group of 50 mice was treated one to three times weekly with heated air-refined (oxidized) petroleum asphalt [bitumen class 2]. No carcinoma was observed at the site of treatment in the 32 mice surviving more than seven weeks (10 autopsied). In a complementary group (10 males and 10 females), the air-refined asphalt, diluted in toluene (one volume of toluene to 10 of melted asphalt) was applied three times weekly for up to two years (284 applications). Topical squamous-cell carcinomas developed in nine of 20 mice autopsied. Of 15 toluene-treated control animals, one developed a skin papilloma (Simmers, 1965a). [The Working Group noted the high mortality in the early part of the study and that data on subsequent survival were incomplete.]

In a third experiment, a mixture of 'aromatics' and 'saturates' [a fraction of a class 1 bitumen], isolated by fractionation of a steam-refined asphalt from a California crude petroleum, was applied three times weekly (about 33.4 mg per application) to the intrascapular non-shaved skin of 25 male and 25 female C57 black mice. The number of applications ranged from 72 to 242 because of differential survival. Of 30 mice studied microscopically, 13 showed skin papillomas (13) and cancers (seven epidermoid carcinomas, five baso-squamous cancers and one sebaceous-gland carcinoma). Other tumours found included one epidermoid carcinoma of the anus and two leiomyosarcomas (one subcutaneous and one intestinal) (Simmers, 1965b). [The Working Group noted that no control was used.]

One group of 100 and three groups of 50 male and female black C57 mice were treated by skin application with road petroleum asphalts [bitumens class 1] obtained by steam distillation of crudes from Venezuela, Mississippi and California, and by steam-vacuum distillation of one Oklahoma crude, respectively. Each mouse received one drop of asphalt liquefied with acetone on the neck skin twice weekly for up to two years. One skin carcinoma was observed in the group treated with the Mississippi sample, and one skin papilloma was observed in the groups treated with the Oklahoma and the Mississippi samples. No skin tumour was found in the groups treated with the Venezuela or California samples or in 200 untreated mice (Hueper & Payne, 1960). [The Working Group noted the inadequate reporting of the experiment.]

A group of 50 (25 male and 25 female) C57 black mice received skin paintings twice weekly on the nape of the neck with a heated sample of an air-blown asphalt [bitumen class 2] used for roofing purposes [dose unspecified]. Treatment was continued for up to two years; one skin carcinoma was reported (Hueper & Payne, 1960). [The Working Group noted that no control was used and that data on survival were not reported.]

Two cracking-residue (destructive thermal distillation) bitumens (BN-5 and BN-4) [class 8] and four residual bitumens (straight distillation) (BN-5, BN-4, BN 3 and BN-2) [class 1] in benzene (40% solutions) were tested for carcinogenicity by weekly skin painting for

19 months (70 applications) to SS-57 white mice. With the cracking-residue bitumen BN-5, nine of 49 survivors at the time of appearance of the first tumour (ninth month) developed tumours at the treatment site: five cornified squamous-cell carcinomas, one fibrosarcoma and three papillomas. In addition, seven animals developed pulmonary adenomas and adenocarcinomas and one, a squamous-cell carcinoma of the forestomach. With the cracking-residue bitumen BN-4, four of the 42 mice alive at the appearance of the first tumour (10th month) had skin tumours, two of which were carcinomas (one cornified, one noncornified) and two papillomas, and all four had pulmonary adenomas. With the residual bitumens, BN-5, BN-4, BN-3 and BN-2, tumours were reported in two (one cornified squamous-cell carcinoma and one sebaceous carcinoma) of 43, none of 30, two (one fibrosarcoma and one papilloma) of 43 and none of 30 mice surviving 9 months, respectively. In addition, lung tumours were observed in 5/43, 1/30, 1/43 and 1/30 mice, respectively. In 23 control mice painted with benzene only, no skin tumour was seen; one mouse developed lung adenomas (Kireeva, 1968). [The Working Group noted that the low incidences of lung tumours could not be evaluated in the absence of adequate data on survival of both treated and control animals.]

The carcinogenic activity of eight road-paving-grade asphalts [bitumens class 1] produced by vacuum distillation from well-defined crude sources was studied. The different asphalts dissolved in benzene (10% solutions) were applied twice weekly to the skin of groups of 24-32 male and female random-bred Swiss albino mice with a calibrated dropper delivering 25 µl of solution, corresponding to 2.5 mg of asphalt per application. Mean survival times were 81 weeks for asphalt-treated and 82 weeks for benzene-treated mice. At the end of the experiment, six of 218 animals treated with the different asphalts developed skin tumours: one was a carcinoma and five 'papillomatous growths'. In 26 control mice treated with benzene only, one 'papillomatous growth' was observed (Wallcave et al., 1971).

Emmett et al. (1981) treated 50 male C3H/HeJ mice with standard roofing petroleum asphalt [bitumen class 2] dissolved in toluene (1:1 w/w). Each animal received 50 mg of the solution on the intrascapular skin twice weekly for 80 weeks. No skin tumour was observed in 26 mice that survived 60 or more weeks or in 37 of a control group of 50 mice treated with toluene only. Of 50 positive-control mice treated with benzo[a]pyrene (0.1% toluene solution, 50 µg/application), 31/39 (79%) surviving at the time of appearance of the first skin tumour had skin tumours (24 malignant, seven papillomas; average latent period of papillomas, 32 weeks).

(b) *Subcutaneous and/or intramuscular injection*

Mouse: A pooled sample of six different steam- and air-blown (oxidized) petroleum asphalts [bitumens classes 1 and 2], suspended in olive oil (1%) was injected s.c. in the interscapular region of 62 (33 males and 29 females) C57 black mice. Each mouse received 0.2 ml per injection twice weekly for 41 weeks, then once weekly. A control group of 60 animals (32 males and 28 females) received similarly olive oil only. In the treatment group, the first sarcoma appeared 36 weeks after the beginning of the experiment. A total of eight sarcomas was observed at the injection site. No injection-site tumour was noted in the control group (Simmers et al., 1959).

In a further study, two groups of 50 (25 male and 25 female) C57 black mice received two s.c. injections (at intervals of three and four months, respectively) of 200 mg per injection of heated steam-refined asphalt [bitumen class 1] or heated air-refined (oxidized) asphalt [bitumen class 2]. For injection, the steam-refined asphalt was heated to 70°C and the air-refined asphalt to 100°C. No skin tumour was observed in 32 autopsied mice from the group receiving steam-refined asphalt. Five malignant tumours (two rhabdomyosarcomas, one sebaceous-gland carcinoma, two not described) were found in the 38 autopsied mice treated with air-refined asphalt (Simmers, 1965a). [The Working Group noted the absence of controls.]

Four groups of 50 (25 males and 25 females) C57 black mice received six i.m. injections every two weeks in the right thigh of 0.1 ml of a tricaprylin dilution (equal parts) of petroleum asphalts [bitumens class 1] obtained by steam distillation of crudes from Mississippi, California and Venezuela, and by steam-vacuum distillation of one Oklahoma crude. After two years, injection-site sarcomas were noted in one mouse in each of the groups treated with samples from crudes from Mississippi, California and Venezuela. No such tumour was observed in the group treated with the sample from Oklahoma crude or in tricaprylin controls (Hueper & Payne, 1960). [The Working Group noted the absence of data on survival.]

Rat: Four groups of 30 Bethesda black rats each received 12 i.m. injections every two weeks into the right thigh of 0.2 ml of a tricaprylin dilution (equal parts) of petroleum asphalts [bitumens class 1] obtained by steam distillation of crudes from Mississippi, California and Venezuela, and by steam-vacuum distillation of one Oklahoma crude. After two years of observation, one, six, two and four rats, respectively, had sarcomas at the site of injection (Hueper & Payne, 1960). [The Working Group noted the absence of a vehicle-control group.]

3.2 Other relevant biological data

(a) Experimental systems

Toxic effects

A number of effects [which may occur as intercurrent respiratory illness] have been reported in tests of chronic inhalation exposure of guinea-pigs, rats and mice to bitumen fumes [bitumens classes 1 and 2], aerosol or smoke (Hueper & Payne, 1960; Simmers, 1964). These effects include patchy regions of emphysema, bronchiolar dilatation, pneumonitis and severe localized bronchitis. Squamous metaplasia was seen rarely.

Skin effects after exposure to samples of eight different bitumens [class 1] were studied in random-bred Swiss albino mice (Wallcave *et al.*, 1971). Mice were given twice-weekly applications of 25 µl of bitumen solution (10% in benzene) to shaved areas of the back for an average of 81 weeks. Epidermal hyperplasia was a general finding. Inflammatory infiltration of the dermis, cutaneous ulceration with abscess formation and amyloidosis of the spleen and kidney were commonly observed.

No data were available to the Working Group on effects on reproduction and prenatal toxicity, or on absorption, distribution, excretion and metabolism.

Mutagenicity and other short-term tests

An extract of a 'road-coating tar' [cutback bitumen class 3] in dimethyl sulphoxide (DMSO) was mutagenic to *Salmonella typhimurium* TA98 in the presence of an Aroclor-induced rat-liver metabolic system (S9). Vapours, particles and aerosols emitted at 550°C, 350°C and 250°C (collected in DMSO) were also weakly mutagenic, both in the presence and absence of S9 in *S. typhimurium* TA98 and/or TA100 (Penalva *et al.*, 1983).

(b) Humans

Toxic effects

Keratoconjunctivitis was frequent among roofers who worked with both bitumens [class 2] and coal-tar pitches; these changes were frequently associated with exposure to coal-tar pitches but not with that to bitumens (Emmett *et al.*, 1977).

In a study of 250 road workers exposed to bitumens [presumably class 1], two burns due to contact with hot bitumen and six instances of chronic eczema were noted. It was the authors' opinion that occupational dermatitis was not a serious problem in these jobs (Chanial & Joseph, 1964).

Zeglio (1950) studied 22 workers who insulated electrical cables and telegraph and telephone lines with bitumens [presumably class 2] heated to 120°C. There were frequent complaints of cough and phlegm, burning of the throat and chest, hoarseness, headache and nasal discharge. Physical examination disclosed bronchitis, rhinitis, oropharyngitis and laryngitis among these workers.

No data were available to the Working Group on effects on reproduction and prenatal toxicity, on absorption, distribution, excretion and metabolism or on chromosomal effects.

3.3 Case reports and epidemiological studies of carcinogenicity in humans

Oliver (1908) reported two cases of scrotal cancer, a rare tumour, in workers exposed to bitumens for 13 and 20 years, respectively. Concomitant exposure to coal-tars and pitches was reported in one case (see also Appendix, p. 243).

Henry (1947) reviewed 3753 cases of skin cancer in 2975 persons[1]. Of these cases, 13 were in 12 persons applying road surface material and nine were in makers of road material. Henry stated that most of these workers were in contact with coal-tars.

A 12-year mortality study was made of 5939 members of a roofers' union in the USA alive in 1960 and with at least nine years of union membership, 5788 of whom were traced through 1971 (Hammond *et al.*, 1976). Expected numbers of deaths were computed on the basis of US life tables specific for age and single calendar year. The standardized mortality ratio (SMR) for all deaths was 103 in the first six years of the study and 110 in the second six years. Lung cancer ratios (SMR) generally increased with increasing time since joining the union: 9-19 years, 22 observed, SMR 92; 20-29 years, 66 observed, SMR 152; 30-39 years, 21 observed, SMR 150; and over 40 years, 12 observed, SMR 247. The following were recorded in the group with 20 or more years' membership in the union: for cancer of the oral cavity, larynx and oesophagus, 31 observed, SMR 195; for cancer of the stomach, 24 observed, SMR 167; for leukaemia, 13 observed, SMR 168; and for cancer of the bladder, 13 observed, SMR 168. In the total cohort, five deaths were from skin cancer other than melanoma (1.18 expected). [No data on smoking were available.]

---

[1]The UK Factory and Workshop Act 1901 required that diseases contracted in premises to which the Act applied be notified to HM Chief Inspector of Factories by the medical practitioners concerned. The list of such notifiable diseases was extended in 1 January 1920 to include 'epitheliomatous ulceration or cancer of the skin due to pitch, tar, bitumen, paraffin or mineral oil, or any compound, or residue of any of these substances, or any product' (Henry, 1947).

In a study by Menck and Henderson (1976) of occupational differences in lung cancer rates in white men in Los Angeles County, USA, mortality data for 1968-1970 and morbidity data from the county Cancer Registry for 1972-1973 were pooled. The 3938 men under study were classified by their last recorded occupation and by the industry in which they were employed. This information was obtained from death certificates for mortality and from hospital records for incident cases. Age-adjusted expected numbers of deaths and incident cases were calculated on the basis of a 2% sample of 1970 census records for Los Angeles County classified by occupation and industry. Among roofers, a significantly increased risk of lung cancer was observed, with a 'standard mortality ratio' of 496 based on six deaths and five incident cases. [No data on smoking were available.]

Milham (1982) conducted a proportional mortality analysis of deaths among white male residents in the State of Washington during 1950-1979. Among roofers and slaters, there were four deaths due to laryngeal cancer (proportionate mortality ratio, PMR, 270) and 53 deaths due to cancer of the bronchus and lung (PMR, 161). [No data on smoking were available.]

Decouflé et al. (1977), in a survey of occupations, looked at cases of cancer and of non-neoplastic diseases recorded at Roswell Park Memorial Hospital, NY, during 1956-1965. Persons employed in clerical jobs were used for comparison with those in other occupations. Relative risks were adjusted for age and smoking habits. A non-significant relative risk of 2.95 was reported for cancer of the buccal cavity and pharynx among roofers and slaters. The relative risk for lung cancer in this trade was not reported.

According to Hammond et al. (1976), roofers in the USA are mainly engaged in applying hot pitches or hot bitumens. In former years, [coal-tar] pitches (see monograph, p. 83) were used more frequently than bitumens [class 2], but today bitumens are more commonly used. Most men work with both substances.

[No study of petroleum refinery workers was considered, although the Working Group recognized that some workers in that industry might be exposed to bitumens.]

4. Summary of Data Reported and Evaluation

4.1 Exposures

World usage of bitumens is estimated to be more than 60 million tonnes per year; exposures occur in a variety of applications. A total of 90-95% of bitumen is used hot (>100°C) in road construction, roofing and flooring. Fumes from these operations contain polynuclear aromatic compounds, although almost all of the mass of material in the fumes remains uncharacterized. Cutback bitumens, blended grades and bitumen emulsions are usually used at ambient or warm temperatures, where the potential for direct skin contact may be greater.

Blended or fluxed bitumens, which represent a relatively small percentage of the total usage, may contain aromatic oils, thermally-cracked petroleum residues or coal-tar products, which contain polynuclear aromatic compounds.

4.2 Experimental data[1]

Steam-refined petroleum bitumens were tested by application to the skin of mice. Skin tumours were produced with undiluted bitumens, with dilutions in benzene and with a fraction of steam-refined bitumen.

When air-refined (oxidized) bitumens were applied to the skin of mice, no tumour was found with undiluted bitumens; but, in one experiment, an air-refined bitumen in solvent (toluene) produced topical skin tumours.

Two cracking-residue bitumens produced skin tumours when applied to the skin of mice.

A pooled mixture of steam- and air-blown petroleum bitumens in benzene produced tumours at the site of application on the skin of mice.

One sample of heated, air-refined bitumen injected subcutaneously into mice produced a few sarcomas at the injection sites.

A pooled mixture of steam- and air-blown petroleum bitumens produced sarcomas at the site of subcutaneous injection in mice. Steam-distilled bitumens injected intramuscularly produced local sarcomas in one experiment in rats.

Both an extract of road-surfacing bitumen and its emissions were mutagenic to *Salmonella typhimurium*.

Overall assessment of data from short-term tests on bitumens[a]

	Genetic activity			Cell transformation
	DNA damage	Mutation	Chromosomal effects	
Prokaryotes		+[b]		
Fungi/green plants				
Insects				
Mammalian cells (*in vitro*)				
Mammals (*in vivo*)				
Humans (*in vivo*)				
Degree of evidence in short-term tests for genetic activity: *inadequate*				Cell transformation: no data

[a]The groups into which the table is divided and the symbol + are defined on pp. 16-18 of the Preamble; the degrees of evidence are defined on p. 18.

[b]Road-surfacing bitumen, including emissions

---

[1]Subsequent to the meeting of the Working Group, the Secretariat became aware of a study showing that solutions of the fumes from two types of roofing bitumens, generated at 232°C or 316°C, produced skin tumours when applied topically to mice (Thayer *et al.*, 1983).

4.3 Human data

No epidemiological study of workers exposed only to bitumens is available. A cohort study of US roofers indicates an increased risk for cancer of the lung and suggests increased risks for cancers of the oral cavity, larynx, oesophagus, stomach, skin and bladder and for leukaemia. Some support for excess risks of lung, oral cavity and laryngeal cancers is provided by other epidemiological studies of roofers. As roofers may be exposed not only to bitumens but also to coal-tar pitches and other materials, the excess cancer risk cannot be attributed specifically to bitumens. Several case reports of skin cancer among workers exposed to bitumens are available; however, exposure to coal-tars or products derived from them cannot be ruled out.

4.4. Evaluation[1]

There is *sufficient evidence*[2] for the carcinogenicity of extracts of steam-refined bitumens, air-refined bitumens and pooled mixtures of steam- and air-refined bitumens in experimental animals.

There is *limited evidence* for the carcinogenicity of undiluted steam-refined bitumens and for cracking-residue bitumens in experimental animals.

There is *inadequate evidence* for the carcinogenicity of undiluted air-refined bitumens in experimental animals.

There is *inadequate evidence* that bitumens alone are carcinogenic to humans.

5. References

Ahonen, K., Engström, B. & Lehtinen, P.U. (1977) *Bitumen Fume Concentration Moisture Isolating with Hot Bitumen: Indoor or Poor Ventilated Basement* (Finn.) (*Report No. 7*), Helsinki, Institute of Occupational Health

American Conference of Governmental Industrial Hygienists (1983) *Threshold Limit Values for Chemical Substances in the Work Environment Adopted by ACGIH for 1983-84*, Cincinnati, OH, p. 10

The Asphalt Institute (1982) *A Brief Introduction to Asphalt and Some of its Uses* (*Manual Series No. 5*), 8th ed., College Park, MD

---

[1]For definitions of the italicized terms, see Preamble pp. 15-16 and 19.

[2]In the absence of adequate data on humans, it is reasonable, for practical purposes, to regard chemicals for which there is sufficient evidence of carcinogenicity in animals as if they presented a carcinogenic risk to humans.

The Asphalt Institute (1983) *Specifications for Paving and Industrial Asphalts (Specification Series No. 2* (SS-2)), College Park, MD

Beggs, T.W. (1981) *Emission of Volatile Organic Compounds from Drum-Mix Asphalt Plants (EPA 600/2-81-026)*, Cincinnati, OH, US Environmental Protection Agency

Bingham, E., Trosset, R.P. & Warshawsky, D. (1979) Carcinogenic potential of petroleum hydrocarbons. A critical review of the literature. *J. environ. Pathol. Toxicol., 3*, 483-563

Brandt, H.C.A. & De Groot, P.C. (1985) Sampling and analysis of bitumen fumes. 3. Laboratory study of emissions under controlled conditions. *Ann. occup. Hyg.* (in press)

Brandt, H.C.A. & Molyneux, M.K.B. (1985) Sampling and analysis of bitumen fumes. 2. Field exposure measurements. *Ann. occup. Hyg.* (in press)

Bright, P.E., Chipperfield, E.H., Eyres, A.R., Frati, E., Hoehr, D., Rettien, A.R. & Simpson, B.J. (1982) *Health Aspects of Bitumens (CONCAWE Report No. 7/82)*, The Hague, The Netherlands, International Study Group for Conservation of Clean Air and Water - Europe

British Standards Institution (1974) *Method for Determination of Penetration of Bituminous Materials (BS 4691)*, London

British Standards Institution (1982a) *Bitumens for Building and Civil Engineering. Part 1. Specification for Bitumens for Road Purposes (BS 3690)*, Part 1, London

British Standards Institution (1982b) *Bitumens for Building and Civil Engineering. Part 2. Specification for Bitumens for Industrial Purposes (BS 3690)*, Part 2, London

British Standards Institution (1983a) *Softening Point of Bitumen (Ring and Ball) (BS 2000)*, Part 58, London

British Standards Institution (1983b) *Viscosity of Cutback Bitumen and Road Oil (BS 2000)*, Part 72, London

Broome, D.C. (1973) Bitumen. In: Hobson, G.D., ed., *Modern Petroleum Technology*, 4th ed., New York, John Wiley & Sons, pp. 804-815

Brown, J. & Fajen, J. (1977a) *Industrial Hygiene Survey Report at Zeric Roofing Corporation (Report No. 48.15)* Cincinnati, OH, National Institute for Occupational Safety and Health

Brown, J. & Fajen, J. (1977b) *Industrial Hygiene Survey Report at Asbestos Roofing Company (Report No. 48.17)* Cincinnati, OH, National Institute for Occupational Safety and Health

Brown, J. & Fajen, J. (1977c) *Industrial Hygiene Survey Report at Asbestos Roofing Company (Report No. IWS-48.16, PB82-151234)*, Cincinnati, OH, National Institute for Occupational Safety and Health

Byrd, R.A. & Mikkelsen, O.G. (1979) *Measurements of Concentrations of Asphalt Fumes, Concentrations of Organic Vapours and Chemical Analysis of Asphalt Fumes*, Hvidovre, Denmark, Association of Danish Asphalt Industries

Chanial, G. & Joseph, J.Y. (1964) Bitumen dermatoses in the Lyonnais area (Fr.). *Arch. Mal. prof. Méd. Trav. Sécur. soc.*, 25, 453-454

Chipperfield, E.H. (1984) *IARC Review on Bitumen Carcinogenicity. Bitumen Production, Properties and Uses in Relation to Occupational Exposures (IP 84-006)*, London, Institute of Petroleum

Claydon, M.F., Christian, F., Eyres, A.R., Guelfo, G., Lentge, H., Mitchell, J.G., Molyneux, M.K.B., Moore, J.P., Rousseaux, G., Sanderson, J.T. & Webb, C.L.F. (1984) *Review of Bitumen Fume Exposures and Guidance on Measurement*, The Hague, International Study Group for Conservation of Clean Air and Water - Europe

Decouflé, P., Stanislawczyk, K., Houten, L., Bross, I.D.J. & Viadana, E. (1977) *A Retrospective Survey of Cancer in Relation to Occupation (DHEW (NIOSH) Publ. No. 77-178)*, Cincinnati, OH, National Institute for Occupational Safety and Health, p. 186

Emmett, E.A., Stetzer, L. & Taphorn, B. (1977) Phototoxic keratoconjunctivitis from coal-tar pitch volatiles. *Science*, 198, 841-842

Emmett, E.A., Bingham, E.M. & Barkley, W. (1981) A carcinogenic bioassay of certain roofing materials. *Am. J. ind. Med.*, 2, 59-64

Erdman, J.G. & Ramsey, V.G. (1961) Rates of oxidation of petroleum asphaltenes and other bitumens by alkaline permanganate. *Geochim. cosmochim. Acta*, 25, 175-188

Evans, J.V. (1978) Asphalt. In: Kirk, R.E. & Othmer, D.F., eds, *Encyclopedia of Chemical Technology*, 3rd ed., Vol. 3, New York, John Wiley & Sons, pp. 284-327

Federal Office for Workers' Safety (1982) *11 Februar Verordnung über gefährliche Arbeitsstoffe, Bundesgesetzblatt I* (Regulation for hazardous substances in the workplace), Dortmund

Futagaki, S.K. (1981) *Petroleum Refinery Workers Exposure to PAHs at Fluid Catalytic Cracker, Coker and Asphalt Processing Units (Contract No. 210-78-0082) (PB83-131755)*, Cincinnati, OH, National Institute for Occupational Safety and Health

Gross, D., Konetzke, G.W. & Schmidt, E. (1979) On the industrial hygienic situation of mastic asphalt finishers with special reference to the carcinogenic action of tar, asphalt and bitumen (Ger.). *Z. ges. Hyg.*, 25, 655-659

Hammond, E.C., Selikoff, I.J., Lawther, P.L. & Seidman, H. (1976) Inhalation of benzpyrene and cancer in man. *Ann. N.Y. Acad. Sci.*, 271, 116-124

Henry, S.A. (1947) Occupational cutaneous cancer attributable to certain chemicals in industry. *Br. med. Bull.*, 4, 389-401

Hervin, R.L. & Emmett, E. A. (1976) *Health Hazard Evaluation Determination (Report No. 75-102-304)*, Cincinnati, OH, National Institute for Occupational Safety and Health

Hueper, W.C. & Payne, W.P. (1960) Carcinogenic studies on petroleum asphalt, cooling oil and coal tar. *Arch. Pathol.*, *70*, 106-118

Institute of Petroleum (1979) *Bitumen Safety Code*, London

International Labour Office (1980) *Occupational Exposure Limits for Airborne Toxic Substances*, 2nd (rev.) ed. (*Occupational Safety and Health Series No. 37*), Geneva, pp. 46, 47

Kawahara, F.K., Santner, J.F. & Julian, E.C. (1974) Characterization of heavy residual fuel oils and asphalts by infrared spectrophotometry using statistical discriminant function analysis. *Anal. Chem.*, *46*, 266-273

Kinsey, J.S. (1976) An evaluation of control systems and mass emission rates from dryer-drum hot asphalt plants. *J. Air Pollut. Control Assoc.*, *26*, 1163-1165

Kireeva, I.S. (1968) Carcinogenic properties of coal-tar pitch and petroleum asphalts used as binders for coal briquettes. *Hyg. Sanit.*, *33*, 180-186

Lawther, P.J. (1971) Pollution at work in relation to general pollution. *R. Soc. Health J.*, *91*, 250-253

von Lehmden, D.J., Hangebrauck, R.P. & Meeker, J.E. (1965) Polynuclear hydrocarbon emissions from selected industrial processes. *J. Air Pollut. Control Assoc.*, *15*, 306-312

Malaiyandi, M., Benedek, A., Holko, A.P. & Bancsi, J.J. (1982) *Measurement of potentially hazardous polynuclear aromatic hydrocarbons from occupational exposure during roofing and paving operations*. In: Cooke, M., Dennis, A.J. & Fisher, G.L., eds, *Polynuclear Aromatic Hydrocarbons: Physical and Biological Chemistry, 6th International Symposium*, Columbus, OH, Battelle Press, pp. 471-489

Menck, H.R. & Henderson, B.E. (1976) Occupational differences in rates of lung cancer. *J. occup. Med.*, *18*, 797-801

Milham, S., Jr (1982) *Occupational Mortality in Washington State 1950-1979 (Contract No. 210-80-0088)*, Cincinnati, OH, National Institute for Occupational Safety and Health, p. 1257

National Institute for Occupational Safety and Health (1977) *Criteria for a Recommended Standard... Occupational Exposure to Asphalt Fumes (DHEW (NIOSH) Publication No. 78-106)*, Washington DC, US Government Printing Office

National Institute for Occupational Safety and Health (1980a) *Projected Number of Occupational Exposures to Chemical and Physical Hazards*, Cincinnati, OH, p. 11

National Institute for Occupational Safety and Health (1980b) *Worker Exposure to Polyaromatic Hydrocarbons at Selected Petroleum Refinery Process Units, Coastal States Petrochemical Company Refinery, Corpus Christi, Texas (Report No. IWS-90.16)*, Cincinnati, OH, pp. 19-36

National Institute for Occupational Safety and Health (1981) *National Occupational Hazard Survey* (microfiche), Cincinnati, OH, pp. N10, O10, A11, B11, D11

National Institute for Occupational Safety and Health (1983) *NIOSH Manual for Analytical Methods*, Vol. 8, Cincinnati, OH

Okawa, M.T. & Apol, A.G. (1977) *Health Hazard Evaluation Determination (Report 76-54-436)*, Cincinnati, OH, National Institute for Occupational Safety and Health

Oliver, T. (1908) Tar and asphalt workers' epithelioma and chimney-sweeps cancer. *Br. med. J.*, *ii*, 493-494

Penalva, J.M., Chalabreysse, J., Archimbaud, M. & Bourgineau, G. (1983) Determining the mutagenic activity of a tar, its vapors and aerosols. *Mutat. Res.*, *117*, 93-104

Pfeiffer, J.P. & van Doormaal, P.M. (1936) The rheological properties of asphaltic bitumens. *J. Inst. Petr. Technol.*, *22*, 414-440

Priha, E., Koistinen, P., Liius, R., Rantanen, S. & Roto, P. (1980) *Moisture Insulation with Bitumen in Construction Work (Publication No. 29)*, Helsinki, Institute of Occupational Health

Puzinauskas, V.P. (1980) *Exposures of Paving Workers to Asphalt Emissions (Research Report 80-1)*, College Park, MD, Asphalt Institute

Puzinauskas, V.P. & Corbett, L.W. (1975) *Report on Emissions from Asphalt Hot Mixes (Research Report 75-1A)*, College Park, MD, Asphalt Institute

Puzinauskas, V.P. & Corbett, L.W. (1978) *Differences Between Petroleum Asphalt, Coal-tar Pitch and Road Tar (Research Report 78-1)*, College Park, MD, Asphalt Institute

Reed, L.D. (1983) *Health Hazard Evaluation (Report No. HETA-82-067-1253)*, Cincinnati, OH, National Institute for Occupational Safety and Health

Simmers, M.H. (1964) Petroleum asphalt inhalation by mice. Effects of aerosols and smoke on the tracheobronchial tree and lungs. *Arch. environ. Health*, *9*, 727-734

Simmers, M.H. (1965a) Cancers from air-refined and steam-refined asphalt. *Ind. Med. Surg.*, *34*, 255-261

Simmers, M.H. (1965b) Cancer in mice from asphalt fractions. *Ind. Med. Surg.*, *34*, 573-577

Simmers, M.H., Podolak, E. & Kinosita, R. (1959) Carcinogenic effects of petroleum asphalt. *Proc. Soc. exp. Biol. Med.*, *101*, 266-268

Tharr, D.G. (1982) *Health Hazard Evaluation (Report No. HETA-81-432-1105)*, Cincinnati, OH, National Institute for Occupational Safety and Health

Thayer, P.S., Harris, J.C., Menzies, K.Y. & Niemeier, R.W. (1983) *Integrated chemical and biological analysis of asphalt and pitch fumes*. In: Waters, M.D., Sandhu, S.S., Lewtas, J., Claxton, L., Chernoff, N. & Nesnow, S., eds, *Short-term Bioassays in the Analysis of Complex Environmental Mixtures III*, New York, Plenum Press, pp. 351-366

Todd, A.S. & Timbie, C.Y. (1979) *Industrial Hygiene Survey. Preliminary Survey of Wood Preservative Treatment Facility at McArthur Lumber and Post Company, Inc.,*

McArthur, OH (*Report No. PB81-230682*), Cincinnati, OH, National Institute for Occupational Safety and Health

Todd, A.S. & Timbie, C.Y. (1981) *Industrial Hygiene Report. Comprehensive Survey of Wood Preservative Treatment Facility at Cascade Pole Company, McFarland Cascade, Tacoma, WA* (*Report No. PB82-174160*), Cincinnati, OH, National Institute for Occupational Safety and Health

Todd, A.S. & Timbie, C.Y. (1983) *Industrial Hygiene Surveys of Occupational Exposure to Wood Preservative Chemicals* (*Contract No. 210-78-0060*), Cincinnati, OH, National Institute for Occupational Safety & Health

US Environmental Protection Agency (1978) *Supplement No. 8 for Compilation of Air Pollutant Emission Factors*, 3rd ed. (including Supplements 1-7) (*NTIS PB 288 905*), Research Triangle Park, NC, pp. 8.1-5, 8.2-4, 8.2-5

US Food and Drug Administration (1980) Food and drugs. *US Code Fed. Regul., Title 21*, Part 176.170, 149

US Occupational Safety and Health Administration (1983) Occupational exposure to coal tar pitch volatiles; modification of interpretation. *US Code Fed. Regul., Title 29*, Part 1910; *Fed. Regist., 48 (No. 15)*, 2764-2768

Virtamo, M., Riala, R., Schimberg, R., Tolonen, M., Lund, G., Peltonen, Y. & Eronen, R. (1979) *Bituminous Products in Road Paving Operations* (Finn.) (*Publication No. 20*), Helsinki, Institute of Occupational Health

Wallcave, L., Garcia, H., Feldman, R., Lijinsky, W. & Shubik, P. (1971) Skin tumorigenesis in mice by petroleum asphalts and coal-tar pitches of known polynuclear aromatic hydrocarbon content. *Toxicol. appl. Pharmacol., 18*, 41-52

Witherspoon, P.A. (1962) Colloidal nature of petroleum. *Trans. N. Y. Acad. Sci., 25*, 344-361

Yanysheva, N.Y., Kireyeva, I.S. & Serzhantova, N.N. (1963) The content of 3,4-benzpyrene in petroleum bitumens (Russ.). *Gig. Sanit., 28*, 71-73

Yen, T.F., Erdman, J.G. & Pollack, S.S. (1961) Investigation of the structure of petroleum asphaltenes by X-ray diffraction. *Anal. Chem., 33*, 1587-1594

Yen, T.F., Erdman, J.G. & Saraceno, A.J. (1962) Investigation of the nature of free radicals in petroleum asphaltenes and related substances by electron spin resonance. *Anal. Chem., 34*, 694-700

Zeglio, P. (1950) Changes in respiratory tract after bitumen vapours (Ital.). *Rass. Med. Ind., 19*, 268-273

GLOSSARY

AIR BLOWING. The process by which compressed air is blown into a feedstock, which may include VACUUM RESIDUE, PROPANE-PRECIPITATED ASPHALT mixed with ATMOSPHERIC RESIDUE, and/or a FLUX OIL at 200-280°C. This process dehydrogenates the residue resulting in oxidation and polymerization. See OXIDIZED BITUMENS.

AIR-BLOWN ASPHALTS. See OXIDIZED BITUMENS.

AIR-REFINED BITUMENS. Penetration bitumens produced by partial blowing

ASPHALT. A mixture of bitumen and mineral materials

ASPHALT COLD MIXES. ASPHALT mixtures made using CUTBACK BITUMENS or BITUMEN EMULSIONS, which can be laid at ambient temperatures

ASPHALTENES. Highly polar aromatic materials (molecular weight 2000-5000, number average) forming the hard core of micelles in the bitumen colloidal system. They can be precipitated with heptane.

ASPHALT MIXES. Mixtures of graded mineral aggregates (sized stone fractions, sands and fillers) with a controlled amount of PENETRATION BITUMEN

ATMOSPHERIC DISTILLATION. Distillation at atmospheric pressure

ATMOSPHERIC RESIDUE. Residue of ATMOSPHERIC DISTILLATION

BASE OILS. Petroleum-derived products consisting of complex mixtures of straight and branch-chained paraffinic, naphthenic (cycloparaffin) and aromatic hydrocarbons, with carbon numbers of 15 or more and boiling-points in the range of 300-600°C.

BITUMEN EMULSIONS [class 5]. Fine dispersions of bitumens in aqueous solutions. These are known as EMULSIFIED ASPHALTS in the USA.

BITUMEN ENAMEL. An external coating for protecting steel pipes. The term can also be used for bitumen paints (formulated CUTBACK BITUMENS or BITUMEN EMULSIONS).

BITUMEN MACADAM. A type of ASPHALT mix with a high stone content and containing 3-5% by weight of bitumen

BITUMEN PRIMER. A CUTBACK BITUMEN made to treat bare metal surfaces giving a bond between the metal and an enamel

BLENDED BITUMENS [class 6]. Bitumens blended with fluxes of high boiling-point (>350°C)

BRIQUETTE. See BRIQUETTING.

BRIQUETTING. The process by which fine materials (e.g., coal dusts, metal tailings) are mixed with a bitumen (or other) binder to form conveniently handled blocks or pellets

CATALYTIC AIR-BLOWN BITUMENS. Special grades of OXIDIZED BITUMENS produced using catalysts in AIR BLOWING

COLLOID MILLS. High-speed shearing devices in which hot bitumen can be dispersed in an aqueous solution to produce a BITUMEN EMULSION

CRACKING-RESIDUE BITUMENS. See THERMAL BITUMENS.

CRUDE OIL. See CRUDE PETROLEUM.

CRUDE PETROLEUM. A naturally-occurring mixture, consisting predominantly of hydrocarbons but also containing sulphur, nitrogen or oxygen derivatives of hydrocarbons, which can be removed from the earth in a liquid state

CUTBACK BITUMENS [class 3]. Bitumens produced by mixing PENETRATION BITUMENS or OXIDIZED BITUMENS with volatile diluents such as WHITE SPIRIT, KEROSENE or GAS OIL

CYCLICS (NAPHTHENE AROMATICS). These comprise the naphthene aromatic compounds with the lowest molecular weight (500-900, number average). They constitute 50-60% by weight of the total bitumen and represent the major proportion of the dispersion medium for the ASPHALTENES and adsorbed resins. This fraction is derived using solvent precipitation and adsorption chromatography.

DRUM-MIXER. An ASPHALT mixing device in which mixtures of MINERAL AGGREGATE and bitumen can be made in a rotating drum

EMULSIFIED ASPHALTS. See BITUMEN EMULSIONS.

FILLER. Fine mineral matter (e.g., limestone dust, Portland cement or stone dust) employed to give body to a bituminous binder or to fill the voids of a sand

FLEXIBLE PAVEMENTS. Road surfacings made from layers of ASPHALT mixes

FLUX. Heavy petroleum distillate used in making BLENDED or FLUXED BITUMENS

FLUXED BITUMENs. See BLENDED BITUMENS.

FLUX OILS. Less viscous forms of FLUX

GAS OIL. A liquid petroleum distillate with a viscosity and boiling-range between those of KEROSENE and lubricating oil

HARD BITUMENS [class 4]. Bitumens produced using extended VACUUM DISTILLATION with some AIR BLOWING or AIR BLOWING from PROPANE-PRECIPITATED ASPHALT. They have low penetration values and high softening-points.

KEROSENE. A petroleum distillate consisting of hydrocarbons with carbon numbers predominantly in the range of C_9 through C_{16} and boiling in the range of 150-290°C

LAKE ASPHALT. Most commonly met form of NATURAL ASPHALT, occurring in Trinidad

MASTIC ASPHALT. An ASPHALT made using a very fine MINERAL AGGREGATE with a PENETRATION BITUMEN. These materials can be poured and levelled by hand.

MECHANICAL LAYING MACHINE. Machine fitted to receive ASPHALT mixtures and to spread them as a compacted layer or course of controlled width and height

MINERAL AGGREGATE. A controlled combination of stone fractions and FILLER

MODIFIED BITUMENS [class 7]. Products or specialized applications made by incorporating polymers, elastomers or other products into PENETRATION or OXIDIZED BITUMENS

MOPPER. A worker who spreads hot bitumen on a roof with a mop

NATURAL ASPHALT. Naturally-occurring mixture of bitumens and mineral matter formed by oil seepages in the earth's crust

OXIDIZED BITUMENS [class 2]. Bitumens produced by extensive AIR-BLOWING. Also referred to as AIR-BLOWN ASPHALTS or ROOFING ASPHALTS in the USA.

PENETRATION BITUMENS [class 1]. Bitumens classified by penetration value. Similar in properties to VISCOSITY-GRADED ASPHALTS in the USA and Australia, VACUUM RESIDUES, STRAIGHT-RUN BITUMENS and STEAM-REFINED BITUMENS. Because some penetration bitumens are produced using VACUUM DISTILLATION and AIR BLOWING, whereas others are obtained by VACUUM DISTILLATION alone, the term STRAIGHT-REDUCED BITUMENS does not describe all penetration bitumens.

PENETRATION TEST. Specification test in which the indentation of a bitumen in tenths of a millimetre at 25°C is measured using a specified needle with a loading of 100 g.

PROPANE-PRECIPITATED ASPHALT. See SOLVENT PRECIPITATION.

PUG-MILL. Mixer used to combine stone materials and bitumen in an asphalt-mixing plant. The mixing is effected by high-speed stirring with paddle blades at elevated temperatures.

RAKERMAN. Worker at the back of a finishing machine who adjusts minor imperfections in the levelled asphalt course, usually by means of a hand rake

RESIDUAL BITUMEN. Bitumen obtained as a VACUUM RESIDUE

RESINS (POLAR AROMATICS). Very adhesive fractions with molecular weights of 800-2000 (number average) which act as dispersing agents for ASPHALTENES in the bitumen colloidal system. This fraction is derived using solvent precipitation and adsorption chromatography.

ROAD OILS. Term sometimes used in the USA for CUTBACK BITUMENS

ROCK ASPHALT. Naturally-occurring form of ASPHALT, usually a combination of bitumen and limestone. Found in south-eastern France, Sicily and elsewhere

ROOFING ASPHALTS. See OXIDIZED BITUMENS.

ROOFING FELT. A sheet material generally supplied in rolls and used for waterproofing roofs

ROOFING KETTLE. A heated vessel used to raise the temperature of roofing bitumens to working level. They may be fitted with various types of heating devices and lids.

ROTARY DRUM DRYER. A device in an asphalt-mixing plant used to dry and heat stone materials

SATURATES. These comprise predominently straight- and branched-chain aliphatic hydrocarbons together with alkyl naphthenes and alkyl aromatics. Their molecular weight is similar to that of CYCLICS. This fraction is derived using solvent precipitation and adsorption chromatography.

SCREED. Levelling device at the rear of a finishing machine. Can also apply to boards or metal edges used to level MASTIC ASPHALT

SKIP-HOIST. A device for transfer of ASPHALT MIXES from a PUG-MILL to storage

SOFTENING-POINT TEST. Specification test in which the temperature is measured in °C at which a bitumen (in the form of a disc under given loading conditions) softens and extends a fixed distance

SOLVENT EXTRACTS. Aromatic by-products obtained from the refining of BASE OILS

SOLVENT PRECIPITATION. The process by which a hard product, PROPANE-PRECIPITATED ASPHALT, is separated from a vacuum residue by solvent precipitation (usually with propane). PROPANE-PRECIPITATED ASPHALT is truly a bitumen by the definitions applied in this monograph. In the USA, this process is called 'solvent deasphalting' and the product, SOLVENT-REFINED ASPHALT.

SOLVENT-REFINED ASPHALT. Term used in the USA for PROPANE-PRECIPITATED ASPHALT

STEAM-REFINED BITUMENS. VACUUM RESIDUES that have been subjected to STEAM STRIPPING

STEAM STRIPPING. Injection of steam into a residue which aids VACUUM DISTILLATION

STRAIGHT-REDUCED BITUMENS. VACUUM RESIDUES used as bitumens. STEAM STRIPPING may have been used in their production.

STRAIGHT-RUN BITUMENS. Similar to STEAM-REFINED BITUMENS and STRAIGHT-REDUCED BITUMENS

SURFACE DRESSING. Process used to seal road surfaces; a thin film of bitumen, CUTBACK BITUMEN or BITUMEN EMULSIONS is spread, covered with a single or double layer of chippings, and then rolled.

SURFACE TREATMENT. May include SURFACE DRESSING and other techniques, such as spraying with minor amounts of BITUMEN EMULSION to bind surfaces together

THERMAL BITUMENS [class 8]. Bitumens produced by thermal cracking

VACUUM DISTILLATION. Distillation of ATMOSPHERIC RESIDUE under vacuum yielding GAS OIL for the manufacture of waxy distillates, lube oil, distillate feedstocks for catalytic cracking and hydrocracking and VACUUM RESIDUE

VACUUM RESIDUE. Residue obtained by VACUUM DISTILLATION

VISCOSITY. Resistance to flow. For bitumen products, test methods include vacuum-capillary and orifice-type viscometers.

VISCOSITY-GRADED ASPHALTS. See PENETRATION BITUMENS.

WHITE SPIRIT. A distillate petroleum product free of rancid or objectionable odours, boiling-range 150-200°C; sometimes described as 'Stoddard solvent'

COAL-TARS AND DERIVED PRODUCTS[1]

1. Chemical and Physical Data

1.1 Synonyms and trade names

Coal-tars (general)

Chem. Abstr. Services Reg. No.: 8007-45-2

Chem. Abstr. Name: Coal tar

Synonyms: Pix carbonis; crude coal tar

Trade Names: Carbo-Cort; Polytar Bath; Supertah; Estar; Lavatar; Syntar; Zetar

High-temperature coal-tars[2]

Chem. Abstr. Services Reg. No.: 65996-89-6

Chem. Abstr. Name: Tar, coal, high-temp.

Synonyms: Coal tar; tar decanter sludge

---

[1]This monograph deals with coal-tar products, including crude coal-tars and the higher boiling (above 205°C) primary distillation fractions and residues derived therefrom. Excluded are products derived by further processing of these primary distillation fractions, e.g., individual chemicals, such as naphthalene and anthracene, or the tar acids, e.g., phenol, cresols and xylenols, derived from low-temperature coal-tars. However, primary distillation fractions that have undergone only mild processing, such as centrifugation or alkali washing, which are then used to produce creosote and other saleable products, are included. A glossary of terms used in this monograph is given on p. 156.

[2]High-temperature coal-tars are of two main types: coke-oven tar and continuous vertical-retort (CVR) tar. Coke-oven tar is by far the most important; CVR tars are no longer commercially significant in most countries.

Low-temperature coal-tars

Chem. Abstr. Services Reg. No.: 65996-90-9

Chem. Abstr. Name: Tar, coal, low-temp.

Synomyms: Coal tar; coal oil

The important common distillation fractions and residues from coal-tars made by the high-temperature and low-temperature processes, as well as information on their main components, are given in Table 1.

Table 1. Properties of primary distillation fractions and residues from coal-tars[a]

Chemical Abstracts Name	Chemical Abstracts Services Registry Number	Distillation range (°C)	Main components
A. Products derived from high-temperature (>700°C) coal-tars[b,c]			
Light oil/overheads	65996-78-3	<180	Mainly toluene, xylenes, benzene and indene-naphtha
Carbolic oil		180-205	Mainly higher alkylbenzenes, phenol and alkylphenols, indene, xylene and naphthalene
Naphthalene oil		200-230	Mainly naphthalene and methylnaphthalene plus smaller proportions of alkylbenzenes, thionaphthalene and alkylphenols; usually alkaline washed to recover the acids; pure naphthalene may be recovered by crystallization and centrifugation
Creosote oil/wash oil	8001-58-9	230-290	Mainly alkylnaphthalenes, naphthalene, diphenylacenaphthene, fluorene, plus small amounts of higher phenols
Light anthracene oil		260-310	Mainly anthracene, phenanthrene and carbazole, with small amounts of fluorene and pyrene; an anthracene-rich fraction may be recovered by crystallization and centrifugation
Heavy anthracene oil/base oil		>310	Mainly polynuclear aromatic compounds of higher molecular weight
Medium-soft pitch	65996-93-2	Residue	40-50% of polynuclear aromatic compounds with 4-7 rings
B. Products derived from low-temperature (<700°C) coal-tars			
Light oil		90-160	Primarily C_6-C_{10} hydrocarbons
Middle oil		160-240	Primarily phenols, hydrocarbons and aromatic nitrogen bases
Heavy/wax oil		240-360/ 235-450	Both contain hydrocarbons, heavy oil and phenols
Pitch		Residue	Principally polynuclear aromatic compounds and phenols

[a]Based on McNeil (1983)

[b]Similar products can be made from coal-tars produced by the continuous vertical-retort process, but this process is not currently commercially significant in most countries.

[c]Additional information on other components is given in the text under the heading '*High-temperature coal-tar distillate fractions and their mixtures*', p. 90.

COAL-TARS AND DERIVED PRODUCTS

1.2 Description

For the purpose of this monograph, the materials with boiling-points of above 205°C, described in Section 1.1 above, and the products derived from them have been categorized into the following eight classes:

Class 1: Coal-tars
    1.1 High-temperature
    1.2 Low-temperature
    1.3 Unspecified or other

Class 2: Naphthalene oils
    2.1 High-temperature
    2.2 Unspecified or other

Class 3: Creosote oils/wash oils
    3.1 High-temperature
    3.2 Unspecified or other

Class 4: Anthracene oils
    4.1 High-temperature, light
    4.2 High-temperature, heavy
    4.3 Unspecified or other

Class 5: Middle oils

Class 6: Heavy oils/wax oils
    6.1 Heavy oil
    6.2 Wax oil

Class 7: Pitches
    7.1 High-temperature
    7.2 Low-temperature
    7.3 Unspecified or other

Class 8: Coal-tar-derived materials not otherwise classified (not sufficiently described to permit assignment to other classes)

This classification is based on the practices commonly used in the industry to produce saleable products. Although the techniques of processing may vary, all of the products of tar distillation may be categorized using the above classification.

Throughout this monograph, when the names of crude coal-tars or derived products are used as headings for separate discussions, the class to which the material belongs, determined to the best of the Working Group's ability, is given in square brackets. In separate discussions of formulated products in which the coal-tar or derived distillates or pitches used to make the products are described, the appropriate classes of the coal-tar or derived products are given.

Coal-tars [class 1]

Coal-tars are by-products of the destructive distillation of coal, called carbonization or coking. The composition and properties of a coal-tar depend mainly on the temperature of the carbonization and, to a lesser extent, on the nature of the coal used as feedstock (Kloffner et al., 1981; Novotny et al., 1981). Coal-tars are usually viscous liquids or semisolids, black or almost black in colour, with a characteristic, naphthalene-like odour (Wilson & Wells, 1950; Hawley, 1981; Windholz, 1983). In general, coal-tars are complex combinations of hydrocarbons, phenols and heterocyclic oxygen, sulphur and nitrogen compounds. Over 400 compounds have been identified in coal-tars, and probably as many as 10 000 are actually present (Trosset et al., 1978; McNeil, 1983). Table 2 presents data for some main components of selected coal-tars and illustrates that the relative proportions of these main components is quite different in tars made by low-temperature processes compared with those made by high-temperature processes (e.g., the former have a higher content of phenols and tar acids and a lower content of medium-soft pitch). Information on other components is presented later in the discussion of high-temperature coal-tars (see below).

Table 2. Some constituents of coal-tars produced by various processes[a]

	Average weight % of dry tar					
	High-temperature processes				Low-temperature processes	
	Coke-oven tars			CVR[b] tars	Conventional low-temperature tars	Lurgi tars
	UK	FRG	USA	UK	UK	UK
Benzene	0.25	0.4	0.12	0.22	0.01	0.02
Toluene	0.22	0.3	0.25	0.22	0.12	0.05
ortho-Xylene	0.04	-	0.04	0.06	0.05	0.05
meta-Xylene	0.11	0.2	0.07	0.13	0.10	0.07
para-Xylene	0.04	-	0.03	0.05	0.04	0.03
Ethylbenzene	0.02	-	0.02	0.03	0.02	0.04
Styrene	0.04	-	0.02	0.04	0.01	0.01
Phenol	0.57	0.5	0.61	0.99	1.44	0.97
ortho-Cresol	0.32	0.2	0.25	1.33	1.48	1.14
meta-Cresol	0.45	0.4	0.45	1.01	0.98	1.83
para-Cresol	0.27	0.2	0.27	0.86	0.87	1.51
Xylenols	0.48	-	0.36	3.08	6.36	5.55
High-boiling tar acids	0.91	-	0.83	8.09	12.89	11.95
Naphtha	1.18	-	0.97	3.21	3.63	3.02
Naphthalene	8.94	10.0	8.80	3.18	0.65	2.01
α-Methylnaphthalene	0.72	0.5	0.65	0.54	0.23	0.63
β-Methylnaphthalene	1.32	1.5	1.23	0.68	0.19	1.05
Acenaphthene	0.96	0.3	1.05	0.66	0.19	0.57
Fluorene	0.88	2.0	0.64	0.51	0.13	0.62
Diphenylene oxide	1.50	1.4	-	0.68	0.19	0.57
Anthracene	1.00	1.8	0.75	0.26	0.06	0.32
Phenanthrene	6.30	5.7	2.66	1.75	1.60	0.28
Carbazole	1.33	1.5	0.60	0.89	1.29	0.22
Tar bases[c]	1.77	0.73	2.08	2.09	2.09	2.50
Medium-soft pitch	59.8	54.4	63.5	43.7	26.0	33.1

[a]From McNeil (1983)
[b]CVR, continuous vertical-retort process
[c]Tar bases include quinoline, isoquinoline, methylquinolines, pyridine, α- and β-picolines, ethylpyridine and lutidines; only a minority of the basic constituents in CVR and low-temperature tars have been identified.

High-temperature coal-tars [class 1.1]

High-temperature coal-tars are the condensation products obtained by the cooling of the gas evolved in the high-temperature (>700°C) carbonization of coal. High-temperature coal-tars are of two main types: coke-oven tars and continuous vertical-retort (CVR) tars.

Coke ovens for blast furnaces use the highest temperature (1250-1350°C); a slightly lower temperature (1000-1100°C) is applied in continuous vertical-retorts for the manufacture of domestic heating coke and gas (McNeil, 1983).

The aromaticity of the tars increases and the content of paraffins and phenols decreases when the carbonization temperature increases (McNeil, 1983). Thus, coke-oven tars contain relatively small amounts of aliphatic hydrocarbons, whereas CVR tars contain a higher proportion of straight-chain or slightly branched-chain paraffins (about 20% in the lower-boiling fractions of tar to 5-10% in the higher distillate oils). Coke-oven tars contain about 3% of phenolic compounds in the fractions distilling at up to 300°C. CVR tars, on the other hand, contain 20-30% of phenols (McNeil, 1983).

Both aromatic and heterocyclic rings occur in substituted and unsubstituted forms. The aromatic compounds in CVR tars are mostly alkyl derivatives, whereas coke-oven tars consist predominantly of compounds containing unsubstituted rings (McNeil, 1983).

Both theoretical and analytical results indicate that the polynuclear aromatic hydrocarbon (PAH) profile of tars is relatively independent of the starting material, being mainly a function of temperature (Kleffner et al., 1981). PAH concentrations found in high-temperature coal-tars are as shown in Table 3.

Table 3. Concentrations (%) of polynuclear aromatic hydrocarbons (PAH) found in high-temperature coal-tars[a]

PAH	Concentration (%)	PAH	Concentration (%)
Naphthalene	10	Benzo[e]pyrene	0.50
Anthracene	1.5	Benzo[a]fluoranthene	0.30
Phenanthrene	5.0	Benzo[b]fluoranthene	0.30
Pyrene	2.1	Benzo[j]fluoranthene	0.30
Fluoranthene	3.3	Benzo[k]fluoranthene	0.40
Tetracene (naphthacene)	0.25	Dibenz[a,j]anthracene	0.10
Benz[a]anthracene	0.65	Dibenz[a,h]anthracene	0.10
Chrysene	1.1	Picene	0.15
Triphenylene	0.13	Benzo[ghi]perylene	0.55
Perylene	0.25	Anthanthrene	0.18
Benzo[a]pyrene	0.55	Indeno[1,2,3-cd]pyrene	0.50

[a]From Kleffner et al. (1981)

Table 4. Concentrations of polynuclear aromatic hydrocarbons (PAHs) found in coke-oven tars[a]

PAH	Concentration (%)
Naphthalene	1.0-5.0
Biphenyl	0.1-1.0
Fluorene	0.0-1.0
Phenanthrene	8.7
Anthracene	5.5
Fluoranthene	1.0-5.0
Pyrene	9.9
Chrysene and/or triphenylene	2.4
Benzo[a]pyrene	0.4
Benzo[e]pyrene	0.4
Perylene	0.1-1.0

[a]From Novotny et al. (1981)

The idea that the carbonization temperature is the most important parameter affecting PAH composition is also supported by a study by Novotny et al. (1981). They found that coal-tar samples derived from coals of four different geographical origins were similar in their major constituents. The main PAHs found in coke-oven tars in that study were as shown in Table 4.

Table 5. Polynuclear aromatic hydrocarbons (PAHs) found in high-temperature coal-tars

PAH	Type of coal-tar	PAH concentration (mg/kg)	Reference
Anthracene	Pharmaceutical Crude	2880-4350 11 000	Lijinsky et al. (1963) Mariich & Lenkevich (1978)
Benz[a]anthracene	Pharmaceutical Samples from Detroit	6240-6980 210-602	Lijinsky et al. (1963) Colucci & Begeman (1965)
Benzo[b]chrysene	Pharmaceutical	800-930	Lijinsky et al. (1963)
Benzo[b]fluoranthene	Crude Samples from Detroit	Detected 209-492	Kruber & Oberkobusch (1952) Colucci & Begeman (1965)
Benzo[j]fluoranthene	Pharmaceutical Samples from Detroit	450-630 62-205	Lijinsky et al. (1963) Colucci & Begeman (1965)
Benzo[k]fluoranthene	Pharmaceutical Samples from Detroit	1070-1080 86-608	Lijinsky et al. (1963) Colucci & Begeman (1965)
Benzo[ghi]perylene	Pharmaceutical Samples from Detroit	1230-1890 317-664	Lijinsky et al. (1963) Colucci & Begeman (1965)
Benzo[a]pyrene	Pharmaceutical Experimental Crude Samples from Detroit	1760-2080 149 500 93-354	Lijinsky et al. (1963) Tomkins et al. (1980) Zorn (1978) Colucci & Begeman (1965)
Benzo[e]pyrene	Pharmaceutical	1850-1880	Lijinsky et al. (1963)
Chrysene	Pharmaceutical Crude Samples from Detroit	2130-2860 5600 99-728	Lijinsky et al. (1963) Mariich & Lenkevich (1973) Colucci & Begeman (1965)
Dibenz[a,c]anthracene	Crude	Detected	Lang et al. (1959)
Dibenz[a,h]anthracene	Pharmaceutical Crude	230-300 Detected	Lijinsky et al. (1963) Lang et al. (1959)
Dibenz[a,j]anthracene	Crude	Detected	Lang et al. (1959)
Dibenzo[a,i]pyrene	Crude	Detected	Schoental (1957)
Fluoranthene	Pharmaceutical Crude Samples from Detroit	17 700-17 800 21 500 94-820	Lijinsky et al. (1963) Mariich & Lenkevich (1973) Colucci & Begeman (1965)
Fluorene	Crude	13 700	Mariich & Lenkevich (1973)
Indeno[1,2,3-cd]pyrene	Crude Samples from Detroit	Detected 135-457	Lang et al. (1959) Colucci & Begeman (1965)
Perylene	Pharmaceutical	700-760	Lijinsky et al. (1963)
Phenanthrene	Pharmaceutical Crude	13 600-17 500 43 000	Lijinsky et al. (1963) Mariich & Lenkevich (1973)
Pyrene	Pharmaceutical Crude Samples from Detroit	7950-10 500 12 300 144-1045	Lijinsky et al. (1963) Mariich & Lenkevich (1973) Colucci & Begeman (1965)

In laboratory coking experiments, Masek (1966) observed that the formation of benzo[a]pyrene started at 700°C and increased with temperature. At 1100°C, the highest temperature investigated, the benzo[a]pyrene content of the coal-tar was 0.2%.

Modern, computerized capillary gas chromatography/mass spectrometry makes detailed PAH analysis possible. Thus, Novotny et al. (1981) were able to make the structural assignments of 40 neutral PAHs, and Romanowski et al. (1983) recorded more than 200 different PAHs in a high-temperature coal-tar. More than 140 components of a high-temperature coal-tar distillate were separated and identified or characterized by their mass spectra by Borwitzky and Schomburg (1979). Data from several studies on coal-tars are summarized in Table 5.

The differences in PAH concentrations given in various reports are the result at least partially of analytical difficulties. This was especially true of earlier studies.

Several nitrogen-containing compounds have been identified in high-temperature coal-tars, and carbazole is one of the main ones. In coke-oven tar, the average percentage by weight is given as 1.33 in the UK, 1.5 in the Federal Republic of Germany and 0.6 in the USA. A percentage of 0.89 is reported for CVR tars (McNeil, 1983). A Ukrainian coal-tar was found to have a carbazole content of 1.40% (Mariich & Lenkevich, 1973). High-temperature coal-tars contain several other aromatic nitrogen compounds (see Table 6). In addition, N-methylcarbazole, pyridine, acridine, picolines, methyl- and dimethylquinolines, benzoquinolines, benzocarbazoles, benzindoles and methylbenzindoles (Novotny et al., 1981; McNeil, 1983) have been identified. Kosuge et al. (1982) isolated five aza-polynuclear aromatic hydrocarbons (phenaleno[1,9-gh]quinoline, pyrido[3,2-c]carbazole, quino[3,2-c]carbazole, benzo[h]naphtho[2,1,8-def]quinoline and benzo[de]naphtho[1,8-gh]quinoline) and six aminopolynuclear aromatic hydrocarbons (1-aminofluoranthene, 1-aminophenanthrene, 3-aminophenanthrene, 1-aminopyrene, 2-aminopyrene and 5-aminobenz[a]anthracene) from a basic extract of a coal-tar (exact type not given, but probably a coke-oven tar).

Table 2 gives typical average percentages by weight of phenol and cresols in coal-tars. High-temperature coal-tars also contain quinones (e.g., di-isopropylhydroquinone and tert-butylhydroquinone) and several polyaromatic phenols, such as naphthols, methylnaphthols, fluorenols, pyrenols and fluoranthenols (Novotny et al., 1981). Collin and Zander

Table 6. Some heteronuclear nitrogen compounds in high-temperature coal-tars

Compound	Average weight % of dry tar	Reference
Acridine	0.6	Collin & Zander (1982)
Quinoline	0.3 0.23	Collin & Zander (1982) Mariich & Lenkevich (1973)
Phenanthridine	0.2	Collin & Zander (1982)
Isoquinoline	0.2	Collin & Zander (1982)
2-Methylquinoline	0.2	Collin & Zander (1982)
7,8-Benzoquinoline	0.2	Collin & Zander (1982)
Indole	0.2 0.15	Collin & Zander (1982) Mariich & Lenkevich (1973)
2-Methylpyridine	0.02	Collin & Zander (1982)

(1982) have reported a 2,3-benzodiphenylene oxide content of 0.2% in high-temperature coal-tars. Sulphur-containing aromatic hydrocarbons in high-temperature coal-tars include thiophene (McNeil, 1983), benzo[b]thiophene (0.3%; Collin & Zander, 1982), dibenzothiophene (0.3%; Collin & Zander, 1982) and phenanthro[4,5-bcd]thiophene (Novotny et al., 1981).

Low-temperature coal-tars [class 1.2]

Low-temperature coal-tars are the condensation products obtained by cooling of the gas evolved in low-temperature (<700°C) carbonization of coal. They are black viscous liquids more dense than water (US Environmental Protection Agency, 1979) and are less aromatic than high-temperature coal-tars. The content of aromatic hydrocarbons (usually these are alkyl-substituted) is only 40-50%. Low-temperature coal-tars also contain 30-35% of non-aromatic hydrocarbons and about 30% of alkali-extractable phenolic compounds in their distillate oils (Considine, 1974; McNeil, 1983).

The total nitrogen content of a coal-tar prepared in a laboratory at 600°C from a German coal was 1.05%, about 50% of which could be separated by acid extraction. The non-basic nitrogen compounds identified were carbazole and its alkyl and aryl derivatives. In the basic fraction, 94 nitrogen compounds were detected. The aza compounds were found to be predominantly basic nitrogen compounds, and appeared to be derived schematically by the replacement of one benzene ring by a pyridine ring (Burchill et al., 1983a).

Coal-tars produced in synthetic natural-gas processes in which low-temperature pyrolysis of coal is employed, e.g., the Lurgi gasification process, are similar in composition and properties to other low-temperature coal-tars (McNeil, 1983). Table 2 gives some components of selected low-temperature and Lurgi-process coal-tars.

More information on the composition of low-temperature coal-tars is given in the IARC monograph on coal gasification (IARC, 1984a).

Distillate fractions (tar oils) [classes 2, 3, 4, 5 and 6]

The distillate fractions (collectively termed tar oils) obtained by fractional distillation of crude coal-tars have been commonly described in the past as creosote. This term is now more frequently regarded as referring to the fractions or blends of fractions specifically used for timber preservation. To compound this confusion in terms further, in a publication by the US Environmental Protection Agency (1981) it is stated that the term 'creosote' refers in EPA documents to coal-tar, creosote and coal-tar neutral oil.

Creosote has also been called brick oil, coal-tar creosote, coal-tar oil, creosote from coal-tar, creosote oil, creosotum and dead oil. However, in current terminology, 'creosote oil' is used to describe a distillation fraction which is only one of the fractions used to make creosote.

The term 'creosote' in this monograph refers only to the products made from coal-tars and not to 'creosote, wood' (Chem. Abstr. Services Reg. No. 8021-39-4).

High-temperature coal-tar distillate fractions and their mixtures

The higher-boiling (>205°C) distillate fractions of high-temperature coal-tars are as follows.

Naphthalene oil [class 2.1] (boiling range, 210-220°C, solidifying-point, 65-75°C) (Collin & Zander, 1982) contains naphthalene, phenols, benzothiophene and other aromatic compounds.

Creosote oil/wash oil [class 3.1] (boiling range, 230-290°C) contains, besides small amounts of naphthalene, naphthalene derivatives, indole, diphenyl, acenaphthene, diphenylene oxide, fluorene, phenol derivatives (2%) and quinoline base (4-6%) (Collin & Zander, 1982). A benzo[a]pyrene concentration of 0.01% was found in a wash-oil sample (Masek, 1973).

Anthracene oil [class 4.1 and 4.2] (boiling range, 300-450°C) is a semisolid, greenish-brown crystalline material. Anthracene oil from the primary distillation of coal-tars is obtained in two fractions. The lower-boiling fraction (light anthracene oil) has a high content of phenanthrene, anthracene and carbazole; the higher-boiling fraction (heavy anthracene oil) has a high content of fluoranthene and pyrene (Collin & Zander, 1982). Benzo[a]pyrene concentrations ranged between 0.01 and 0.06% in samples of anthracene oil [classes 4.1 and 4.2] (Masek, 1973). The analysis of a commercial anthracene oil [class 4.3] is shown in Table 7.

Table 7. Components of an anthracene oil [class 4.3] (mg/g)[a]

Anthracene	34.8[b]
Fluoranthene	64.7[c]
Pyrene	125[c]
Benz[a]anthracene	7.3[c]
Benzo[e]pyrene	0.02[b]
Dibenz[a,h]anthracene and dibenz[a,c]anthracene	0.03[b]

[a]From Lehmann et al. (1984)
[b]Determined by gas chromatography
[c]Determined by high-performance liquid chromatography

Creosote has a varying composition but can be described generally as a yellow-dark-green-brown oily liquid (Hawley, 1981), consisting of aromatic hydrocarbons, e.g., anthracene, naphthalene and phenanthrene derivatives, some tar acids (e.g., phenol, cresols and xylenols) (McNeil, 1983) and tar bases (e.g., pyridine and lutidine derivatives) (Considine, 1974).

One source (Nestler, 1974) has reported 162 compounds which can be expected to be found in creosote. This was based on a compilation of compounds identified from coal carbonization and assumed that creosote, as a bulk distillate of coal-tars, would contain those compounds whose boiling-points fall within the boiling range of creosote. Of these 162 compounds, only a limited number (less than 20) have been identified at levels exceeding 1% but these constitute the major portion of creosote. PAHs (mostly unsubstituted) generally account for at least 75% of creosote (Lorenz & Gjovik, 1972).

The major components of some creosotes have been reported by several sources. These data are presented in Table 8.

One source (Lijinsky et al., 1963) has reported the presence of the following additional polynuclear aromatic compounds (PACs) in creosote: benzo[j]fluoranthene, benzo[k]fluoranthene, benzo[e]pyrene, benzo[b]chrysene and perylene.

Table 8. Major components of some creosotes

Component	Creosote				
	A[a]	B[b]	C[c]	D[d]	E[e]
Acenaphthene	9.0	14.7	3.1	9.0	-
Anthracene	2.0	-[f]	1.5	7.0	-
Benz[a]anthracene	-	-	-	-	0.16-0.26
Benzo[a]pyrene	-	-	-	-	0.04-0.06
Benzofluorenes	2.0	1.0	-	-	-
Biphenyl	0.8	1.6	-	1.0	-
Carbazole	2.0	1.2	2.4	-	-
Chrysene	3.0	2.6	-	-	-
Dibenz[a,h]anthracene	-	-	-	-	0.01-0.04
Dibenzofuran	5.0	7.5	1.1	4.0	-
9,10-Dihydroanthracene	-	-	0.2	-	-
Dimethylnaphthalenes	2.0	2.3	-	-	-
Fluoranthene	10.0	7.6	3.4	3.0	0.49-0.93
Fluorene	10.0	7.3	3.1	9.0	-
Methylanthracenes	4.0	3.9[g]	-	-	-
Methylfluorenes	3.0	2.3	-	-	-
1-Methylnaphthalene	0.9	1.7	3.0[h]	12[h]	-
2-Methylnaphthalene	1.2	2.8	3.0[h]	12[h]	-
Methylphenanthrenes	3.0	-[i]	-	-	-
Naphthalene	3.0	1.3	15.8	18.0	-
Phenanthrene	21.0	17.4[j]	10.7	16.0	-
Pyrene	8.5	7.0	2.2	1.0	-

[a]Probably a creosote [classes 3, 4, 5 and 6]; results of replicate runs were within ±0.7% (Lorenz & Gjovik, 1972)
[b]Creosote [classes 3, 4, 5 and 6]; results of replicate runs were within ±0.7% (Lorenz & Gjovik, 1972)
[c]Data are averages of nine creosote samples (class 3) (Stasse, 1954; Nestler, 1974)
[d]Typical creosote [classes 3, 4, 5 and 6] used as a wood preservative for the impregnation of railway sleepers (Andersson et al., 1983)
[e]Range from three different creosotes for wood impregnation [classes 3, 4 and 5] (Lehmann et al., 1984)
[f]Data included under phenanthrene
[g]Data include methylphenanthrenes
[h]Methylnaphthalenes
[i]Data included under methylanthracenes
[j]Data include anthracene

Table 9. Polynuclear aromatic hydrocarbons (PAHs) found in saleable mixtures of primary distillation fractions of high-temperature coal-tars

PAH	Product	PAH concentration (mg/kg)
Benz[a]anthracene	Three road tars	8400-13 100
	Protective paint	5800
Benzo[a]pyrene	Three road tars	5300-7600
	Protective paint	13 100
	Coal-derived fuel oil[b]	82.3[b]
Dibenz[a,h]anthracene	Three road tars	not detected-1300
	Protective paint	2600
Fluoranthene	Three road tars	17 400-30 900
	Protective paint	9000
Pyrene	Three road tars	22 000-35 800
	Protective paint	20 900

[a]From Lehmann et al. (1984), unless otherwise indicated
[b]Tomkins et al. (1980)

In a sample of US commercial creosote, benzo[a]pyrene was present at a concentration of about 0.3% (Black, 1982). According to analyses conducted by the Institute of Occupational Health in Helsinki, the concentration of benzo[a]pyrene in creosotes available in Finland was 1 g/kg (Schimberg, 1981).

Other saleable mixtures of primary distillation fractions of high-temperature coal-tars include, for example, carbon-black oil, road tar and protective paint (see Table 15). Table 9 summarizes the concentrations of some PAHs in various such products.

Low-temperature coal-tar distillate fractions and their mixtures

No quantitative data were available to the Working Group regarding the chemical composition of low-temperature coal-tar distillate fractions and their mixtures.

Coal-tar pitch [class 7]

Coal-tar pitch is a dark-brown-black, shiny, amorphous residue produced during the distillation of coal-tars (Hawley, 1981). Pitch is composed of many different compounds which interact to form eutectic mixtures; consequently it does not show a distinct melting or crystallization point. Rather, it is characterized by its softening point (the temperature at which a given viscosity is reached). Depending on the depth of distillation, pitches with different softening points can be obtained (Considine, 1974), e.g., medium-soft pitch or very hard pitch.

Pitch contains PAHs and their methyl and polymethyl derivatives, as well as heteronuclear compounds. With increasing molecular weight of the fractions, only qualitative information is available about pitch constituents. The ring systems become larger, more heteronuclear atoms are present, and the number and length of alkyl chains decrease (Considine, 1974; McNeil, 1983). Benzo[a]carbazole, benzo[b]carbazole and benzo[c]carbazole were detected by Bender *et al.* (1964) in a coal-tar pitch made by an unspecified process [class 7.3]. Benzo[a]pyrene concentrations of 100-350 µg/kg were found in lignite-tar pitches [class 7.3] used in the German Democratic Republic (Schunk, 1979).

High-temperature coal-tar pitches [class 7.1]

The least complex of the PAHs in high-temperature coal-tar pitches are the compounds with four rings; e.g., chrysene, fluoranthene, pyrene and benz[a]anthracene. The compounds with four to seven rings constitute about 40-50% of a medium-soft coke-oven pitch. PAHs found in high-temperature pitches in various studies are summarized in Table 10.

Some 20 nitrogen compounds (including quinoline, indole, 4-azafluorene, 7,8-benzoquinoline, 2,3-benzoquinoline, 5,6-benzoquinoline, carbazole, methylcarbazole, azafluoranthene, azapyrene, phenanthro[bcd]pyrrole, benz[c]acridine, dibenzoquinoline, benzocarbazole and 2,3-benzocarbazole) were identified in a coke-oven pitch by Burchill *et al.* (1983b). The concentrations ranged from 62-2690 mg/kg.

The following PAHs were found in old coal-tar pitches in the form of the bulk tear-off dust of pitch roofs: phenanthrene, anthracene, fluoranthene, pyrene, benz[a]anthracene, chrysene, benzo[b]fluoranthene, benzo[k]fluoranthene and benzo[a]pyrene (Reed, 1983).

Table 10. Polynuclear aromatic hydrocarbons (PAHs) found in high-temperature coal-tar pitches

PAH	Use of coal-tar pitch	PAH concentration (mg/kg)	Reference
Anthracene	Not given	1.5	Gorman & Liss (1984)[a]
Anthracene and/or phenanthrene	Roofing	34 000	Hervin & Emmett (1976)
Benz[a]anthracene	Not given Roofing Roofing	8.4 8900-12 500 169 000-324 000	Gorman & Liss (1984) Wallcave et al. (1971) Malaiyandi et al. (1982)
Benz[a]anthracene and chrysene	Roofing	16 000	Hervin & Emmett (1976)
Benzo[b]fluoranthene	Not given	22.4	Gorman & Liss (1984)
Benzo[ghi]fluoranthene	Electrodes	400-3400	Arrendale & Rogers (1981)
Benzo[k]fluoranthene	Not given Roofing Roofing	11.2 1670-4500 12 000	Gorman & Liss (1984) Malaiyandi et al. (1982) Hervin & Emmett (1976)
Benzo[ghi]perylene	Roofing	754-3980	Malaiyandi et al. (1982)
Benzo[a]pyrene	Not given Electrodes Roofing Roofing	16 4500-13 200 8400-12 500 4290-11 000	Gorman & Liss (1984) Arrendale & Rogers (1981) Wallcave et al. (1971) Malaiyandi et al. (1982)
Benzo[a]pyrene and benzo[e]pyrene	Roofing	10 000	Hervin & Emmett (1976)
Benzo[e]pyrene	Not given Roofing	14.3 5400-7000	Gorman & Liss (1984) Wallcave et al. (1971)
Chrysene	Not given Electrodes Roofing Roofing	19.6 2600-17 000 42 500-88 000 7400-10 000	Gorman & Liss (1984) Arrendale & Rogers (1981) Malayandi et al. (1982) Wallcave et al. (1971)
Dibenz[a,c]anthracene	Not given	Detected	Lang et al. (1959)
Dibenz[a,h]anthracene	Not given Roofing	Detected 317-1680	Lang et al. (1959) Malaiyandi et al. (1982)
Dibenz[a,j]anthracene	Not given	Detected	Lang et al. (1959)
7,12-Dimethylbenz[a]anthracene	Roofing	1070-2950	Malaiyandi et al. (1982)
Fluoranthene	Electrodes Roofing Roofing	5200-38 800 19 800-32 500 28 000	Arrendale & Rogers (1981) Malaiyandi et al. (1982) Hervin & Emmett (1976)
Fluorene	Electrodes	800-4000	Arrendale & Rogers (1981)
Indeno[1,2,3-cd]pyrene	Not given Roofing Roofing	Detected Not detected-2460 7300-9300	Lang et al. (1959) Malaiyandi et al. (1982) Wallcave et al. (1971)
Perylene	Roofing	2090-5840	Malaiyandi et al. (1982)
Phenanthrene	Not given Electrodes	5.4 7500-40 300	Gorman & Liss (1984) Arrendale & Rogers (1981)
Pyrene	Not given Electrodes Roofing Roofing	14.3 4500-34 900 17 500-23 600 20 000	Gorman & Liss (1984) Arrendale & Rogers (1981) Malaiyandi et al. (1982) Hervin & Emmett (1976)
Total PAH content	Roofing	265 000-500 000	Malaiyandi et al. (1982)

[a] The Working Group noted the large difference between the PAH levels reported by Gorman & Liss and those reported by other authors.

COAL-TARS AND DERIVED PRODUCTS

Low-temperature coal-tar pitches [Class 7.2]

No data were available to the Working Group regarding the concentration of individual PACs. One source (McNeil, 1983) has reported that low-temperature coal-tar pitches contain hydroxy and polyhydroxy derivatives of PAHs.

1.3 Chemical and physical properties

Coal-tars [class 1]

Table 11 presents the average composition and properties of coal-tars produced by various commercial processes in the UK, the Federal Republic of Germany and the USA.

Table 11. Average properties of some commercial coal-tars produced by various processes[a]

Property	[Class 1.1] High-temperature processes				[Class 1.2] Low-temperature processes	
	Coke-oven tars			CVR[b] tars,	Conventional low-temperature tars,	Lurgi tars,
	UK	FRG	USA	UK	UK	UK
Yield (L/t)[c]	33.6	26.8	--	70.9	95.5	12.7
Density at 20°C (g/cm^3)	1.169	1.175	1.180	1.074	1.029	1.070
Water (wt%)	4.9	2.5	2.2	4.0	2.2	2.8
Carbon (wt%)	90.3	91.4	91.3	86.0	84.0	84.2
Hydrogen (wt%)	5.5	5.25	5.1	7.5	8.3	7.7
Nitrogen (wt%)	0.95	0.86	0.67	1.21	1.08	1.09
Sulphur (wt%)	0.84	0.75	1.2	0.90	0.74	1.39
Ash (wt%)	0.24	0.15	0.03	0.09	0.10	0.02
TI[d] (wt%)	6.7	5.5	9.1	3.1	1.2	0.7

[a]From McNeil (1983)
[b]CVR, continuous vertical-retort process
[c]L/t, litres per tonne
[d]TI, toluene-insoluble components

Solubility: Coal-tars are completely or nearly completely soluble in benzene and nitrobenzene; partially soluble in acetone, carbon disulphide, chloroform, diethyl ether, ethanol, methanol, petroleum ether and sodium hydroxide solution; and slightly soluble in water (Windholz, 1983).

High-temperature coal-tar distillation fractions and residues [classes 2.1, 3.1, 4.1, 4.2 and 7.1]

Descriptions of typical fractions from the distillation of high-temperature coal-tars were not available to the Working Group. Therefore, information from selected UK manufacturers has been used.

Table 12 gives properties of the distillation fractions of high-temperature coal-tar from one UK company (Thomas Ness Ltd, 1984).

Table 12. Properties of the distillation fractions of high-temperature tar from one UK company[a]

Property	Naphthalene oil [class 2.1]	Creosote oil/wash oil[b] [class 3.1]	Light anthracene oil[c] [class 4.1]	Heavy anthracene oil/base oil [class 4.2]	Pitches [class 7.1] Medium-soft	Pitches [class 7.1] Hard (electrode)
Specific gravity (15.5°C)	1.030	1.062	1.092	1.155	1.292 (20°C)	-
Water (% v/v)	0.5	0.2	0.2	0.1	Trace	0.15 (% w/w)
Tar acids (% v/v)	0.5	-	3.5	-	-	-
Flash-point (°C, method of Pensky Martens)	170	99	99	143	-	-
Crystallizing-point (°C)	60	-	-	-	-	-
Softening-point (°C, method of Ring & Ball)	-	-	-	-	80.0	103.0
Viscosity (method of Redwood I at 37.8°C; sec)	-	40 (21°C)	43	35	-	-
Volatile matter (% w/w)	-	-	-	-	63.0	-
Toluene insolubles (% w/w)	-	-	-	-	25.0	37.0
Quinoline insolubles (% w/w)	-	-	-	-	4.8	15.0
Ash (% w/w)	-	-	-	-	0.35	0.15
Distillation (% v/v, °C) 5 / 10 / 50 / 90 / 95	208 / 211 / 218 / 228 / 235	- / - / - / Up to 300 / -	226 / 239 / 306 / 345 / 353	296 / 316 / 373 / - / -	- / - / - / - / -	Water/up to 110°C, 0.15; 110-170°C, nil; 170-230°C, nil; 230-270°C, 0.4%w/w; 270-300°C, 0.4%w/w; 300-360°C, 4.5%w/w

[a]From Thomas Ness Ltd (1984)
[b]After alkaline washing to remove the tar acids
[c]After centrifugation to recover an anthracene-rich portion

Low-temperature coal-tar distillation fractions [classes 5, 6.1, 6.2, 7.2]

Another UK company has reported the properties of some low-temperature distillation fractions and their mixtures (Coalite & Chemical Products Ltd, 1984); these are given in Table 13.

Table 13. Properties of the distillation fractions of low-temperature tar and their mixtures from one UK company[a]

Property	Creosote oil [class 5]	Refined tar suitable for blending with bitumen [class 8; probably classes 6 and 7.2]	Medium-soft pitch [class 7.2]
Specific gravity (20°C)	1.024	1.10	1.14-1.20
Water (% v/v)	0.2	0.5 (% w/w max)	0.5 (% w/w max)
Tar acids (% v/v)	14 (in distillate up to 315°C)	-	-
Flash-point (°C, method of Pensky Martens)	132	175	-
Softening-point (°C, method of Ring & Ball)	-	-	73-77
Viscosity (EVI, °C)	-	54 ± 1.5	-
Toluene insolubles (% w/w)	0.2	6.0 (% w/w max)	6.0 (% w/w max)
Ash (% w/w max)	-	-	0.8
Distillation (% v/v, °C)			
2	230	-	(Nil below 300°C)
5	-	322	
45	315	-	
50	-	380	
70	-	405	
85	355	-	

[a]From Coalite & Chemicals Products Ltd (1984)

1.4 Technical products and impurities

Coal-tars, pharmaceutical grade

Coal-tar is available in the USA in a US Pharmacopeia (USP) grade with a maximum residue on ignition of 2.0%. Coal-tar ointment, USP (obtained by combining coal-tar with Polysorbate 80 [a sorbitan mono-oleate polyoxyethylene derivative] and blending with zinc oxide paste) and coal-tar topical solution, USP (made by combining coal-tar with Polysorbate 80 and diluting with ethanol to an ethanol content of 81.0-86.0%) are also available in the USA (US Pharmacopeial Convention, 1980).

Coal-tar gained admission to the *British Pharmacopoeia* in 1898. The official preparations of coal-tar in the UK are: *pix carbonis*, *pix carbonis praeparata* and *liquor picis carbonis*. *Pix carbonis* is crude commercial tar. *Pix carbonis praeparata* is commercial coal-tar heated at 50°C for one hour to remove some of the lighter oil fractions. *Liquor picis carbonis* is made by macerating prepared coal-tar 20% and quillaia 10% in 90% ethanol for seven days and filtering (Rook *et al.*, 1956).

Reported concentrations of individual PAHs in pharmaceutical-grade coal-tars are included in Table 5, p. 88.

Creosote for wood preservation

The American Wood-Preservers' Association (AWPA) standards, P1-78 for new creosote and the creosote in use for land and fresh water and P2-68 for creosote and creosote solutions, are given in Table 14.

Table 14. The American Wood-Preservers' Association (AWPA) standards P1-78 and P2-68 for creosotes[a]

Property	PI-78		P2-68 (Grade A)	
	New	In use	New	In use
Specific gravity (38°C)	1.05	1.05	1.06-1.11	1.06-1.11
Water (% v/v max)	1.5	3.0	3.0	3.0
Xylene insolubles (% w/w max)	0.5	1.5	-	-
Benzene insolubles (% w/w max)	-	-	2.0	3.0
Distillation (% w/w)				
up to 210°C	<2	<2	<5	<5
up to 235°C	<12	<12	<25	<25
up to 270°C	10-35	10-35	-	-
up to 315°C	40-65	40-65	>36	>36
up to 355°C	60-77	60-77	>60	>60

[a]From American Wood-Preservers' Association (1983); the AWPA standards also include other requirements not shown here.

The specifications established by some European state railways for the distillation fractions of creosote used on railway sleepers are given in Table 15 (Schulz, 1983).

High-temperature coal-tar pitches [class 7.1]

Coal-tar pitches used in roofing, damp-proofing, and waterproofing are covered by ASTM Standard Specification D450-78 (American Society for Testing and Materials, 1983). Three grades of coal-tar pitch are defined by several properties, including a so-called total bitumen content (soluble in carbon disulphide) of 72-85%, a maximum ash content of 0.5%, and a maximum water content of 0%.

Pitches used in other applications (e.g., electrode binders and tar-based pipeline coatings) are covered by some national specifications, but are mainly sold to user specifications based on performance (McNeil, 1983).

Table 15. Specifications of some European state railways for the distillation fractions of creosote used on railway sleepers[a]

Country	Railway	Distillation fractions (%)												
		150 °C	170 °C	200 °C	205 °C	210 °C	230 °C	235 °C	270 °C	300 °C	315 °C	355 °C	360 °C	375 °C
Belgium	SNCB					≤2[c]		≤12[c]	20-40[c]		45-65[c]	65-82[c]		
Federal Republic of Germany	DB							≤10[c]		20-40[c]		55-75[c]		
	BP							≤15		≤30		65-85		
France	SNCF	≤0.5[b]		≤4[b]										
	PTT		≤1[b]											
United Kingdom	BSI				≤6			≤40[b]			≤60[b]			
	BSI				≤5		≤40				≤65[b]		≤85[b]	
	GPO				≤5		5-30				≤78	73-85		
							10-30				40-78	73-85		
											≤78			
Italy	FS	≤2[c]						≤25[c]						
	ENEL	≤3		≤10				≤25	≤20					
The Netherlands						≤5[b]		≤15[b]	≤20[b]			75-85[b]		
Austria	ÖBB			≤6[c]				≤15[c]		≤30[c]		65-85		
	Post			≤6[c]				≤15[c]		≤30[c]		≤75		
Denmark, Norway, Sweden and Finland						≤4		≤20						
Switzerland	SBB					≤5[b]		≤10			≤30	60-90		≤90[c]
Spain								5-25[b]	≤20[b]		60-80[b]	≤75[b]		≤90
USA	AWPA					≤2[b]		≤12[b]	20-40[b]		45-65[b]	65-82[b]		
	AWPA					≤2[b]		≤12[b]	20-40[b]		45-65[b]	65-75[b]		

[a]From Schultz (1983); the specifications also include other requirements not shown here
[b]vol %
[c]wt %

2. Production, Use, Occurrence and Analysis

2.1 Production and use

Coal-tars

(a) *Production*

When coal is carbonized to make coke and/or gas, crude coal-tar is one of the by-products. The commonly used carbonization processes have been described in previous IARC monographs (IARC, 1984a). The low-temperature processes used to produce solid smokeless fuels for industrial and home heating have also been described (Codd et al., 1971). A process for the production of tar from lignite, as practised in the 1920s in the UK, was outlined by Kennaway (1924). An unusual blast furnace used to make iron in Scotland, using splint coal as well as coke, generated tar as a by-product (blast-furnace tar) (Berenblum, 1930). The latter two tars were described as being low-temperature tars.

One source (McNeil, 1983) has estimated total world production of crude coal-tars at 18 million tonnes, approximately 14 million tonnes of which were from coke ovens. The production of crude coke-oven tars by several countries during 1975 and 1981, as estimated by this source, is shown in Table 16.

Production of coal-tars world-wide is closely linked with steel production, because of the need for coke in steel making. As the pattern of steel production has altered, corresponding changes in coal-tar availability have occurred. Thus, tar production has decreased in the USA and most countries in western Europe, and increased in Asia, South America

Table 16. Estimated production of crude coke-oven tar by country or region (thousands of tonnes)[a]

Country	Production volume		
	1975	1980	1981
Benelux (Belgium, the Netherlands and Luxembourg)	390	280	240
Canada	-	-	223
Czechoslovakia	250	-	-
France	460	430	426
German Democratic Republic	1000	-	-
Germany, Federal Republic of	1350	1120	1089
Italy	300	308	298
Japan	2000	-	-
Poland	760	-	-
Spain	200	163	162
South Africa	-	382	292
United Kingdom	719	450	403
USA	-	2375	-
USSR	3500	-	-

[a]From McNeil (1983)

and Australia. A new coal-tar industry has developed in South Africa as a consequence of the coal-based production of petroleum substitutes. An additional factor in the rapid decline in UK coal-tar production during the 1960s and 1970s was the discovery and exploitation of natural gas, which entirely displaced gas manufacture by coal carbonization.

In 1913, US production of coal-tar as a by-product of coke production was 436 million litres (US Tariff Commission, 1919). In recent years, US production of coal-tar by coke-oven operators has been declining gradually; in 1981 and 1982, productions of 2.1 million tonnes (472.2 million gallons) and 1.4 million tonnes (316.4 million gallons), respectively, of coal-tar were reported (US International Trade Commission, 1982, 1983).

In the Federal Republic of Germany, coal-tar production declined from 2.1 million tonnes in 1960 to 1.3 million tonnes in 1978 (Schecker, 1984).

Total UK production of coal-tar was 561 thousand tonnes in 1981 (403 tonnes from coke ovens and 158 thousand tonnes from low-temperature ovens). This is down dramatically from the 3 million tonnes produced in 1955 (approximately 60% of which was continuous vertical-retort tar and 1% was low-temperature tar) (McNeil, 1983).

In Japan, coal-tar distillation was carried out industrially for the first time in 1901. Production of coal-tar in Japan in 1982 was 2.58 million tonnes.

(b) Uses and processing

Coal-tar is suitable for burning as a fuel in the steel industry in open-hearth furnaces and blast furnaces because of its availability, its low sulphur content and its high heating value (Perch & Muder, 1974). Use of crude coal-tar as a fuel is not believed to be significant in Europe; an estimated 99% of the coal-tars produced in the UK and the Federal Republic of Germany is distilled.

Both high-temperature and low-temperature coal-tars are reportedly used topically in the treatment of psoriasis and other chronic skin diseases. Coal-tar products are available in the USA in many pharmaceutical vehicles, including creams, ointments, pastes, lotions, bath and body oils, shampoos, soaps and gels. Shampoos are the most important of these products. Coal-tar products reported to be present in these formulations include: coal-tars (levels of 0.18-10%); coal-tar solution (2-48.5%); coal-tar extract (5%); tar distillate (3-25%); coal-tar fraction (1.25%); and acetyl alcohol-coal-tar (4%) (Hoover, 1975; Baker, 1982; Gilbertson, 1982). Coal-tar extract may be used in the USA at levels of 45-55 mg/g in topical neomycin sulphate-hydrocortisone ointment (US Food and Drug Administration, 1982a).

Coal-tar, USP grade, is approved for use in the USA in denatured alcohol, formulae 38-B and 38-F (US Department of the Treasury, 1983).

Several surface-coating formulations in the USA are reported to contain coal-tar at varying concentrations (National Institute for Occupational Safety and Health, 1983a). It is believed to act as the binder and filler in these products and is frequently used as a modifier for epoxy-resin surface coatings.

The US use patterns for crude coke-oven coal-tars in 1980 and 1982 are presented in Table 17.

Table 17. US use patterns for coke-oven coal-tars in 1980 and 1982[a]

Use	1980 volume		1982 volume	
	million litres	%	million litres	%
By coke-oven operators	734	35.2	347	28.4
Refined	492	23.6	-	-
Directly burned as fuel	227	10.9	-	-
Other	15	0.7	-	-
Sold to tar distillers for refining	1352	64.8	873	71.6
TOTAL	2086	100.0	1220	100.0

[a]From US Department of Energy (1981, 1983)

The majority of crude coal-tars produced in the USA are distilled into refined chemicals and bulk products. Table 18 summarizes the available data on the US production of the various products obtained by distillation of coal-tars.

In 1981, UK production of coal-tar pitch totalled 195 thousand tonnes and creosote-oil production amounted to 172 thousand tonnes (McNeil, 1983).

In Japan, an estimated 2.57 million tonnes of crude coal-tars were used in 1982. About 20% of this was burned as fuel; the remaining 80% was distilled to give the following products: pitch, 50%; creosote oil, 35%; naphthalene, 10%; and minor products (tar acids, anthracene, road tar), 5%.

Table 18. US production of coal-tar products in 1982[a]

Product	Production
Crude light oil[b]	358 million litres
Intermediate light oil[b]	12 million litres
Light-oil distillates:	
Benzene[b,c]	64
Solvent naphtha	>9.1 tonnes
Toluene[b]	5 million litres
Xylene[b]	0.96 million litres
Pyridine, crude bases	>2.3 tonnes
Naphthalene, crude	106 thousand tonnes
Methylnaphthalene	>2.3 tonnes
Crude tar-acid oils	>9.1 tonnes
Cresylic acid, crude	>9.1 tonnes
Creosote-oil distillate (100% creosote basis)[c]	137 million litres
Creosote in coal-tar solution (100% solution basis)[c]	168 million litres
Other distillate products:	
Carbon-black oil	>4.5 tonnes
Creosote tar-acid oil	>4.5 tonnes
Crude coal-tar solvent	>4.5 tonnes
Crude tetralin	>4.5 tonnes
Priming and refractory oil	>4.5 tonnes
Tars:	
Road tar	>4.5 tonnes
Crude tar for other uses	>15.9 tonnes
Refined tar for other uses	44 million litres
Pitch of tar	2770 thousand tonnes
Pitch emulsion	>2.3 tonnes
Refined anthracene	>2.3 tonnes

[a]From US International Trade Commission (1983)

[b]Produced by coke-oven operators

[c]See the IARC monograph on benzene (IARC, 1982a) for further information on these two products

History of refining

Following a demonstration by Accum in 1818 that tar could be distilled to give a volatile oil, the London and Westminster Chartered Gas, Light and Coke Company established its own tar works at Poplar for the production of pitch and refined tar for supply to the Royal Navy.

The first method of tar processing to be established was distillation in cast-iron pot-stills to produce two products: a crude light oil, used as a solvent for making paints and varnishes and as a rubber solvent for Mackintosh's waterproofing process (see IARC, 1982b), and a refined tar for use as a protective paint for treating ship's cordage and for making lamp-black.

Tar was first distilled in wrought-iron, riveted vessels, directly fired in brick settings, in batches of 10-30 tons. Later versions of pot-stills were equipped with a fractionation column; some were designed for continuous operation. Although the introduction of continuous pipe-stills marked an important development in tar distillation, a number of pot-stills of improved design are still in operation in some countries.

Distillation of high-temperature coal-tars

Nowadays, continuous stills with capacities of 100-700 tonnes per day (McNeil, 1983) are employed for the distillation of coal-tars made by high-temperature processes [class 1.1]. There are a number of different types of continuous tar-stills in operation, but all operate on the same principles. They include a tube-still furnace through which tar flows continuously, flash chambers in which water and volatiles are separated from the crude tars, and one or more fractionating columns in which the tar vapours are separated into a series of distillation oils of increasing boiling range (Collin & Zander, 1982).

The most common types of tar-stills found in western European countries include: (1) The *Aberhalden* design, which employs a pipe-still furnace and a pitch-flash tower; (2) the *Koppers* design, having coparate columns for each oil fraction; (3) the *Wilton* design, in which heated tar is pumped to the bottom section of the pitch column, and the mixture of tar and hot pitch is pumped through the main furnace oil to the upper part of the pitch column where the distillate oils flash-off; vapours from the pitch column are fed to the fractionating column; and (4) the *Castrop-Rauxel* design, which is capable of producing a wide range of narrow-band tar-oil fractions without secondary fractionating (McNeil, 1983).

Carbolic oil, naphthalene oil [class 2.1], creosote oil [class 3.a], light anthracene oil [class 4.1], heavy anthracene oil [class 4.2], and pitch [class 7.1] are also obtained. See also the section: 'High-temperature coal-tar distillate fractions and their mixtures', p. 90.

Distillation of low-temperature coal-tars

Production of low-temperature coal-tars is carried out in the UK at the present time by two processes: the Coalite and the Rexco processes. The Coalite process has been operated for a considerably longer time and hence the tar has been investigated and developed more fully. In addition, in South Africa the Lurgi process, used as the first step in a Fischer-Tropsch synthesis of hydrocarbons, produces a tar similar to that produced by the Coalite process.

The distillation of Coalite-process coal-tar is as follows. The crude tar, on initial separation, contains 12% of water, it is settled until the water content falls to 4%. The resulting coal oil is fed to a heating section which contains both radiant and convection banks, into a flash dehydrator where the water is removed and then into an atmospheric fractionating column. At this stage light oil and middle oil [class 5] are removed from an offtake point in the column. The residue from this atmospheric column is known as 'reduced crude' and it can either be withdrawn and utilized as a heavy fuel or as a road-construction material, or it can be further processed by feeding it to a vacuum fractionating column. Heavy oil and wax oil [class 6] are removed from offtakes in the column, and pitch [class 7.2] is removed on a continuous basis by means of a barometric leg.

Blending and other secondary processes

The only primary fractions that are sold by tar distillers without further processing are creosote oil and pitch. Even these can be subjected to additional process stages. Tar acid and tar-base concentrates are removed from the primary fractions which contain them, and chemical feedstocks present in relatively high proportion, such as anthracene and naphthalene, are extracted for refining. The remaining portions of the fractions are usually blended in varying proportions to meet product specifications, which are generally given in terms of physical characteristics such as specific gravity, viscosity, flash-point and distillation range.

Figure 1 illustrates the secondary processes and blending operations generally employed for high-temperature tars, and Figure 2 presents operations for low-temperature tars. The figures also show the main uses to which the saleable products are put. [In these figures, rectangles are used to identify processes, capital letters to identify saleable products, and parentheses to identify uses.] The most important blended products are listed in Table 19 for high-temperature tars and in Table 20 for low-temperature tars.

Direct uses of primary distillation fractions and residues from high-temperature tars

The principal uses for these fractions and residues are as follows: (1) *naphthalene oils*, naphthalene production; (2) *creosote/wash oils*, benzene recovery; (3) *anthracene oils*, anthracene paste production; (4) *medium-soft pitches*, briquetting of smokeless solid fuel (briquetting pitch), impregnation of electrodes (impregnation pitch), and impregnation of fibre (fibre-pipe pitches); and (5) *hard (electrode) pitches*, binder in the manufacture of electrodes and graphite.

Blended products from primary distillation fractions of high-temperature coal-tars

Creosote oils are obtained by mixing strained naphthalene oil, wash oil and strained or light anthracene oil, and are mainly used as wood preservatives.

Fluxing oils are obtained by mixing naphthalene oil, wash oil and light anthracene oil, and are used for extending the workability of road binders.

Carbon-black oils are obtained by mixing light anthracene oil and heavy anthracene oil, and are used as a feedstock for production of carbon blacks.

Road tars are mixtures of pitch, creosote oil, light anthracene oil and heavy anthracene oil, and are used for road and paving construction. Road tars may also contain bitumen (15-85%). The composition of such mixtures of tars and bitumens is chosen for the particular application.

Fig. 1. Processing of high-temperature tars

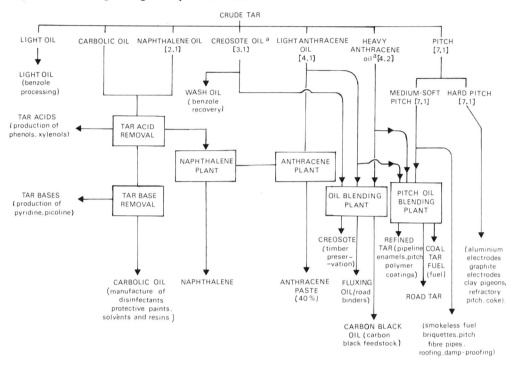

Fig. 2. Processing of low-temperature tars

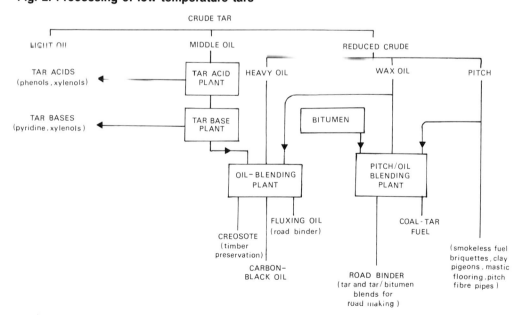

Table 19. Products obtained from high-temperature coal-tar fractions

Fraction	Saleable products (components from first column)
1. Light oil/overheads	Overheads (1)
2. Carbolic oils	Carbolic oils (2)
3. Naphthalene oils [class 2.1]	Naphthalene (3)
4. Creosote oils/wash oils [class 3.1]	
5. Light anthracene oils [class 4.1]	Wash oils (4)
6. Heavy anthracene oils/base oils [class 4.2]	Creosote (3, 4, 5)
7. Pitches [class 7.1]	Fluxing oils (3, 4, 5) Anthracene pastes (5) Carbon-black oils (5, 6) Refined tars (5, 6, 7) Road tars (4, 5, 6, 7) Coal-tar fuels (4, 5, 6, 7) Medium-soft (briquetting) pitches (7) Pitches for roofing and damp-proofing (7) Hard (electrode) pitches (7) Protective paints (2, 7) Pitch/polymer coatings (7 or 2, 5, 6, 7 with aromatic solvents) Pitch refractories (5, 6, 7)

Table 20. Products obtained from low-temperature coal-tar fractions

Fraction	Saleable products (components from first column)
1. Light oils	Light oils (1)
2. Middle oils [class 5]	Creosote (2, 3, 4)
3. Reduced crudes[a]	Fluxing oils (2, 3, 4)
4. Heavy oils [class 6.1]	Carbon-black oils (5)
5. Wax oils [class 6.2]	Coal-tar fuels (4, 5, 6)
6. Pitches [class 7.2]	Road tars (2, 4, 6) Tar/bitumen blends (2, 3, 6 and bitumen) Medium-soft pitches (6)

[a][*], starting material for classes 6.1, 6.2 and 7.2

Coal-tar fuels are mixtures of pitch, creosote oil, light anthracene oil and heavy anthracene oil, and are used for heating industrial premises.

Protective paints are solutions of pitches in coal-tar oils and various solvents, and are used as anti-corrosion paints.

Pitch/polymer coatings are blends incorporating the use of pitches as extenders for epoxy, urethane and other resins, and are used as damp-proofing and protective coatings in corrosive environments.

Direct uses of primary distillation fractions from low-temperature coal-tars

Carbon-black oils can be made directly from wax oils.

COAL-TARS AND DERIVED PRODUCTS

Pitches are used for smokeless-fuel briquetting and for the manufacture of clay-pigeon targets, plastic flooring tiles and pitch-fibre pipes.

Blended products from primary distillation fractions from low-temperature coal-tars

Industrial fuel oils

Fluxing oils are blends of washed middle oil, reduced crude and heavy oil, and are used to extend the workability of road binders.

Tar/bitumen blends are blends of washed middle oil, reduced crude and pitch with bitumen and are used for road making. Low-temperature tars are more compatible with bitumen than are high-temperature tars. See also the monograph on bitumens in this volume.

Road tars are blends of washed middle oil, heavy oil and pitch, and are used for road making. Road tar may also contain bitumen. The composition of the product depends on the purpose of use.

Creosote oils are blends of middle oil, reduced crude and heavy oil, and are used for timber preservation. They have lower specific gravity and greater fluidity at low temperatures than high-temperature creosotes.

Creosote

(a) *Production*

The various methods of blending primary distillation fractions to produce creosote are described above under '*Blending and other secondary processes*'.

The commercial production of creosote from coal-tar by batch distillation was reported in the UK during the first half of the eighteenth century.

Creosote has been produced commercially in the USA since 1917 (US Tariff Commission, 1919). At present, production data are available for creosote for use only as a wood preservative. These data are given on both 'creosote, distillate as such (100% creosote basis)' and 'creosote in coal tar solutions (100% solution basis)'. Creosote, distilled as such (100% creosote basis), is presently produced in the USA by five companies, who reported a total production of 147 000 tonnes (36.3 million gallons) in 1982 (US International Trade Commission, 1983), down from the 331 000 tonnes (81.9 million gallons) produced by six companies in 1981 (US International Trade Commission, 1982). 'Creosote in coal tar solutions (100% solution basis)' is presently produced in the USA by five companies, who reported a total production of 179 000 tonnes (44.3 million gallons) in 1982 (US International Trade Commission, 1983), down from the 247 000 tonnes (61.1 million gallons) produced by five companies in 1981 (US International Trade Commission, 1982).

The 1981 production of creosote in the UK is estimated to have been 172.5 thousand tonnes, down from 279.6 thousand tonnes in 1975 (McNeil, 1983).

The 1982 production of creosote in Spain by four companies is estimated to have been in the range of 10-50 thousand tonnes.

Commercial production of creosote in Japan started prior to 1933. Total production in 1982 by four companies is estimated to have been 505 thousand tonnes.

(b) *Use*

It is believed that the creosote used in the USA finds application almost exclusively in wood preservation, where it accounts for an estimated 60% of all wood preservatives in use (Layman, 1982). The estimated annual usage by application method is as follows: approximately 98% by the pressure-treatment process; almost 2% by the non-pressure process; and 0.2% by brush, dip, spray or soak (US Department of Agriculture, 1980a).

In 1981, the total estimated US usage of all creosote types (100% creosote basis) used as a wood preservative is estimated to have been 505 000 tonnes. This includes 248 000 tonnes (62.5 million gallons) of 'creosote in coal tar solutions (100% creosote basis)', 182 000 tonnes (45.9 million gallons) of 'creosote distilled as such (100% creosote basis)', and 75 000 tonnes (18.9 million gallons) of 'creosote in petroleum solutions (100 % creosote basis)'. Creosote and its solutions were used to preserve 4.2 million m^3 of wood products in 1981. The breakdown of this usage by the products treated was as follows (million m^3): railway sleepers, 2.83; poles, 0.54; pilings, 0.24; lumber and timber, 0.22; railway-switch sleepers, 0.18; fence posts, 0.10; and unidentified end-use, 0.10. Creosote and its solutions were used to treat more than 96% of all the sleepers and switch sleepers in the USA in 1981 (American Wood-Preservers' Association, 1982).

Creosote is registered in the USA for use as an animal or bird repellent, animal dip, miticide, fungicide, herbicide and insecticide. However, it is believed that currently creosote is used in only limited quantities and in only three of these applications: as an animal or bird repellent, as an insecticide (ovicide), and as an animal dip (US Department of Agriculture, 1980b).

Creosote is reported to be used at a level of about 12% in a tap-hole refractory cement (used to close openings in furnaces or ovens after material has been withdrawn) (Chrostek, 1980).

It has been reported that creosote can be used as a frothing agent in mineral flotation (Aplan, 1980). Coal-tar creosote can also be used as a feedstock for the production of carbon blacks (see IARC, 1984b) by either the furnace or the lampblack process (McNeil, 1983).

In the UK, the 1981 usage of creosote is estimated by one source to have been as follows: wood preservative, 46%; feedstock for carbon blacks, 36.5%; component in road-tar manufacture, 16.5%; and production of horticultural winter wash oils and disinfectant emulsions, 1% (McNeil, 1983).

In Finland, 19 343 tonnes of creosote were used to treat 211 200 m^3 of wood in 1982 (The Finnish Wood Preserving Association, 1982).

Coal-tar pitches

(a) *Production*

The processes for production of coal-tar pitches have been described above under '*Distillation of high-temperature coal-tars*' and '*Distillation of low-temperature pitches*'.

In the past, coal-tar pitches were allowed to cool and solidify in open bays before being broken up by explosive charge. However, because loading of the resultant lump pitch created a dust nuisance, the practice has been discontinued. Pitches are presently handled either in liquid form or in conditioned form produced by cooling the molten pitch in water-cooled conveyor belts or through nozzles into cold water.

In 1982, US production of coal-tar pitches was reported to be 2.8 million tonnes, and production of refined tar (for uses other than road tar) was 48.6 thousand tonnes (US International Trade Commission, 1983).

(b) Use

Roughly 2 million tonnes of coal-tar pitches are believed to be used annually in the USA. The major use is as the binder for aluminium smelting electrodes (see IARC, 1984a). Pitches are also used in roofing, surface coatings, for pitch coke production and for a variety of other applications.

One source (McNeil, 1983) has estimated that 550 thousand tonnes of coal-tar pitch were used in the USA in 1979 for binding aluminium smelting electrodes. For the Söderberg electrode, about 30% of a paste electrode is medium-hard coke-oven pitch; prebaked electrodes use only about 18% of binder pitch.

Although bitumen is a more usual roofing material, coal-tar pitch is also used for this purpose, especially in the USA. When used for roofing, the coal-tar pitch is heated and applied at approximately 200°C (Hittle & Stukel, 1976). US usage of roofing pitch for flat-topped buildings has been estimated at 200 thousand tonnes (McNeil, 1983).

Coal-tar pitch is also used in surface coatings. US production of such products is believed to be more than 100 thousand tonnes per year. Black varnishes (soft pitches fluxed usually with heavy anthracene oil) are used to some extent as protective coatings for industrial steelwork and timber buildings and as antifouling paints for marine applications. Pipe-coating enamels, made by fluxing a coke-oven pitch with anthracene oil, are used to protect buried oil, gas and water pipes from corrosion (McNeil, 1983). About 75% of all underground petroleum, gas and municipal water pipelines in the USA have been coated with coal-tar enamels (Larson, 1978).

Coal-tar pitch is used to impregnate and strengthen the walls of brick refractories; a soft pitch or refined tar seeps into the brick and is fired to produce a pitch coke which prolongs the life of the oven. A very hard pitch is also used as a binder for foundry cores. The pitch carbonizes upon contact with hot metal to strengthen the core (Considine, 1974). These are believed to be minor uses in the USA (annual usage, 5% or less for each).

Target pitch (a very hard pitch) is used with a clay or limestone filler to produce brittle clay pigeons used for target practice (Considine, 1974). Less than 5% of the US usage of pitch is used for this purpose annually.

An unknown quantity of pitch coke is produced in the USA by carbonizing coke-oven pitch; it is used as the carbon component of electrodes, carbon brushes and shaped carbon and graphite articles (Considine, 1974).

Western European usage of coal-tar pitch in 1979 as a binder for electrodes has been estimated at 440 thousand tonnes. Of this total, the UK is believed to have supplied 155 thousand tonnes (65 thousand tonnes domestically and 90 thousand tonnes for export) and the Federal Republic of Germany most of the remainder (McNeil, 1983).

In 1980, the USSR and the Federal Republic of Germany each produced about 150 thousand tonnes of pitch coke made from coal-tar pitch (McNeil, 1983).

About 45 thousand tonnes of the low-temperature pitch made annually in the UK is used as road binder. Tar/bitumen blends, which may be polymer-modified, are used as surface-dressing binders in the UK and France. Road surfaces containing binders of bitumen and 20-25% coal-tar pitch have improved, longer-lasting antiskid properties (McNeil, 1983).

In 1981, 56 thousand tonnes of coal-tar pitch were used in the UK and France in the production of smokeless, precarbonized briquettes containing 8-10% of medium-soft coke-oven pitch (temperature about 90°C). These briquettes were formerly made from western European pitch but recently this market has been satisfied by imports from eastern Europe (McNeil, 1983).

The Netherlands is the largest European consumer of coal-tar pitch for roofing purposes (McNeil, 1983).

In 1981, about 4 thousand tonnes of pitch were used in the UK to produce about 19 thousand tonnes of pipe-coating enamel, most of which was exported to the Middle East (McNeil, 1983).

About 10 thousand tonnes per year of refined tar (made by fluxing a high-temperature pitch to a low softening-point with strained anthracene or heavy oils) are used in the UK as an extender for resins (e.g., epoxy and polyurethane). Such formulations produce abrasion-resistant, waterproof films, which are used for coating storage tanks, marine pilings and bridge decks. They are highly resistant to petroleum-based fuels (Hoiberg, 1965; McNeil, 1983).

In 1981, about 2.75 thousand tonnes (down from 30 thousand tonnes in 1964) of soft coal-tar pitch were used in the UK to impregnate paper tubes to produce pitch-fibre pipes at a temperature of about 165°C. These pipes are used for the transport of sewage and effluents and for irrigation (McNeil, 1983).

Other end uses for coal-tar pitch in western Europe also vary in importance from region to region. Road tars, produced by fluxing of medium-soft pitch with high-boiling tar oils, still have an important market in Europe, although it has declined considerably in recent years. Usage in the UK is about 35 thousand tonnes per year (McNeil, 1983), down from 700 thousand tonnes in 1930 and 100 thousand tonnes in 1975 (McNeil, 1983).

An estimated 970 thousand tonnes of coal-tar pitch were used in Japan in 1982. The use pattern was as follows: feedstock for pitch coke, 35%; binder for briquettes, 25%; binder for foundry cores, 15%; electrode binder, 10-15%. Use as a binder for refractories made up most of the remaining quantity, and use as a fuel for coke ovens accounted for the balance. One source (McNeil, 1983) has reported that the practice of using coal-tar pitch as fuel has been phased out in other regions of the world.

In 1971, the most recent year in which the US production volume of road tar was reported, 152 million litres were produced (US Tariff Commission, 1973). In 1981, US production of refined tar (for uses other than road tar) was 41.7 million litres (US International Trade Commission, 1982).

2.2 Legislation

Seven countries limit occupational exposure to coal-tar pitch volatiles by regulation or recommended guideline. Their standards are listed in Table 21.

In the Federal Republic of Germany, coal-tars, pitches and primary distillation fractions blended with bitumen are subject to regulations to limit occupational exposure (Federal Office for Workers' Safety, 1982).

Coal-tars are recognized as being carcinogenic by three countries (the Federal Republic of Germany, Japan and Switzerland). Coal-tar pitch volatiles are recognized as being carcinogenic by seven countries (Australia, Belgium, the Federal Republic of Germany, Italy, the Netherlands, Switzerland and the USA) (International Labour Office, 1980).

The US Environmental Protection Agency has identified coal-tars as a hazardous constituent; materials containing coal-tars and some constituents of coal-tars (e.g., naphthalene) are classified as hazardous wastes (US Environmental Protection Agency, 1982).

Coal-tar products for use in the treatment of dandruff, seborrhoea and psoriasis have been reviewed by a US Food and Drug Administration advisory panel; the panel recommended a Category I classification (generally recognized as safe and effective for the claimed therapeutic indication) for coal-tar shampoos and a Category III classification (insuffi-

Table 21. National occupational exposure limits for coal-tar pitch volatiles[a]

Country	Year	Concentration (mg/m^3)	Interpretation[b]	Status
Australia	1978	0.2	TWA	Guideline
Belgium	1978	0.2	TWA	Regulation
Italy	1978	0.2	TWA	Guideline
Netherlands	1978	0.2	TWA	Guideline
Switzerland	1978	0.2	TWA	Regulation
USA				
OSHA[c]	1983	0.2	TWA	Regulation
ACGIH	1983-1984	0.2	TWA	Guideline
NIOSH[d]	1977	0.1	TWA	Guideline
Yugoslavia	1971	0.2	Ceiling	Regulation

[a]From National Institute for Occupational Safety and Health (NIOSH) (1977); International Labour Office (1980); American Conference of Governmental Industrial Hygienists (ACGIH) (1983); and US Occupational Safety and Health Administration (OSHA) (1983a)

[b]TWA, time-weighted average

[c]The OSHA regulation is for coal-tar pitch volatiles (benzene-extractable fraction), which has been interpreted to include all the fused polynuclear hydrocarbons which volatilize from the distillation residues of coal, petroleum (excluding asphalt), wood and other organic matter (e.g., anthracene, benzo[a]pyrene, phenanthrene, acridine, chrysene and pyrene).

[d]The NIOSH guideline refers to the cyclohexane-extractable fraction of coal-tar products (coal-tar, coal-tar pitch, creosote or mixtures of these substances).

cient data available to permit final classification) for other topical dosage forms (US Food and Drug Administration, 1982b).

The Bureau of Alcohol, Tobacco and Firearms of the US Department of the Treasury (1983) has approved coal-tars USP for use as denaturants for prescribed formulae 38-B and 38-F for denaturing alcohol used as a solvent in various cosmetic, pharmaceutical and biocidal preparations.

In September 1978, the US Environmental Protection Agency (EPA) (1978) published a notice of rebuttable presumption against registration (RPAR) on 'creosote' (which refers to coal-tars, creosote and coal-tar neutral oil used as wood preservatives). In August 1984, the EPA proposed to cancel all non-wood uses of creosote on the basis of data that show that it causes cancer and mutagenic effects in test animals.

The EPA has identified creosote as a toxic waste and requires that persons who generate, transport, treat, store or dispose of it comply with the regulations of the federal hazardous waste management programme. Both creosote and the bottom sediment sludge from the treatment of waste-waters from wood-preserving processes involving use of creosote and/or pentachlorophenol are included in the list of hazardous wastes (US Environmental Protection Agency, 1982). In addition, the EPA has proposed a rule which would require that notification be given whenever discharges containing 0.454 kg of either creosote or bottom sediment sludge are made into waterways (US Environmental Protection Agency, 1983).

2.3 Occurrence

(a) *Natural occurrence*

Coal-tars and derived products do not occur as such in Nature.

(b) *General population and occupational exposure*

[The Working Group noted that recent reports have indicated that one of the principal methods for determining occupational exposure to airborne particulates and volatiles from coal-tars and derived products may not be reliable (see Section 2.4 Analysis).]

Coal-tar products

One source has reported that approximately 2500 workers at 50 plant locations in the USA are potentially exposed to emissions of coal-tar pitch volatiles resulting from coating metallic pipes with hot-applied (about 250°C) coal-tar enamels (Larson, 1978).

Occupational exposure to coal-tars and pitch which takes place at coke-oven, coal gasification, foundry and aluminium production work is described in the previous volume of *IARC Monographs* (IARC, 1984a). Therefore, these exposures to coal-tar itself are not discussed further in this monograph.

Exposure of the general population to coal-tar through dermatological preparations can occur. Among conditions treated, psoriasis alone is estimated to affect about 2% of the US population. Coal-tar preparations are usually prescribed in concentrations of 1-10% in ointment or paste. In 1971, Shabad *et al.* analysed three coal-tars used for ointments (one from the USSR, one from Switzerland and one from the USA). The respective benzo[a]pyrene concentrations were 5190, 5020 and 225 µg/g. Fysh et al. (1980) observed very large differences in the benzo[a]pyrene content of coal-tar shampoos. They estimated that if 5 ml of the shampoo with the highest benzo[a]pyrene content were applied to the scalp, 8 mg of benzo[a]pyrene would be present. Hirohata et al. (1973) examined five tar-containing skin preparations and found levels of benzo[a]pyrene varying from none detected to 14.5 µg/g. Swallow and Curtis (1980) analysed seven skin preparations containing coal-tars available in New Zealand. Levels of benzo[a]pyrene varied from 1-360 µg/g. The lower extreme was found in a preparation which may contain wood tar, since the name of this preparation was 'Pinetarsol'. In addition, benz[a]anthracene, chrysene, benzo[b]fluoranthene, benzo[k]fluoranthene, benzo[j]fluoranthene, benzo[e]pyrene, perylene, dibenz[a,j]anthracene, indeno[1,2,3-cd]pyrene, dibenz[a,h]anthracene and benzo[ghi]perylene were detected. The sum of these polynuclear aromatic hydrocarbons (PAHs) ranged from 8-2160 µg/g.

Creosote

(a) Occupational exposure

The major source of occupational exposure to creosote is during its use as a wood preservative[1]. A recent study by the National Institute for Occupational Safety and Health (1983b) reported on the occupational exposure to creosote (measured as coal-tar pitch volatiles using method P & CAM 217 (Todd & Timbie, 1983) and an alternative ultraviolet spectroscopy method) in the air during various operations at several wood-preserving facilities using the pressure-treatment process. The results, combined with those from a report of the US Department of Agriculture (1980a), for the major work operations are given in Table 22.

Table 22. Occupational exposure to creosote[a] at various work operations in US wood-preserving facilities

Operation	Concentrations in air (mg/m^3)		
	NIOSH gravimetric	UV spectrometric	Reference
Treating operator	0.01-0.20 0.007-1.343	- <0.002-0.120	US Department of Agriculture (1980) National Institute for Occupational Safety & Health (1983b)
Locomotive operator	0.01-0.09 0.013-0.159	- 0.003-0.032	US Department of Agriculture (1980) National Institute for Occupational Safety & Health (1983b)
Switchman	<0.01-0.28 0.045-0.35	- 0.013-0.019	US Department of Agriculture (1980) National Institute for Occupational Safety & Health (1983b)
Sample borer	0.04-0.05	-	US Department of Agriculture (1980)
Forklift operator	0.010-0.020	0.016-1.211	National Institute for Occupational Safety & Health (1983b)

[a]Probably a blend of classes 2.1, 3.1 and 4.1

---

[1]It should be noted that, in addition to creosote, workers in the wood-preserving industry are usually exposed to pentachlorophenol (IARC, 1979a) and inorganic arsenics (IARC, 1980)

These data show large differences between the results obtained by the gravimetric and ultraviolet spectrometric methods. The reports also indicate that the highest exposures occurred mainly during transfer tasks and cylinder unloading of the treated product.

The US Department of Agriculture (1980a) has estimated that commercial thermal and dip-treatment workers (about 100 persons) have consistently high inhalation exposure to creosote. Commercial pressure-treatment workers (about 4000 persons) were estimated to have occasional high inhalation exposure to creosote. Skin contact was classified as minimal, except among maintenance workers, who were estimated to have occasional high exposure.

Air samples taken at a US plant where railway sleepers were impregnated with creosote [a blend of classes 2.1, 3.1 and 4.1] were found to contain creosote-derived PAHs in the range of 0.05-650 µg/m^3. Naphthalene, fluorene and phenanthrene were the major chemicals identified (Andersson et al., 1983).

In another US wood-preserving plant, where railway sleepers and telephone poles were pressure treated with a 70/30 mixture of creosote and coal-tar solution [a blend of classes 2.1, 3.1, 4.1 and 1.1], levels of particulate polynuclear organic material (PPOM) varied from 0.07-0.55 mg/m^3. No benzo[a]pyrene was detected in air samples. Total polynuclear aromatic components (PACs) ranged from 0.004-0.11 mg/m^3 (Markel et al., 1977).

About 500 workers are employed by about 80 wood-preserving plants in Finland. In addition, more than 500 people are occupationally exposed to creosote while installing creosoted items. The Institute of Occupational Health conducted industrial hygiene surveys on exposure to creosote [classes 2.1, 3.1, 4.1 and 4.2] in two Finnish pressure-treating plants and in one railway-switch assembly shop in 1980-1982. The levels of coal-tar pitch volatiles (CTPVs) (cyclohexane-soluble fraction) varied from 0.13-0.39 mg/m^3 and benzo[a]pyrene concentrations from 0.04-0.07 µg/m^3 in the pressure-treating plants (Korhonen, 1980). In the railway-switch assembly shop the total concentration of particulate PAHs ranged from none detected to 36.3 µg/m^3. The concentrations of individual PAHs in the filter sample with the highest PAH level were as shown in Table 23 (Korhonen & Mulari, 1983).

Table 23. Concentrations of polynuclear aromatic hydrocarbons (PAHs) in the air of a railway-switch assembly shop

PAH	Concentration (µg/m^3)
Phenanthrene	7.6
Anthracene	0.3
Fluoranthene	9.0
Pyrene	7.7
Benzo[a]fluorene	0.8
Benzo[b]fluorene	0.8
Benz[a]anthracene	3.0
Triphenylene/chrysene	3.5
Benzo[b]fluoranthene and benzo[j]fluoranthene	1.0
Benzo[k]fluoranthene	0.7
Benzo[a]pyrene	1.0
Benzo[e]pyrene	0.6
Perylene	<0.1
ortho-Phenylpyrene	0.2
Benzo[ghi]perylene	0.2
Dibenz[a,h]anthracene	0.3

Employee exposure to CTPVs in the air has been determined at a US refractory cement plant that uses creosote [probably classes 2.1, 3.1 and 4.1] in the compounding of a tap-hole cement mix. The benzene-soluble fraction of these CTPVs was determined to be in the range of 0.03-1.42 mg/m^3. The concentrations of five PAHs were also determined (μg/m^3): benzo[a]pyrene, 0.30-4.98; chrysene, 0.31-3.26; pyrene, <0.13-0.87; benz[a]anthracene, 0.22-2.41; and fluoranthene, 0.10-1.00 (Chrostek, 1980).

(b) Water and sediment

Two studies (US Department of Agriculture, 1980a) have reported the retention of residual creosote in marine pilings after 25, 40 and 59 years of service to be in the range of 280-380 kg/m^3. This compares with the current specifications for new marine pilings, which require a creosote level of 380 kg/m^3 for pilings in direct contact with tides or wave actions (American Wood-Preservers' Association, 1983). A study of the loss of creosote into marine waters reported that about 20% of the creosote initially impregnated into marine pilings is expelled during the first year as a result of wood hydration. Another study reported that only one of three Douglas fir marine piles studied had a detectable loss of creosote. This loss averaged 0.23 kg/linear metre per year during the last four years of the eight-year study (US Environmental Protection Agency, 1978).

Samples taken from creosote-treated southern pine pilings exposed to marine-borer attack in the sea for 9.5 years were analysed for 18 creosote components, and the results were compared with those of samples of unexposed treated wood and the original creosote. The pilings retained 93% of the original creosote composition for the chemicals studied. The total amount of creosote and of aromatic hydrocarbons in the creosoted wood had decreased, but the relative amounts of the major components remained unchanged (Lorenz & Gjovik, 1972).

Creosote was found to be present in an oily leachate of a river in Michigan near a former (used from 1902-1949) wood-preserving facility. In addition, the presence of PAHs in sediment samples was believed to have been due to creosote contamination (Black, 1982).

Coal-tar pitch [class 7]

(a) Occupational exposure

Benzo[a]pyrene levels in the ambient air near emission sources associated with the use of coal-tar products have been summarized by Bridbord *et al.* (1976) as follows (μg/m^3): pavement tarring, 78; coal-tar pitch-working area, 75; roof-tarring area, 14; and the Boston Sumner tunnel, 0.07. Eight-hour average exposures to benzo[a]pyrene for coke-oven workers ranged from 7-18 μg/m^3.

Sawicki *et al.* (1962) took samples near roof-tarring [class 7.1] and pavement-tarring [classes 3.1, 4.1, 4.2 and 7.1] operations. The concentrations of individual PAHs varied from none detected to 200 μg/m^3 during roof-tarring, and from none detected to 3700 μg/m^3 during pavement-tarring operations.

Employees involved in roofing operations (including tear-off of old roofs) in Missouri were exposed to particulate polynuclear organic material (PPOM) as cyclohexane solubles at levels of <0.01-1.88 mg/m^3, to PAHs at levels of <0.001-0.247 mg/m^3, and to benzo[a]pyrene plus benzo[e]pyrene at levels of <0.002-0.019 mg/m^3 (Hervin & Emmett, 1976).

Another study found that pitch [class 7.1] workers at a US roofing site inhaled as much as 53 mg of benzo[a]pyrene in seven hours, with concentrations in air ranging from none detected to 135 mg/m^3 (Hammond et al., 1976).

Two US studies deal with the tear-off operation of old coal-tar pitch [class 7.1] roofs. In 1981, the exposures resulting from the tear-off were investigated at two locations. Total dust concentrations ranged from 1.8-6.2 mg/m^3, and about 30% of the dust was respirable. Exposures to cyclohexane solubles during tear-off procedures were found to range from none detected to 0.51 mg/m^3; no cyclohexane solubles were found in the respirable samples. Concentrations of individual PAHs ranged from none detected to 26 µg/m^3. Exposure to PAHs was highest during a power-cutter operation, when the following PAH levels were observed (µg/m^3): fluoranthene, 26; pyrene, 18; benz[a]anthracene, 14; chrysene, 9; and benzo[a]pyrene, 11 (Tharr, 1982).

A worker operating the felt machine using hot coal-tar pitch [class 7.1] had an exposure to cyclohexane solubles of 0.11 mg/m^3. The airborne concentrations of fluoranthene, pyrene, benz[a]anthracene, chrysene and benzo[a]pyrene were 17, 11, 2.6, 2.5 and 0.65 ug/m^3, respectively (Tharr, 1982).

At three US sites where coal-tar pitch [class 7.1] roofing operations were performed, concentrations of 11 PAHs and total PAHs in ambient air and personal samples of applicators and kettlemen are given in Table 24 (Malaiyandi et al., 1982). In all the earlier studies, only the particulate PAHs were investigated; in this study, the gaseous PAHs were also taken into consideration. Significant amounts of PAHs were found to escape from filters.

During a 1982 study, worker exposure to dusts from the roof tear-off operation [coal-tar pitch of class 7.1 was used in the old roofing material] ranged from 0.3-1.1 mg/m^3 (benzene-soluble fraction). The mean airborne concentrations of individual PAHs were (µg/m^3): phenanthrene, 7.6; anthracene, 2.4; fluoranthene, 11.3; pyrene, 10.1; benz[a]anthracene, 5.3; chrysene, 5.9; and benzo[a]pyrene, 5.5 (Reed, 1983).

Hittel and Stukel (1976) studied experimentally the particle size distribution and chemical composition of fumes from roofing pitch [class 7.1]. At 200°C, the mean diameter of fume droplets was found to be 5.5 µm. Phenanthrene and/or anthracene were the principal constituents of the fumes (36.4%). Other main components included fluoranthene (11.8%), carbazole and methyl phenanthrene (9.6%), fluorene (9.1%), pyrene (8.5%) and trimethyl naphthalene (7.6%).

At a US metal products plant, touch-up spray painters using coal-tar paints [class 8] were exposed to CTPVs (benzene extractables) at a time-weighted average concentration of 0.48 mg/m^3 in 1979. Sheer-press operators in the general workroom were exposed to CTPVs [class 8] at time-weighted average levels ranging from none detected to 0.08 mg/m^3 (McQuilkin, 1980).

Air samples were collected in 1975 at a US plant manufacturing a plastic pipe covering. The covering was made by coating a continuous plastic sheet with a hot mixture of coal-tar [class 8], bitumen and powdered polyvinyl chloride. The samples were analysed using the solvent extraction/gravimetric method, and the concentration in air of the benzene-soluble fraction ranged from 0.18-4.41 mg/m^3. Benzo[a]pyrene analyses were performed on seven samples. All samples were below the detection limit, except one, which contained 1.9 µg/m^3 (Gunter & Ligo, 1976).

Table 24. Concentrations of polynuclear aromatic hydrocarbons (PAHs) found in the work environment during coal-tar pitch roofing operations[a]

Site description	Source of samples	Volume of air sample (litres)	PAH[b] concentration (μg/m³)											
			F	P	B[a]A	Chry	DMBA	Per	B[k]F	B[a]P	B[ghi]P	DB[ah]A	IP	Total
Site VI														
Coal-tar Pitch [class 7.1]	Ambient air	810	0.33	0.35	0.59	0.40	0.07	0.01	0.06	0.01	0.07	0.04	ND[c]	2
	Applicator	590	97.40	36.70	64.80	26.40	0.81	0.62	0.96	0.40	ND[c]	ND[c]	ND[c]	228
Roofing	Kettleman	720	144.00	110.00	257.00	798.00	1.37	3.49	2.78	4.22	3.24	0.35	0.98	1325
Site VII														
Coal-tar Pitch [class 7.1]	Ambient air	660	0.33	0.34	1.59	0.79	0.05	0.08	0.06	0.06	0.11	0.04	0.05	4
	Applicator	510	154.00	162.00	145.00	77.60	1.22	1.93	2.03	1.22	1.48	0.80	0.38	548
Roofing	Kettleman	540	51.40	188.00	61.90	32.00	0.88	0.83	1.00	0.62	0.28	0.11	0.11	337
Site VIII														
Coal-tar Pitch [class 7.1]	Ambient air	690	0.70	0.28	3.78	0.74	0.02	0.01	Trace	0.02	0.01	ND[c]	Trace	6
	Applicator	340	53.00	44.80	109.00	28.10	0.28	0.57	0.60	0.93	0.28	ND[c]	0.08	237
Roofing	Kettleman	440	87.30	87.30	523.00	152.00	16.30	6.97	5.19	11.30	4.16	1.49	1.81	897

[a]From Malaiyandi et al. (1982)

[b]F, fluorene; P, pyrene; B[a]A, benz[a]anthracene; Chry, chrysene; DMBA, 7,12-dimethylbenz[a]anthracene; Per, perylene; B[k]F, benzo[k]fluoranthene; B[a]P, benzo[a]pyrene; B[ghi]P, benzo[ghi]perylene; DB[ah]A, dibenz[a,h]anthracene; IP, indeno[1,2,3-cd]pyrene

[c]ND, not detectable

Industrial hygiene surveys conducted at eight US plants undertaking protective coating of pipelines with hot-applied (about 250°C) coal-tar enamel indicated that the mean exposure to CTPVs in the coating operations was 1.9 mg/m^3. The enamels were formulated from refined tar pitch [class 7.1]. Coating operators (mean, 6.5 mg/m^3) and kettle tenders (mean, 2.3 mg/m^3) had the heaviest exposure. The aerosols produced consisted of a high proportion of respirable particulates (85% of particles less than 10 μm). The major PAHs in total emission samples were (μg/m^3): fluoranthene, 3036; methyl phenanthrene, 2321; pyrene, 1429; benz[a]anthracene, 429; chrysene, 295; methylpyrene, 170; and benzo[a]pyrene, 36 (Larson, 1978).

Exposure to CTPVs (benzene solubles) during the carbon-carbon impregnation and densification processes of a US fibre plant using coal-tar pitch [class 7.1] was detected at levels of 0.03-3.01 mg/m^3 in several breathing zone and area samples. The following quantities of individual PAHs were also detected (μg/sample): benz[a]anthracene, 1.08-73.3; benzo[a]pyrene, 0.50-42.7; chrysene, 0.99-40.3; fluoranthene, 0.77-55.5; and pyrene, 0.95-101 (Lewis, 1979).

A study of worker exposures during the transfer of coal-tar pitch [class 7.1] from river barges to an ocean barge has been reported (Gorman & Liss, 1984). Total particulates found in personal and area samples ranged up to 9.52 mg/m^3 for the nine samples obtained, and a respirable fraction was detected in four out of five samples analysed. Individual PAHs were detected at airborne concentrations in the following ranges (μg/m^3): benzo[k]fluoranthene, 0.02-12.88; benzo[b]fluoranthene, 0.05-34.76; chrysene, 0.32-26.58; benzo[e]pyrene, 0.09-38.85; and benzo[a]pyrene, 0.11-38.85. The ranges of airborne concentrations found in the transfer of the coal-tar pitch and in the transfer of a petroleum pitch were reported to have been 0.46-44.99 μg/m^3 for pyrene and 0.11-34.76 ug/m^3 for benz[a]anthracene. (Separate data for the coal-tar pitch transfer only were not reported.)

Skin exposure may be considerable for coal-tar and pitch workers because, owing to heat stress, they often wear few clothes and thus large portions of the body may be directly exposed to CTPVs. In the skin oil of nine roofing workers (also exposed to bitumen), PAHs were detected at levels of 0.048-36 ng for the 36-cm^2 area of forehead sampled; analysis of skin oil from two men with no exposure to roofing materials showed no detectable amount of PAHs (Wolff et al., 1982).

Workers' clothes and underwear were found to be contaminated with benzo[a]pyrene in a Czechoslovak pitch-manufacturing plant. The concentration of benzo[a]pyrene in clothing increased with time despite frequent laundering. This was taken to indicate that normal laundering did not remove benzo[a]pyrene effectively, but dry cleaning was found to be quite effective (Kandus et al., 1972; Jach & Masek, 1973).

Benzo[a]pyrene exposures were measured (area and personal samples) in a unit producing refractory bricks (Table 25). The pitch [class 7.1] is kept at 220°C and the vessel is opened when dipping the bricks. The volatile compounds fluoranthene and pyrene were found at higher concentrations than in the coke-oven plants (Blome, 1981, 1983). Table 25 also presents data on the benzo[a]pyrene concentrations found during: (i) the production of silicon carbide from quartz and coke using pitch [class 7.1] as a binding material; and (ii) melting and moulding of pitch [class 7.1] used for embedding lenses in the optical industry and grinding and polishing of the lenses, as well as removal of the pitch (Blome, 1981, 1983).

(b) Soil

One source has reported that soil polluted by coal-tar pitch contained benz[a]anthracene at a level of 2500 mg/kg, benzo[a]pyrene at 650 mg/kg and chrysene at 600 mg/kg (Fritz & Engst, 1971).

Table 25. Benzo[a]pyrene concentrations at workplaces where pitch [class 7.1] was used[a]

Workplace/process	Benzo[a]pyrene (µg/m³)[b]	
	Area samples	Personal samples
Silicon carbide production (mixing, pressing, heating)		
1980	0.22-5.90	
1982	0.10-1.14	
1983	0.05-0.44	0.20-0.47
Refractory brick production		
old unit	0.07-21.90	3.20-4.50
new unit (with better ventilation)	0.31-0.99	0.73-1.80
Optical industry		
melting and moulding pitch	0.12-0.95	
grinding and polishing lenses	<0.05-2.50	
removing pitch	<0.05-19.70	

[a]From Blome (1983)
[b]Other PAHs were also detected in all workplaces/processes

(c) Water

PAHs were found at a level of 0.41 µg/l in a sample of water taken from the outlet of a water-storage tank. The level in a sample taken from the inlet was 0.029 µg/l. The tank interior had a commercial coal-tar coating which was about five years old but which appeared to be in 'good condition' (Alben, 1980).

2.4 Analysis

(a) Coal-tars and derived products

Owing to the complexity of their composition, the presence of coal-tars and derived products is generally detected by the presence of their specific constituents, especially CTPVs and PAHs, in the air and other media. The National Institute for Occupational Safety and Health (1974, 1977) and Coulter and Petro (1972) have reviewed and evaluated sampling and analytical methods for the detection of coal-tar products in airborne particulates and vapours, including levels of benzene-soluble material and PAHs.

(i) Analysis of CTPVs

In most studies of US occupational exposure to coal-tars and derived products in air, the National Institute for Occupational Safety and Health (NIOSH) Sampling and Analytical Method P & CAM 217 for coal-tar pitch volatile cyclohexane extractables has been used. However, in recent years, other workers (Todd & Timbie, 1980, 1983) have indicated that this method may not be reliable. An alternative analytical procedure described by NIOSH uses ultraviolet spectroscopy to measure the absorption at 252 nm of the cyclohexane extract of the filtered materials (Coulter & Petro, 1972; National Institute for Occupational Safety and Health, 1977; US Occupational Safety and Health Administration, 1983b).

(ii) Analysis of PACs

The most important information on occupational exposure to coal-tars and pitch deals with exposure to PACs. These are found partly in the gaseous phase and partly as particulates. A good sampling method, therefore, includes both an effective filter for particulate PACs and an absorption tube for PAC vapours. After extraction, PACs can be analysed by high-performance liquid chromatography (HPLC) or gas chromatography. In addition, detailed identification of individual PACs requires prior fractionation steps and the use of mass spectrometry. Owing to the extreme complexity of coal-tar composition, analysis is usually limited to the determination of concentrations of the main PAHs. A review of methods of analysis for PACs in environmental samples has been published (IARC, 1979b).

Malaiyandi et al. (1982) (see Table 24) collected the PAHs resulting from coal-tar pitch in the work atmosphere of roofers by collecting air samples using a fibre-glass filter, silver frit membrane and Tenax GC resin sampling train assembly. The filters and resin were Soxhlet extracted with cyclohexane, and the extracts were concentrated and then analysed by HPLC.

Air samples recovered using a similar filter assembly can be analysed by gas chromatography-mass spectrometry (GC-MS) after separate extraction of the filters and the resin (Anderson et al., 1978).

PAHs resulting from coal-tar contamination have been detected in water samples after adsorption on XAD-2 resin. Eluted PAHs were analysed by GC-MS (Alben, 1980).

Coal-tar contamination of skin has been detected using ultraviolet-induced fluorescence because the PAHs in coal-tar products are strongly luminescing species. Visual detection has been used in the past, but Vo-Dinh and Gammage (1980) have developed a luminescence detector for this purpose.

(b) Creosote

The American Wood-Preservers' Association (AWPA) reported the use of AWPA Standard A6-83 (a method for the determination of oil-type preservatives and water in wood) for the determination of creosote levels in preserved wood. This method involves the refluxing and extraction of creosote-containing wood borings or shavings using toluene, xylene or a mixed toluene-xylene solvent. Creosote content is calculated as the difference between the sample weight before and after extraction and is expressed as pounds per cubic foot of treated wood (American Wood-Preservers' Association, 1983).

Black (1982) has described a qualitative method for the identification of creosote in river-water. This method involves obtaining a 'fingerprint match' between a sample of commercial creosote and suspect contaminated river-water using HPLC.

The complexity of creosote precludes its quantitative analysis when it is present at low concentrations in terms of a pure chemical (Nestler, 1974). Instead, the presence of creosote is considered to be implied by the quantitative analysis for, and detection of, its specific PAH constituents. For instance, the US Environmental Protection Agency (1982) reported that the presence of creosote was assumed if phenanthrene and carbazole were detected at a ratio of between 1.4:1 and 5:1.

3. Biological Data Relevant to the Evaluation of Carcinogenic Risk to Humans

Studies on samples of crude coal-tars obtained directly from gas-works, coal-gasification plants or coke-oven plants have been summarized in previous monographs on coal gasification and coke production (IARC, 1984a).

Data relating to timber-preservation oils derived from wood tar have been excluded from this monograph. These products are fundamentally different from coal-tar creosote, typically containing a very high proportion of tar acids, and are similar in their toxicological properties to cresol.

3.1 Carcinogenicity studies in animals

Coal-tars [class 1]

(a) *Skin application*

Mouse: In the earliest report on mice, coal-tar [class 1.3, unspecified] was applied with a brush to the skin of 259 random-bred mice every third or fourth day; 67 mice survived 100 days, and 35 (52%) developed skin papillomas; 16 of these mice later developed skin carcinomas, two with lung metastases. One mouse had a spindle-cell sarcoma (Tsutsui, 1918). Similar findings were reported in subsequent studies (Fibiger & Bang, 1920; Bloch & Dreifuss, 1921).

A group of 45 mice [age, sex and strain not specified] received weekly skin applications [dose not specified] on the dorsal region of an ether extract of a Scottish blast-furnace tar from splint coal [similar to class 1.2]. 'Warts' began to develop after 20 weeks of treatment; after 44 weeks all surviving animals had developed skin tumours. A total of 25 of the original 45 animals developed skin tumours, seven of which were found to be malignant. In addition, 50 mice received weekly applications of untreated tar [class 1.2] for 32 weeks and 100 mice received twice-weekly applications for 38 weeks. Skin tumours started to occur after 15 weeks; mortality was reported to be high, and a total of 26 animals developed skin tumours, three of which were malignant (Berenblum, 1930). [The Working Group noted the absence of concurrent controls.]

Three groups of 60 mice [age, sex and strain not specified] received twice-weekly applications [dose not specified] of three different samples of untreated Scottish blast-furnace tars from splint coal [similar to class 1.2] for 14 weeks followed by weekly applications for 56 weeks, by which time all the animals had died. 'Warts' appeared after 16 weeks; seven, ten and eight animals in the respective groups developed skin tumours, of which a total of three were found to be malignant. In addition, the same treatment schedule was followed in two groups of 60 mice receiving untreated English blast-furnace tar [class 1.2] or an ether extract of it. 'Warts' occurred in the group treated with the untreated tar after 21 weeks. Skin tumours developed in eight animals; there were no malignant tumours. In the ether-extract group, 'warts' started to occur after 12 weeks; 24 animals developed skin tumours, nine of which were malignant (Bonser, 1932). [The Working Group noted the absence of concurrent controls and that survival was not specified.]

Benzene (Soxhlet) extracts (1:1) of two coal-tars [both class 1.1] from different sources were tested in two groups of 30 mice (16 males, 14 females) [strain not specified]. Each animal received a single drop of the test material applied with a dropper to the skin twice weekly for 22 weeks. After eight weeks of treatment, there were 22 survivors in one group and 26 in the other. The reported mean doses of tar applied after four months of treatment were 0.6 g and 0.2 g. At the end of the experiment [probably after seven months], skin tumours were found in six of the 22 survivors (27%) in one group and eight of 26 survivors (31%) in the other (Górski, 1959). [No benzene control group was reported.]

Heated high-temperature coal-tar [class 1.1] (obtained from a processor of coke-oven tars) was applied twice weekly [dose not specified] to the nape of the neck of 50 C57BL mice (25 of each sex) for up to two years. Four of these animals developed papillomas and 22 developed carcinomas of the skin (Hueper & Payne, 1960). [The Working Group noted the absence of concurrent controls and that no survival data were reported.]

Shustova and Samoilovich (1971) studied the carcinogenic activity of a neutralized residual tar [class 1.3, unspecified] produced at a coke factory and used as a road-surfacing material, which contained, as the coal-tar, predominantly aromatics, naphthalene (4%), anthracene, heterocyclics, compounds containing oxygen (phenol (3%), cresols), sulphur (thiophene, thionaphthene) and nitrogen (pyridine, carbazole, pyrrole). The test substance contained benzo[a]pyrene at a concentration of 94 mg/kg. A group of 50 (C57BL x CBA)F_1 mice, two-and-a-half to three months old [sex not specified], was painted on the interscapular skin, with the neutralized tar diluted with benzene (100 g of tar and 20 g of benzene), thrice weekly for 10 months. After three months of treatment, three mice had skin tumours; after five months, 50%; and after six months, 100%. Among 40 mice examined, histological analysis revealed 27 with squamous-cell carcinomas (five of which had metastases), 10 with papillomas and one with sarcoma. In a control group of 50 mice that received benzene only, no skin tumour was found.

Pharmaceutical coal-tars

Twelve white mice were painted twice weekly for 41 weeks with undiluted *liquor picis carbonis* B.P. (a 20% solution of coal-tar [class 1.3, unspecified] in ethanol). Within this time seven mice developed skin papillomas, four of which later developed into squamous carcinomas of the skin. The solution contained 0.1% benzo[a]pyrene (Berenblum, 1948).

A tar ointment [class 1.3, unspecified] (consisting of 0.02% of a steroid hormone, 1% salicyclic acid and 1.5% coal-tar [class 1.3] in a lanolin base) was applied to the skin of seven male and 11 female (C57 x CBA)F_1 mice five times weekly for 1.5 months and then thrice weekly for a further 10.5 months (total, 186 applications). Each application of 0.1 g of the ointment contained about 23 µg of benzo[a]pyrene. The total dose of tar was 18 g and of benzo[a]pyrene 4.3 mg. All surviving mice were killed at 1.5 years. Of 17 mice surviving at the time of appearance of the first tumour, 16 developed skin tumours, 13 of which were carcinomas (three metastasizing). All of the 13 animals surviving at 12 months had skin tumours (Linik, 1970; Shabad *et al.*, 1970, 1971). [No control was reported.]

C57 x CBA mice [number and sex not specified] were painted with two coal-tars used for ointments [class 1.3, unspecified] twice or thrice weekly for 10-12 months [dose not specified]. The animals were observed for 1.5 years. In the group treated with one coal-tar (origin USSR) containing benzo[a]pyrene at a level of 5190 µg/g, 18/19 survivors had skin tumours [type unspecified]. In the group treated with the other coal-tar (origin CIBA), containing benzo[a]pyrene at a level of 5020 µg/g, 20/21 survivors had skin tumours [type unspecified] (Shabad et al., 1971). [The Working Group noted the absence of concurrent controls.]

Another coal-tar preparation (*Pix lithanthracis*) [class 1.3, unspecified] was tested in 100 female 10-week-old NMRI (Wiga colony) mice, that received two drops of a 5% solution in dimethyl sulphoxide thrice weekly for four weeks and then one drop weekly until spontaneous death. The longest survival time was 16 months. Cornifying and often metastatic squamous-cell carcinoma of the skin developed in 54%. All animals surviving 12 months had a skin carcinoma (Hilfrich & Mohr, 1972).

Rat: A group of 24 'Pie' rats, one month old [sex not specified], were painted 'strongly' with crude coal-tar [class 1.3, unspecified] thrice weekly for one to two months then twice a week until their spontaneous death. During the first six months, 15 rats died; seven died during the following six months; and the last two rats died between the 13th and 15th month of treatment. No skin tumour was observed. However, the six rats surviving more than 10 months developed single or multiple epidermoid lung carcinomas (Möller, 1925). [Data on control rats were not reported.]

Rabbit: In a historical study of the first experimental documentation of chemical carcinogenesis (Yamagiwa & Ichikawa, 1915), coal-tar [class 1.3, unspecified] was painted on the ears of rabbits. Of 39 rabbits painted on the inner ear surface, 17 survived 70 days, and 12 of these had papillomas. Of 26 rabbits painted on the outer ear surface, six survived 70 days, one with a papilloma. Of 30 rabbits painted on open ear wounds, two survived 70 days and no skin tumour was found.

(b) *Intramuscular administration*

Mouse: A dichloromethane extract of a coal-tar fume condensate [class 1.1] was diluted in olive oil 1:3 v/v. This olive oil solution was injected twice weekly (six injections of 0.15 ml) into the muscle of the right thigh of 50 male and 50 female C57BL mice. After two years of observation, 50 mice developed sarcomas at the injection site (Hueper & Payne, 1960). [The Working Group noted that there were no controls.]

Creosotes

(a) *Skin application*

Mouse: A full-strength creosote oil [class 3.2, unspecified] and an anthracene oil [class 4.1] were tested in groups of 50 male and female albino white mice by twice-weekly applications for at least 25 weeks. In the group receiving creosote oil, 10 of the 19 animals surviving were found to have papillomas and nine had large keratinizing carcinomas. In the group receiving anthracene oil, eight of the 20 mice surviving at 25 weeks were found to have papillomas and six had large keratinizing carcinomas (Woodhouse, 1950). [The Working Group noted that there was no concurrent control group; however, pine oil and linseed oil produced no skin tumour under similar experimental conditions.]

In another study, one drop of a 20% or 80% solution of a creosote [a blend of classes 2.1, 3.1 and 4.1] in toluene was applied three times weekly to the skin of groups of 10 female C57Bl mice for the lifespan of the animal or until skin tumours developed. All mice exposed to both dilutions developed papillomas and seven of eight effective mice in each of the two groups developed epidermoid carcinomas, some of which metastasized to pulmonary or regional lymph nodes. The time to papilloma appearance in 50% of the animals was 26 and 21 weeks in those receiving 20% and 80% solutions of creosote, respectively. Control animals receiving toluene did not develop any skin neoplasm. A 'light creosote oil' derived from the residual oil of the naphthalene-recovery operation [class 2.1] was tested in male C57Bl mice by applying one drop of a 50% solution in toluene three times a week for life. Of 11 mice surviving 20 or more weeks of treatment, all developed skin tumours within 45 weeks (Poel & Kammer, 1957).

Undiluted creosote oil [class 3.1] was applied twice weekly (one drop of a 2% solution in acetone) for up to 70 weeks to a group of 30 female Swiss mice. The effective number of mice surviving at the appearance of the first tumour was 26. Of these, 13 were found to have 23 skin tumours, of which 16 were considered carcinomas. The average latent period for development of the tumours was 50 weeks (Lijinsky et al., 1957). [The Working Group noted that a specific control group was lacking.] The carcinogenic effect was considered almost identical in potency to that of repeated applications of a 0.01% solution of 7,12-dimethylbenz[a]anthracene in mineral oil, as reported by Saffiotti and Shubik (1956). In addition, a 2% solution of the basic fraction of creosote oil in acetone was applied twice weekly for 70 weeks to a group of 30 mice following a single painting with one drop of a 1% solution of 7,12-dimethylbenz[a]anthracene in mineral oil. At 56 weeks, 12 mice survived; none of these had skin tumours or developed them up to 70 weeks (Lijinsky et al., 1957). [The Working Group noted that the basic fraction alone was not tested.]

Undiluted creosote [a blend of classes 2.1, 3.1, 4.1 and 4.2] was applied twice weekly in amounts of 25 µl to the skin of 30 random-bred female mice. At the end of the experiment (28 weeks), the mice had an average of 5.4 papillomas, and 82% had carcinomas. Similar application of creosote to another group of 30 mice for only four weeks followed by 32 weeks of observation resulted in no skin tumour. A third group treated with creosote for four weeks followed by twice-weekly applications of 25 µl of 0.5% croton oil in benzene developed 2.8 papillomas per mouse at 28 weeks and 4.4 papillomas per mouse at 36 weeks; 46% had carcinomas at week 44. A control group given croton oil alone also developed skin tumours at lower incidences (0.1 papilloma per mouse at 28 weeks), but, by 44 weeks, eight of 26 surviving mice given croton oil alone had 13 papillomas and one carcinoma (Boutwell & Bosch, 1958).

An increased frequency of lung adenomas was also obtained in mice treated with the same creosote [a blend of classes 2.1, 3.1, 4.1 and 4.2] used by Boutwell and Bosch (1958). A group of 24 albino mice treated topically with 25 µl of creosote twice weekly for six months (starting at three weeks of age) showed, after two additional months of observation, 139 lung adenomas (5.8 tumours per mouse). Nine lung adenomas were observed in 19 control mice (0.5 tumour per mouse). In addition, a group of 29 mice born and kept in creosote-impregnated wooden cages received skin painting with 25 µl creosote twice weekly for five months (starting after weaning). At eight months of age they showed 315 lung adenomas (10.8 per mouse). In a second experiment, a different lot of albino mice, aged two months, were treated by skin application of 25 µl of the same creosote twice weekly for four weeks (nine applications). After 10 months, 37 lung adenomas in 23 mice were observed (1.6 tumours per mouse); 15 lung adenomas were found in 50 similar mice treated with croton oil or benzene (0.3 tumour per mouse) (Roe et al., 1958).

A basic fraction of creosote [distillation range including classes 2.1 and 3.1 (Cabot et al., 1940)] was tested in mice by Sall and Shear (1940) for modifying effects on benzo[a]pyrene-induced tumorigenesis. A group of 100 female strain A mice was divided into five subgroups and painted three times a week (124-145 times) with a 1% solution of the basic fraction in benzene, 0.02% or 0.05% benzo[a]pyrene in benzene, or 0.05% benzo[a]pyrene in benzene containing 1% of the basic fraction of creosote. No tumour was produced in any of the mice receiving the basic creosote fraction alone. However, the latent period for tumour appearance was shorter in the groups receiving benzo[a]pyrene and the basic fraction of creosote as compared with that for mice receiving benzo[a]pyrene alone, suggesting that the basic fraction of creosote decreased the latency of the tumorigenic effects of benzo[a]pyrene.

Groups of 20 female, albino, random-bred mice were painted with benzene solutions containing 0.2% or 0.05% benzo[a]pyrene, or 0.2% or 0.05% benzo[a]pyrene together with a creosote oil [distillation range including classes 2.1 and 3.1, the same as in Sall and Shear (1940)], 90% of which distilled at 160-300°C, or various fractions thereof. The test materials were applied to the shaven back by five strokes of a swab dipped once into the test solution. The benzene solution containing 0.2% benzo[a]pyrene was applied thrice weekly for 20 weeks and then twice weekly for six weeks (72 applications in 26 weeks). The solution containing 0.05% benzo[a]pyrene was applied thrice weekly for 38 weeks (114 applications). Animals developing papillomas 4 mm in diameter which did not regress were classified as 'tumour bearing'; time to first appearance of tumours was recorded. Tumours developed in 19/20 animals receiving 0.2% benzo[a]pyrene in benzene, in 12 of these after 13 weeks. Tumours were induced after 22 weeks in 16/20 animals receiving 0.05% benzo[a]pyrene in benzene. A basic fraction of the creosote (1% solution) obtained by aqueous hydrochloric acid extraction increased benzo[a]pyrene tumorigenesis (with respect to both time to appearance of tumours and number of tumours). A phenolic fraction obtained by extraction into aqueous sodium hydroxide exhibited an enhancing effect; this effect was more marked with the lower concentration of benzo[a]pyrene. The phenolic and basic fractions tested alone failed to induce tumours (Cabot et al., 1940).

Pitches [class 7]

(a) *Whole-body exposure*

Mouse: In the first reported experiment with coal-tar pitch, carried out by Barnewitz (1928), an unspecified number ('a considerable number') of white mice were kept in cages with a bedding consisting of peat and coal-tar pitch powder [class 7.3] (1:1). The cages were shaken vigorously twice daily. Papillomas first appeared after eight weeks. Subsequently, 50% of the mice developed skin carcinomas.

(b) *Skin application*

Mouse: In three experiments, a total of 40 effective mice of a 'fairly uniform stock' [sex not specified] were swabbed with a benzene extract of a hard residue from a coke-oven tar [class 7.1] (collected at 377°C) that had been washed with sodium hydroxide and sulphuric acid and then reconcentrated from benzene. The extract was applied on 12 different areas of each mouse, the application sites being rotated so that each area received only three applications, about one month apart, over an 83-day period. Of 23 mice (aged 9-10 months) killed four months after the last administration, 21 had lung tumours; and of 12 mice killed at one to six months and of another group of five surviving mice receiving the same treatment and killed six months after the last administration, 13 had lung tumours. These were confirmed histologically. In two control groups, one of 22 mice aged 8-12 months and one of 16 mice aged 12-16 months, no lung tumour was reported. No skin tumour was found in treated or control mice (Murphy & Sturm, 1925).

Groups of 30 mice (16 males, 14 females) [strain not specified] were treated with benzene (Soxhlet) extracts (1:1) of a hard pitch [class 7.1], a soft pitch [class 7.1] or an anthracene fraction [class 4.3] (all from a single source). Each animal received a single drop of the test material applied with a dropper on the skin twice weekly for 22 weeks. After eight weeks of treatment, the survivors numbered 21, 28 and 24 in the hard-pitch, soft-pitch and anthracene groups, respectively. The reported mean doses of the corresponding products applied after four months of treatment were 1.0, 2.9 and 0.3 g per mouse. At the end of the experiment [probably after seven months], skin tumours were found in eight of the 21 survivors (38%) in the hard-pitch group, 14 of the 28 survivors (50%) in the soft-pitch group and four of the 24 survivors (17%) in the anthracene group (Gorski, 1959). [No benzene control group was reported.]

A group of 49 CC-57 white mice [sex not specified] was treated by weekly skin applications (total, 70) of 40% coal-tar pitch [class 7.1] in benzene over 19 months. The first skin tumour appeared three months after the start of the experiment, at which time 43 mice were still alive. By the end of 12 months, skin tumours had occurred in 37 of these. The majority (29) were squamous-cell carcinomas (Kireeva, 1968).

The carcinogenic activity of two samples of coal-tar pitches [class 7.1] from coke-oven production was tested on the skin of Swiss albino mice. The coal-tar pitches were applied as solutions in benzene (tar content, <9%) twice weekly to the skin with a calibrated dropper delivering 25 µl of solution, corresponding approximately to 1.7 mg of coal-tar pitch per application. The mean survival time for treated animals was 31 weeks, whereas control animals survived 82 weeks. The two coal-tar pitches induced skin tumours in 27/29 mice and in 26/29 mice, which survived with 13 and 18 carcinomas and 27 and 26 papillomas, respectively. One of the 26 benzene-treated mice developed one papilloma (Wallcave et al., 1971).

Emmett et al. (1981) studied the carcinogenic activity of (1) traditional roofing coal-tar pitch [class 7.1]; (2) coal-tar bitumen [class 7.3], with respective softening-points of 65°C and 68.3°C; (3) coal-tar bitumen [class 7.3] from roofing operations; and (4) roofing dust (representative of particulate matter from the site of removal of a 25-year-old roof containing coal-tar pitch). All samples were dissolved in toluene in a 1:1 ratio (w/w) and were applied with either a microlitre pipette or a calibrated glass dropper to the intrascapular shaved skin of male C3H/HeJ mice. Each exposed group consisted of 50 mice, except the group treated with roofing dust (4) which consisted of 20 mice. Mice were treated twice weekly for 80 weeks with 50 mg of the resulting solution. A negative-control group received 50 mg of toluene twice weekly, a positive-control group received 50 mg of benzo[a]pyrene (0.1% toluene solution) under the same conditions. The numbers of tumour-bearing animals were, respectively, 48/49 for (1), 47/48 for (2), 42/45 for (3) and 12/14 for (4). The majority of these skin tumours were malignant (45, 39, 38 and 10 tumours, respectively). In the toluene-treated mice no tumour was observed, whereas in the benzo[a]pyrene positive-control group 31/39 mice developed tumours (24 malignant). In the groups receiving samples containing coal-tar pitch, the average latent period ranged from 16.5 to 21.5 weeks, whereas for the group receiving benzo[a]pyrene the average latent period was 31.8 weeks.

A dichloromethane extract of a sample of coal-tar pitch emissions collected from a conventional tar-pot unit with an external propane burner (temperature, 182-193°C) was studied by skin application for (1) initiating activity, (2) complete carcinogenic activity and (3) promoting activity. The material tested in these studies was reported by Lewtas et al. (1981) to be a 'pitch-based tar' and by Nesnow et al. (1982a,b) to be a 'pine-based pitch'. It has subsequently been reported that the material tested was derived from a coal-tar pitch (Albert et al., 1983; Austin et al., 1984). The same material was tested in several short-term tests, described in section 3.2(a) of this monograph.

(1) For initiating activity, five groups of 40 male and 40 female SENCAR mice, seven to nine weeks of age, were treated with roofing-tar extract dissolved in 0.2 ml of acetone. Single topical applications of 0.1, 0.5, 1.0, 2.0 or 10.0 mg were given; the 10-mg dose was administered as five daily doses of 2 mg. Beginning one week after treatment, doses of 2.0 µg of 12-O-tetradecanoyl-phorbol-13-acetate (TPA) in 0.2 ml of acetone were applied topically twice weekly. A further group of 40 males and 40 females was treated with TPA only. After six months of TPA treatment, the percentages of mice with papillomas (and the mean number of papillomas per mouse) were 12.5 (0.17), 20.5 (0.25), 41.5 (0.60), 36.5 (0.53), 97 (6.0) in the treated groups, respectively, and 6 (0.06) for controls. One year later, the percentages of mice with carcinomas were 7.5, 14, 10, 14 and 35.5 for tar-initiated groups and 2.5 for TPA controls (Lewtas et al., 1981; Nesnow & Lewtas, 1981; Slaga et al., 1981; Nesnow et al., 1982a,b, 1983).

(2) To study its complete carcinogenic activity, five groups of 40 male and 40 female SENCAR mice were treated topically with the same roofing extract dissolved in acetone. Weekly applications of 0.1, 0.5, 1.0, 2.0 and 4.0 mg per mouse were given in 0.2 ml of acetone for 50-52 weeks. The 4-mg dose was administered as two doses of 2 mg. The cumulative percentages of mice with carcinomas (and the mean number of carcinomas per mouse) after one year were 0, 0, 1.5 (0.015), 5.5 (0.05) and 26.5 (0.28), respectively (Nesnow et al., 1983).

(3) For promoting activity, six groups of 40 male and 40 female SENCAR mice were treated topically with a single dose of 50.5 µg benzo[a]pyrene in 0.2 ml acetone. One week later, five groups were treated with roofing-tar extract dissolved in 0.2 ml acetone. Weekly applications (divided into two doses for the highest dose) of 0.1, 0.5, 1.0, 2.0 and 4.0 mg per mouse were given for 34 weeks. The sixth group was treated with 0.2 ml acetone. The percentages of mice with papillomas at 34 weeks were 0, 2.5 (male 5; female 0), 18 (male 20; female 16), 18 (male 23; female 13) and 42.5 (male 55; female 30), respectively. No mouse with a tumour was reported in benzo[a]pyrene-initiated and acetone-treated groups (Nesnow et al., 1983).

3.2 Other relevant biological data

(a) *Experimental systems*

Toxic effects

Coal-tars

Groups of C3H mice were exposed chronically to coal-tar aerosol [class 1.1] for three two-hour periods per week at a concentration of 0.30 mg/l. The most striking change observed was a necrotizing tracheobronchitis. In addition, hyperplasia of the bronchial epithelium was frequently observed, sometimes accompanied by papillary infolding (Horton et al., 1963).

Preparations of pharmaceutical-grade coal-tars

The effects of coal-tars and coal-tar ointments [class 1.3] on the skin of several species have been studied by several investigators (Salfeld et al., 1966: acanthosis in guinea-pigs; Elgjo & Larsen, 1973: hyperplastic epidermis in mice; Ritschel et al., 1975; Sarkany & Gaylarde, 1976; Gloor et al., 1978).

Male, albino New Zealand rabbits were treated with coal-tar in different vehicles by application of ointment disks to shaved areas of the backs to investigate the influence of topical vehicles on tar penetration and retention and on skin irritation. A fair correlation was observed between the depth of penetration and the irritating effects (reddening and swelling of skin) of the tar. However, the vehicle that led to the greatest retention of tar in the skin was also the least irritating (Ritschel et al., 1975).

Coal-tar [class 1.3] applied to the skin of guinea-pigs increased epidermal thickness; the effect varied with the batch of coal-tar (Sarkany & Gaylarde, 1976; Gloor et al., 1978). A similar response was observed in hairless hamsters. Other changes in the skin induced by coal-tar treatment included comedone formation, an increased squame count, and atrophy of the sebaceous glands. There was a high concentration of tar in the follicular ducts and sebaceous glands of exposed hamsters (Foreman et al., 1979).

Combined treatment with coal-tar preparations [class 1.3] and ultraviolet radiation (320-400 nm) caused a marked inhibition of DNA synthesis in vivo in normal and proliferating skin of hairless mice (Stoughton et al., 1978; Walter et al., 1978).

Creosote

No data were available to the Working Group on creosote derived from coal-tars.

Coal-tar pitch

Among young pigs that ate remnants of clay pigeons (the main ingredient of which was coal-tar pitch) [class 7.3], 10-20% died. Degenerative lesions were found in the livers of the young pigs. Adult animals were apparently unaffected (Graham et al., 1940).

Several investigators have reported the effects of feeding coal-tar pitch [class 7.3] to pigs and Peking ducklings. Liver lesions were frequent, ranging from moderate to very severe, and included marked central lobular necrosis (Graham et al., 1940; Luke, 1954; Carlton, 1966; Maclean, 1969).

Coal-tar pitch [class 7.1] solutions (about 10% in benzene) were applied to the shaved areas of both male and female Swiss albino mice twice a week. Each application contained approximately 1.7 mg of pitch. Treatment induced epidermal hyperplasia, often accompanied by inflammatory infiltration of the dermis and occasional ulceration and small abscess formation (Wallcave et al., 1971).

Phototoxic keratoconjunctivitis from coal-tar pitch volatiles [class 7.1] has been produced experimentally in New Zealand white rabbits. Following conjunctival instillation of 10 µl of coal-tar pitch distillate to both eyes, one eye was irradiated with ultraviolet light (330-380 nm). There was no effect in animals treated with ultraviolet light alone. Injection of the lids (vasodilatation), tearing and slight mucous discharge were observed at five and 24 hours after treatment with pitch alone. Eyes exposed to pitch and then irradiated developed injection and oedema of the lids, tearing, photophobia and mucous discharge. These changes were evident until 96 or 120 hours after treatment. All changes were reversible (Emmett et al., 1977).

No data were available to the Working Group on effects on reproduction and prenatal toxicity.

Absorption, distribution, excretion and metabolism

A single topical application of standard coal-tar solution (US Pharmacopeia) to the backs of rats aged 4-6 days 24 hours prior to killing resulted in the induction of aryl hydrocarbon hydroxylase (AHH) in skin and liver. Compared to acetone-treated animals, enzyme levels were induced 15-fold in skin and eight-fold in liver. In addition, topical treatment of pregnant rats with coal-tar resulted in AHH induction in the maternal and prenatal skin and liver (Bickers et al., 1982).

Intraperitoneal injections to rats of creosote oil [class 3.1] in doses of 0.1 or 0.5 g/kg bw resulted in marked induction of 7-ethoxyresorufin O-deethylase in liver and lung microsomes (Söderkvist & Toftgård, 1982).

No data on coal-tar pitch were available to the Working Group.

Mutagenicity and other short-term tests

Preparations of pharmaceutical-grade coal-tars

Each of four therapeutic preparations of coal-tar [class 1.3] tested was mutagenic in *Salmonella typhimurium* TA98 in the presence of an Aroclor induced rat liver metabolic system (S9) (Saperstein & Wheeler, 1979). [The samples were not tested in the absence of S9.]

Three of four coal-tar shampoos [class 1.3] (tested as hexane extracts) were mutagenic in *S. typhimurium* TA100 in the presence but not in the absence of S9. The shampoo with the highest activity contained approximately 50 times more benzo[a]pyrene than the one that was negative (Fysh et al., 1980).

Creosote

Creosote (type P-1, specification of American Wood-Preservers' Association, 1983) [classes 2.1, 3.1 and 4.1] and a coal-tar-creosote mixture (type P-2, class C, specification of American Wood-Preservers' Association, 1983) [classes 2.1, 3.1 and 4.1, or classes 2.1, 3.1, 4.1 and 1.1] were tested for mutagenicity in bacteria and in cultured mammalian cells. Both compounds were mutagenic in *S. typhimurium* TA1537, TA98 and TA100 in the presence of S9; no activity was observed in the absence of S9 or in strain TA1535. The compounds did not induce mutations in *Escherichia coli* strain WP2 (Simmon & Poole, 1978). These creosotes induced a positive concentration-related response in the L5178Y mouse lymphoma cell system ($TK^{+/-}$) in the presence of S9. A much higher concentration was required to induce a positive response in the absence of S9 (Mitchell & Tajiri, 1978).

Bos et al. (1983) reported positive results for creosote (type P-1) [classes 2.1, 3.1 and 4.1] in S. typhimurium TA1537, TA1538, TA98 and TA100 in the presence of S9. The strongest response was obtained with strains TA98 and TA100; no activity was observed with strain TA1535. Fractions of P-1 creosote (separated by thin-layer chromatography) were also mutagenic in S. typhimurium TA98 in the presence of S9 (Bos et al., 1984a).

The urine of rats administered creosote type P-1 [classes 2.1, 3.1 and 4.1] (250 mg/kg bw, i.p.) was mutagenic in S. typhimurium TA98 and/or TA100 when tested in the presence of β-glucuronidase and S9 (higher activity) or with β-glucuronidase only (Bos et al., 1984b).

Coal-tar pitch

A dimethyl-sulphoxide extract of coal-tar pitch (Barrett M-30) [class 7.3] was mutagenic in S. typhimurium TA1537, TA98 and TA100 in the presence but not the absence of S9 (Rao et al., 1979).

Dichloromethane extracts of emissions from a roofing-tar pot (coal-tar pitch-based tar [presumably class 7.3]) were mutagenic in S. typhimurium TA1537, TA1538, TA98 and TA100, in the presence but not the absence of S9. No activity was observed in strain TA1535 (Claxton & Huisingh, 1980; Nesnow & Lewtas, 1981).

The same material was tested further in a number of in-vitro eukaryotic test systems (Casto et al., 1981; Curren et al., 1981; Mitchell et al., 1981). The extract did not induce mitotic recombination in *Saccharomyces cerevisiae* D3 [tested at only one concentration] nor DNA fragmentation (alkaline elution technique) in primary Syrian hamster embryo cells [both tests were performed only in the absence of an exogenous metabolic system]. The extract was mutagenic in BALB/c 3T3 cells (ouabain resistance) and was positive in the L5178Y mouse lymphoma cell system ($TK^{+/-}$), both in the presence and absence of S9. In contrast, it was negative in Chinese hamster ovary (CHO) cells (6-thioguanine resistance) [tested only in the absence of S9]. The extract induced a significant increase in the frequency of sister chromatid exchanges in CHO cells, both in the presence and absence of S9. An observed increase in morphologically-transformed foci in BALB/c 3T3 cells was not statistically significant when tested either in the presence or absence of S9; a significant enhancement of viral transformation was observed in Syrian hamster embryo cells. [The material tested in these studies was reported by Lewtas et al. (1981) to be a 'pitch-based tar' and by Claxton and Huisingh (1980) to be an asphalt tar. It has subsequently been reported that the material tested was derived from a coal-tar pitch (Albert et al., 1983; Austin et al., 1984). Carcinogenicity studies with this material are described in section 3.1 of this monograph.]

(b) *Humans*

Toxic effects

Coal-tars and coal-tar pitch

Information about the effects of coal-tar pitch [various classes] on humans has been derived both from occupational exposures and the therapeutic use of crude coal-tars [class 1.3]. The majority of toxic effects concern the skin and eyes; skin effects include dermatitis, chronic tar-dermatosis, tar or pitch warts, chronic melanosis, folliculitis and pitch acne, as described below.

Chronic tar dermatosis was described in long-term employees of a coal-tar distillery [exposed presumably to all classes]. Changes included alteration in skin texture, with either thickening or glazing and atrophy; numerous simple soft planar warts; rough hyperkeratoses; permanent freckles; and increased capillary telangiectases (Fisher, 1954). Among 241 tar-distillery workers [exposed presumably to all classes], 66 men with a total of 422 tar warts were seen (414 on the head, neck or hands, seven on the scrotum and one elsewhere). No relationship to hair colour was seen. Tar warts were more frequent in workers more susceptible to acute tar erythema. Chronic tar dermatosis did not necessarily precede the development of tar warts (Fisher, 1953).

Hodgson and Whiteley (1970) studied 144 workers in a patent-fuel plant where coal dust and pitch [class 7.3] were fused by steam heat to form fuel ovoids. 'Pitch' warts (clinically and histologically keratoacanthomas) occurred, 26% of which regressed spontaneously. Scrotal proliferative lesions were common (13.5%). Tar keratoses among pitch-tar workers [class 7.3] may occur after exposure to the causal agents has been discontinued (Götz, 1976).

Prolonged contact with tar [presumably all classes] by workers in a tar plant resulted in marked patchy or generalized hyperpigmentation on exposed skin (Morando, 1960).

Crow (1970) described 54 cases of acne due to pitch fumes [class 7.3]. An eruption of pure comedones (morphologically distinguishable from those seen in chloracne) occurred, particularly on the malar region of the face. These generally healed rapidly but sometimes remained and could recur more than ten years later. Folliculitis of the thighs and forearms was common.

Cutaneous photosensitivity from coal-tar pitch has been described by a number of authors (Crow et al., 1961 [class 7.3]; Wulf et al., 1963 [class 1.3]; Hodgson & Whiteley, 1970 [class 7.3]; Wiskemann & Hoyer, 1971 [class 1.3]; Tanenbaum & Parrish, 1975 [class 1.3]; Kaidbey & Kligman, 1977 [class 1.3]). Pitch 'smarts', intense burning sensations, occur on direct exposure to sunlight; erythema usually follows. These phototoxic reactions may be followed by melanosis (Crow et al., 1961 [class 7.3]) and often occur together with ocular photosensitivity (Emmett et al., 1977 [class 7.1]).

Patch testing (to detect allergens that may cause allergic contact dermatitis) of 1664 consecutive patients with eczema revealed that 72 reacted positively to crude coal-tar [class 1.3] [although it was not clear that these patients had coal-tar-induced dermatitis] (Hjorth & Trolle-Lassen, 1963). Allergic contact and photosensitivity reactions to therapeutic tar applications [class 1.3] may cause extensive dermatitis (Starke & Jillson, 1961).

Application of crude coal-tar [class 1.3] to human skin produces an initial hyperplasia during the first two weeks of treatment; prolonged treatment leads to a reduction in overall epidermal thickness but with a hyperkeratotic horny layer (Lavker et al., 1981).

Margolis (1964) described erythema and oedema of the eyelids, forehead and cheeks, tearing, blepharospasm, conjunctivitis, photophobia, severe temporary changes in vision, and corneal changes in 16 workers who had unloaded briquettes of coal-tar pitch [class 7.3]. Susorov (1970) described similar symptoms and photodermatitis as well as headaches and irritation of the upper respiratory passages in 36 workers unloading coke wagons [class 7.3]; the symptoms returned to normal within five days. Burning sensations in the eyes, photophobia and keratoconjunctivitis were described in roofers working with coal-tar pitch [class 7.1]; these symptoms were associated with combined exposure to coal-tar pitch and sunlight but not with bitumens or other roofing materials (Emmett et al., 1977).

There have been studies of the effects of coal-tar pitch on organs other than skin. Morando (1960) described workers exposed to hot pitch or tar [presumably all classes] with abnormal serum protein levels due to apparent hepatic dysfunction, which could have been induced by occupational exposure. Pekker (1967) found that the prevalence of dental caries was higher in workers exposed to products of dry distillation of coal [class 8] than in those with no such exposure; oral mucous membrane disorders, in particular leukoplakia and mucosal oedema, were more frequent in this group than in some other occupational groups. Kapitul'skii (1966) found that pitch-coke electrode manufacturers [class 7.3] exposed to aerosol levels of 23-453 mg/m^3 pitch [class 7.3] and pitch distillates and to heat stress showed a decline in muscular strength and increases in latent period of motor response to sound and chronaxie (a measure of the time taken for excitation of a nervous stimulus) over the course of a work shift.

There are some studies of workers who presumably had long-term exposure to both coal-tar products and bitumens. Hammond et al. (1976) studied a cohort of 5939 members of a roofers' union who had had at least nine years of membership, and compared their mortality with that expected on the basis of US age- and time-specific death rates. They observed 71 deaths due to nonmalignant respiratory disease (standardized mortality ratio, 167), excluding influenza, pneumonia and tuberculosis, in workers with 20 or more years since joining.

Milham (1982) conducted a proportionate mortality analysis of all deaths among white male residents occurring in the State of Washington during 1950-1979. Among roofers, there were seven deaths due to asthma (proportionate mortality ratio, 390; $p < 0.01$).

Creosote

None of the studies described below allowed classification of the creosote to which exposures occurred.

Schwartz (1942) described a temporary eczematous dermatitis of the ankles of workers exposed to wooden block floors newly treated with creosote; photosensitization was not observed; lesions were not seen where creosoted blocks had been treated some months before. Jonas (1943) described 450 subjects with creosote 'burns' from fine sawdust produced by sawing wood saturated with creosote or skin contact with creosoted roof paper. Most had mild erythema in exposed areas, and more severely affected individuals had intense burning and itching followed by desquamation and hyperpigmentation; photosensitization was observed, usually in fair-skinned workers. Heyl and Mellett (1982) described irritant dermatitis and photodermatitis from contact with ammunition boxes impregnated with creosote.

Multiple papillomatous, warty growths were described in a worker exposed to creosote for many years while dipping railway sleepers (Mackenzie, 1898) and in a second worker using timber treated with creosote. Biopsy of a representative lesion from the latter case showed a squamous papilloma (Haldin-Davis, 1935).

Birdwood (1938) described ocular changes (numerous small grey corneal spots and superficial hazy keratitis with permanent visual impairment) in two gardeners who had worked with creosote. Jonas (1943) observed conjunctivitis ranging from hyperaemia to photophobia and pronounced lachrymation in 15% of people observed with cutaneous creosote burns. Permanent corneal scars resulted from chips of creosoted wood. The author also described depression, weakness, severe headache, slight confusion, vertigo and increased salivation and nausea in a small percentage of subjects who developed burns from creosote. Dumber (1962) described headaches, giddiness, nausea, vomiting and copious salivation in 94% of 120 workers who had sprayed warmed creosote (the concentration of creosote vapour in the heated building was up to 0.01 mg/l).

No data were available to the Working Group on effects on reproduction and prenatal toxicity of coal-tars, coal-tar pitch and creosote.

Absorption, distribution, excretion and metabolism

In urine from patients receiving dermal applications of a cream containing 10-20% *coal-tar* [class 1.3], acridine was identified; additional compounds were found on chromatography which were assumed to be coal-tar components (Černiková *et al.*, 1979).

Topical application of crude *coal-tar* solution (US Pharmacopeia) [class 1.3] to human skin caused a two- to five-fold induction of cutaneous aryl hydrocarbon hydroxylase with a maximum response 24 hours after application (Bickers & Kappas, 1978).

No data on creosote or coal-tar pitch were available to the Working Group.

Mutagenicity in urine

Coal-tar preparations

Urine samples (extracts prepared with XAD-2 resin) from psoriatic patients undergoing combined treatment with coal-tar preparations [class 1.3] and ultraviolet light were mutagenic in *S. typhimurium* TA98 in the presence of an Aroclor-induced rat-liver metabolic system (S9); β-glucuronidase had no effect on the response obtained with S9. A time-course study in a patient and in two volunteers without skin lesions showed that mutagenic activity in urine reached a peak six to eight hours after treatment and returned to the background level within 30-72 hours (Wheeler *et al.*, 1981).

Creosote

Urine samples (extracts prepared with XAD-2 resin), collected on 10 consecutive days from three workers employed in a wood preserving plant where P1 creosote [perhaps classes 2.1, 3.1 and 4.1] was used, were not mutagenic in *S. typhimurium* TA98 (tested in the presence of S9 and β-glucuronidase) (Bos *et al.*, 1984b).

Coal-tar pitch

No data were available to the Working Group.

3.3 Case reports and epidemiological studies of carcinogenicity in humans

(a) *Coal-tars and pitches*

Studies of workers exposed to coal-tars in coal gasification and coke production and to pitch in aluminium production are discussed in previous IARC monographs (IARC, 1984a).

(i) *Case reports*

A large number of case reports describing the development of skin cancer in workers exposed occupationally to coal-tars were reviewed, and these are summarized in the Appendix, p. 243. Some were reviewed and evaluated in a previous IARC monograph (IARC, 1984a).

More than 315 cases of scrotal cancer, a rare tumour, were ascribed to coal-tars and pitch and an additional 191 to pitch alone (see Appendix). Most of these cases occurred in patent-fuel workers, pitch loaders, electrical trades and optical lens polishers. [In some series, for example, that of Henry (1947)[1] and the Scottish cases reported by Doig (1970), it appears that mention of exposures to pitch excluded exposures to other coal-derived products.]

(ii) *Descriptive epidemiological study*

Henry (1946) calculated crude cumulative rates of scrotal cancer mortality. All scrotal cancer deaths reported to the Registrar General in England and Wales from 1911-1938 were classified (1631 cases) by occupation and divided by the average population of the 1921 and 1931 censuses for each occupation. Some detailed occupational populations were derived from 'the Board of Trade or a trade itself'. Patent-fuel workers (pitch-briquette makers) had a crude mortality rate for scrotal cancer of 504 per million men based on 24 cases, as compared to a rate of 4.2 per million for the national average and rates of 1 per million or less for most other occupations, including service, transport and retail trades.

(iii) *Analytical epidemiological studies*

Henry et al. (1931) examined in detail 11 429 deaths of men from cancer of the bladder and prostate according to last recorded occupation (as given on the death certificate) out of 13 965 cases that occurred in England and Wales from 1921-1928. Expected numbers of deaths were calculated from death rates for cancers of the bladder and prostate in five-year age groups and applied to the population of workers in each five-year age group according to the census occupation tables. Among 2665 tar-distilling workers, four deaths from blad-

---

[1]The UK Factory and Workshop Act 1901 required that diseases contracted in premises to which the Act applied be notified to HM Chief Inspector of Factories by the medical practitioners concerned. The list of such notifiable diseases was extended on 1 January 1920 to include 'epitheliomatous ulceration or cancer of the skin due to pitch, tar, bitumen, paraffin, or mineral oil, or any compound, or residue of any of these substances, or any product'.

der cancer were observed with 1.2 expected, and one death from prostatic cancer with 1.2 expected; among 1811 patent-fuel workers, two deaths from bladder cancer were observed and 0.5 were expected, and one death from prostastic cancer with 0.4 expected.

A further study in England and Wales of deaths from lung and laryngeal cancer in relation to occupations involving exposure to coal gas (Kennaway & Kennaway, 1936) was evaluated in the IARC monograph on coal gasification (IARC, 1984a).

A 12-year mortality study was made of 5939 members of a roofers' union in USA alive in 1960 and with at least nine years of union membership, 5788 of whom were traced through 1971 (Hammond et al., 1976). Expected numbers of deaths were computed on the basis of US life tables specific for age and single calendar year. The standardized mortality ratio (SMR) for all deaths was 103 in the first six years of the study and 110 in the second six years. Lung cancer ratios (SMR) generally increased with increasing time since joining the union: 9-19 years, 22 observed, SMR, 92; 20-29 years, 66 observed, SMR, 152; 30-39 years, 21 observed, SMR 150; and 40+ years, 12 observed, SMR, 247. The following SMRs were reported in the group with 20 or more years' membership in the union for cancer of the oral cavity, larynx and oesophagus (31 observed, SMR, 195), cancer of the stomach (24 observed, SMR, 167), leukaemia (13 observed, SMR, 168) and bladder (13 observed, SMR, 168). In the total cohort, five deaths were from skin cancer other than melanoma (1.18 expected). [No data on smoking were given.]

In a study by Menck and Henderson (1976) of occupational differences in lung cancer rates among white men in Los Angeles County, USA, mortality data for 1968-1970 and morbidity data from the county cancer registry for 1972-1973 were pooled. The 3938 men under study were classified by their last recorded occupation and the industry in which they were employed. This information was obtained from death certificates for mortality and from hospital records for incident cases. Age-adjusted expected numbers of deaths and incident cases were calculated on the basis of a 2% sample of 1970 census records for Los Angeles County classified by occupation and industry. Among roofers, a significantly increased risk of lung cancer was observed, with a 'standard mortality ratio' of 496 based on six deaths and five incident cases. [No data on smoking were available.]

Milham (1982) conducted a proportional mortality analysis of deaths among white male residents in the State of Washington during 1950-1979. Among roofers and slaters, there were four deaths due to laryngeal cancer (proportionate mortality ratio, PMR, 270, non-significant) and 53 deaths due to cancer of the bronchus and lung (PMR, 161, $p < 0.01$). [No data on smoking were available.]

Decouflé et al. (1977) used cases of cancer and of non-neoplastic diseases recorded at Roswell Park Memorial Hospital, NY, USA, during 1956-1965 in a survey of occupations. Persons employed in clerical jobs were used for comparison with those in other occupations. Relative risks were adjusted for age and smoking habits. A non-significant relative risk of 2.95 was reported for cancer of the buccal cavity and pharynx among roofers and slaters. The relative risk for lung cancer in this trade was not reported.

According to Hammond et al. (1976), roofers in the USA are mainly engaged in applying hot pitch or hot bitumen. In former years, [coal-tar] pitch was used more frequently than bitumen (see monograph in this volume), but today bitumens are more commonly used. Most of the men work with both substances.

In a mortality study in Massachusetts from 1956-1975 of optical lens workers, who are exposed to, among other things, pitch and abrasives when grinding and polishing lenses, a statistically significant excess of all cancers was observed (70 observed, 48 expected) (Wang et al., 1983). Expected deaths were computed on the basis of deaths among people in non-exposed occupations and of similar socioeconomic status. The excess was due to high rates of gastrointestinal cancer; there was a mortality odds ratio of 2.2 for those with medium exposure and 2.5 for long-term exposure. [The mixed exposure makes this study difficult to evaluate in relation to pitch.]

A population-based case-control interview study (McLaughlin et al., 1983) of cancer of the renal pelvis (74 cases and 697 controls) was conducted in the area of Minneapolis-St Paul, USA. After adjustment for age and cigarette smoking, an elevated but non-significant odds ratio was found for occupational exposures to 'petroleum or tar or pitch' (odds ratio, 2.4; 95% confidence interval, 0.9-6.1).

(b) Therapeutic coal-tars

Hodgson (1948) described a case of skin cancer (squamous-cell epitheliomata) in a 62-year-old man who had used tar ointment for over six years to treat pruritus ani. A squamous epithelioma developed on a patch of psoriasis in a 31-year-old man using tar ointment as treatment (Alexander & Macrosson, 1954). A squamous-cell carcinoma on the thigh was reported by Rook et al. (1956) in a road worker aged 60 years who had no relevant occupational exposure but had treated the area with tar ointment for 32 years. Carli (1958) reported an epithelioma in a 64-year-old woman who used tar pomade daily for six months to treat pruritus vulvae. Melanoma developed in a 32-year-old man treated for psoriasis with ultraviolet light and coal-tar for 16 years (Durkin et al., 1978).

Maughan et al. (1980) made a 25-year follow-up study of 426 patients treated with coal-tar and ultraviolet light for atopic dermatitis and neurodermatitis in 1950-1954 at the Mayo Clinic, Minnesota, USA. In an analysis of 260 psoriasis patients followed for 25 years in the same clinic (Pittelkow et al., 1981), the authors stated that the intensity of coal-tar use may have been greater among skin cancer cases.

Stern et al. (1980) performed a case-control study to investigate the relative risk of skin cancer in patients with psoriasis who had been exposed to tar and/or ultraviolet radiation. The majority of the patients had less than 30 months of tar use and less than 100 sunlamp treatments which 'made it impractical to study the effect of these two agents separately'. Halprin et al. (1982) compared the incidence of skin and other cancers in 150 patients with psoriasis with that in a matched control group of diabetics. None of the psoriasis patients had been treated with coal-tar only.

Alderson and Clarke (1983) compiled a cohort of 8405 patients whose records included a diagnosis of psoriasis. No data on treatment were recorded.

[In none of these studies could exposure to coal-tars alone be evaluated.]

(c) Creosote

(i) *Case reports*

O'Donovan (1920) reported three cases of skin cancer in men exposed occupationally to creosote. One of them had been applying creosote to timber for 40 years, and warts had appeared on his hands, legs and behind his ears for seven years.

Cookson (1924) reported a metastasizing squamous epithelioma on the hand of a worker employed for 33 years in carrying creosoted wood.

Henry (1947)[1] reviewed 3753 (3921 sites) cases of cutaneous epitheliomata reported to the British Medical Inspector of Factories from 1920 to 1945 and reported that 35 cases (39 sites), 12 of which were of the scrotum, had exposure to creosote. Of these 35 cases, 14 (15 sites) occurred among workers treating timber with creosote, nine (11 cases) among people handling creosote in storage, and 10 (11 sites) among people using creosote as a releasing agent for brick moulds.

(ii) Descriptive epidemiology study

Henry (1946) also calculated crude mortality rates for scrotal cancer for some groups occupationally exposed to creosote. The crude mortality rate for scrotal cancer during 1911-1938 for brickmakers exposed to 'creosote oil' was 29 per million men, based on nine verified cases, as compared to a rate of 4.2 per million for the national average and rates of one per million or less for groups not exposed to suspected skin carcinogen. In addition, there were several other creosote-exposed groups with increased rates based on fewer cases.

(iii) Analytical epidemiological study

Axelson and Kling (1983) have reported a cohort study on 123 Swedish workers who applied creosote to wood and who had had exposure to creosote and arsenic between 1950 and 1980. Eight workers died from tumours, whereas 6.0 were expected from national statistics. In a subgroup of 21 workers who had been exposed only to creosote for five years or more, three deaths from cancer (leukaemia, pancreas and stomach) were observed as against 0.8 expected. Levels of exposure were not reported.

[The type of creosote involved in these studies was not reported, but the Working Group presumed that exposure in the UK was almost entirely to material derived from coal-tars.]

4. Summary of Data Reported and Evaluation

4.1 Exposure data

Large quantities of crude coal-tars, which contain polynuclear aromatic compounds as major components, are formed as by-products of the destructive distillation of coal, coke-ovens being the major source. The distillate fractions of crude coal-tars alone or in combination with coal-tar pitch, both of which contain polynuclear aromatic compounds,

---

[1]The UK Factory and Workshop Act 1901 required that diseases contracted in premises to which the Act applied be notified to HM Chief Inspector of Factories by the medical practitioners concerned. The list of such notifiable diseases was extended on 1 January 1920 to include 'epitheliomatous ulceration or cancer of the skin due to pitch, tar, bitumen, paraffin, or mineral oil, or any compound, or residue of any of these substances, or any product'.

are mixed to produce various products, such as creosote. Most human exposure to crude coal-tars and derived products takes place during the destructive distillation of coal in foundries and in aluminium production, on which monographs have already been prepared. Members of the general population use topical pharmaceutical coal-tar preparations.

Exposure to polynuclear aromatic compounds in other occupations is especially heavy among workers applying hot coal-tar pitch, such as in roofing, paving, surface coatings and in the production of refractory bricks. There is also potential exposure in coal-tar distillation plants.

Creosote has world-wide use as a wood preservative. In addition to possible inhalation exposures, cutaneous exposure may be considerable.

4.2 Experimental data

Three high-temperature tars, one undiluted and two as benzene extracts, all produced skin tumours, including carcinomas, when applied to the skin of mice. Each of five blast-furnace tars and two extracts of blast-furnace tars produced skin tumours, including carcinomas, after topical application to mice.

Each of five pharmaceutical coal-tar preparations caused skin tumours, including carcinomas, when applied to the skin of mice.

Two unspecified coal-tars both caused skin tumours, including carcinomas, after application to the skin of mice. Lung tumours but no skin tumour were produced in rats after application of a coal-tar to the skin. A fourth, unspecified coal-tar produced tumours when applied to the ears of rabbits.

Intramuscular administration of a coal-tar fume condensate to mice in one experiment gave evidence of sarcoma formation.

Five creosotes or creosote oils all produced skin tumours, including carcinomas, when applied to the skin of mice. One of the creosotes also produced lung tumours in mice after skin application. In two limited studies, a basic fraction of creosote oil was not carcinogenic for the skin of mice.

In two experiments, anthracene oils produced skin tumours (and, in one experiment, carcinomas) when applied to the skin of mice.

Six coal-tar pitches and three extracts of coal-tar pitches all produced skin tumours, including carcinomas, when applied to the skin of mice. An extract of roofing-tar pitch had both initiating and promoting activity in separate experiments.

In one experiment, mice developed skin tumours, including carcinomas, after whole-body exposure to pitch powder.

No data were available to the Working Group on the carcinogenicity of distillation fractions of low-temperature tars or of products derived from them.

Samples of therapeutic coal-tar and extracts of coal-tar shampoos were mutagenic in *Salmonella typhimurium* in the presence of an exogenous metabolic system. Extracts of urine from patients undergoing combined treatment with coal-tar preparations and ultraviolet light were mutagenic in *S. typhimurium*.

Extracts of coal-tar pitch were mutagenic in *S. typhimurium* in the presence of an exogenous metabolic system.

Extracts of roofing-tar emissions were mutagenic in *S. typhimurium* in the presence of an exogenous metabolic system and were mutagenic in two mutation assays and induced sister chromatid exchanges in cultured mammalian cells, both in the presence and absence of an exogenous metabolic system. Viral transformation was enhanced in Syrian hamster embryo cells. A statistically non-significant increase in the number of morphologically transformed foci was observed in BALB/c 3T3 cells. The material did not cause DNA fragmentation in cultured Syrian hamster embryo cells.

Creosote and a coal-tar-creosote mixture were mutagenic in *S. typhimurium* and were positive in the mouse lymphoma L5178Y system, in the presence of an exogenous metabolic system. The urine from rats administered creosote was mutagenic in *S. typhimurium* in the presence of an exogenous metabolic system.

Overall assessment of data from short-term tests on coal-tar derived products[a]

	Genetic activity			Cell transformation
	DNA damage	Mutation	Chromosomal effects	
Prokaryotes		+[b,c,d]		
Fungi/green plants				
Insects				
Mammalian cells (*in vitro*)	−[c]	+[c,d]	+[c]	?[c]
Mammals (*in vivo*)				
Humans (*in vivo*)				
Degree of evidence in short-term tests for genetic activity: *sufficient*				Cell transformation: inadequate

[a]The groups into which the table is divided and the symbols +, − and ? are defined on pp. 16-18 of the Preamble; the degrees of evidence are defined on p. 18.
[b]Pharmaceutical-grade coal-tars [class 1.3]
[c]Coal-tar pitch, including emissions
[d]Creosotes

4.3 Human data

A mortality analysis in the UK from 1946 showed a greatly increased scrotal cancer risk for patent-fuel workers exposed to pitch. Furthermore, a large number of case reports describing the development of skin (including the scrotum) cancer in workers occupationally exposed to coal-tar and/or to pitch have been published.

A cohort study of US roofers indicated an increased risk for cancer of the lung and suggested increased risks for cancers of the oral cavity, larynx, oesophagus, stomach, skin and bladder and for leukaemia. Some support for excess risks of lung, laryngeal and oral-cavity cancer is provided by other epidemiological studies of roofers. Roofers may be exposed not only to a mixture of pitches but also to bitumens and other materials.

One study showed a small excess of bladder cancer in tar distillers and in patent-fuel workers. An elevated risk of cancer of the renal pelvis was also seen in workers exposed to 'petroleum or tar or pitch'.

There have been a number of case reports of skin cancer developing in patients using tar ointments to treat a variety of skin diseases.

A mortality analysis of many occupations indicated an increased risk of mortality from scrotal cancer for creosote-exposed brickmakers. Malignant epitheliomas, about a third of which were of the scrotum, have been reported in several case reports of workers exposed to creosote. The only available cohort study suffered from limitations of size.

4.4 Evaluation[1]

There is *sufficient evidence* for the carcinogenicity in experimental animals of coal-tars, creosotes, creosote oils, anthracene oils and coal-tar pitches.

There is *sufficient evidence* that occupational exposure to coal-tars as it occurs during the destructive distillation of coal is causally associated with the occurrence of skin cancer in humans (IARC, 1984a). The findings of the few studies available on other occupational exposures to coal-tars are consistent with that evaluation.

There is *sufficient evidence* that coal-tar pitches are carcinogenic in humans.

There is *limited evidence* that coal-tar-derived creosotes are carcinogenic in humans.

Taken together, the data indicate that coal-tars and coal-tar pitches are causally associated with cancer in humans and that creosotes derived from coal-tars are probably carcinogenic in humans.

5. References

Alben, K. (1980) Coal tar coatings of storage tanks. A source of contamination of the potable water supply. *Environ. Sci. Technol.*, 14, 468-470

Albert, E., Lewtas, J., Nesnow, S., Thorslund, T.W. & Anderson, E. (1983) Comparative potency method for cancer risk assessment: Application to diesel particulate emissions. *Risk Anal.*, 3, 101-117

Alderson, M.R. & Clarke, J.A. (1983) Cancer incidence in patients with psoriasis. *Br. J. Cancer*, 47, 857-859

---

[1]For definitions of the italicized terms, see Preamble pp. 15-16 and 19.

Alexander, J.O'D. & Macrosson, K.I. (1954) Squamous epithelioma probably due to tar ointment in a case with psoriasis. *Br. med. J., iv*, 1089

American Conference of Governmental Industrial Hygienists (1983) *TLVs, Threshold Limit Values for Chemical Substances in the Work Environment Adopted by ACGIH for 1983-84*, Cincinnati, OH, pp. 16, 42

American Society for Testing and Materials (1983) *1983 Annual Book of ASTM Standards*, Vol. 04.04, Philadelphia, PA, pp. 115-116

American Wood-Preservers' Association (1982) *Wood Preservation Statistics 1981*, Bethesda, MD, J.D. Ferry Associates, pp. 213, 223, 225-227

American Wood-Preservers' Association (1983) *American Wood-Preservers' Association Standards 1983*, Stevensville, MD, pp. P1-78, P2-68

Anderson, L.A., Jr, Ayer, H.E. & Gerhard, M.K. (1978) *Preliminary Industrial Hygiene Characterization Program for the Bruceton Liquefaction Facility* (Final Report, Prepared for the US Department of Energy under Contract No. EW-78-5-22-0221), Cincinnati, OH, University of Cincinnati, pp. 18, 19, 25-27

Andersson, K., Levin, J.-O. & Nilsson, C.-A. (1983) Sampling and analysis of particulate and gaseous polycyclic aromatic hydrocarbons from coal tar sources in the working environment. *Chemosphere, 12*, 197-207

Aplan, F.K. (1980) Flotation. In: Kirk, R.E. & Othmer, D.F., eds, *Encyclopedia of Chemical Technology*, 3rd ed., Vol. 10, New York, John Wiley & Sons, p. 529

Arrendale, R.F. & Rogers, L.B. (1981) Use of capillary column gas chromatography to evaluate the quality of coal tar pitches. *Separation Sci. Technol., 16*, 487-493

Austin, A.C., Claxton, L.D. & Lewtas, J. (1984) Mutagenicity of the fractionated organic emissions from diesel, cigarette smoke condensate, coke oven and roofing tar in the Ames assay. *Environ. Mutagenesis* (in press)

Axelson, O. & Kling, H. (1983) *Mortality among wood preservers with creosote exposure* (Swed.) (Abstract). In: *32nd Nordic Occupational Hygiene Conference*, Solna, Arbetarskyddsstyrelsen (National Board of Occupational Safety and Health), pp. 125-126

Baker, C.E., Jr (1982) *Physicians' Desk Reference*, 36th ed., Oradell, NJ, Medical Economics Company, Inc., pp. 308, 657, 863, 869

Barnewitz, J. (1928) On the experimental tumour production in mice by coal-tar pitch (Ger.). *Dtsch. med. Wochenschr., 54*, 1162-1164

Bender, D.F., Sawicki, E. & Wilson, R.M., Jr (1964) Characterization of carbazole and polynuclear carbazoles in urban air and in air polluted by coal tar pitch fumes by thin-layer chromatography and spectrophotofluorometry. *Int. J. Air Water Pollut., 8*, 633-643

Berenblum, I. (1930) Experimental induction of tumours with blast-furnace tar. *Lancet, ii*, 1344-1346

Berenblum, I. (1948) Liquor picis carbonis (B.P.) a carcinogenic agent. *Br. med. J., ii,* 601

Bickers, D.R. & Kappas, A. (1978) Human skin aryl hydrocarbon hydroxylase induction by coal tar. *J. clin. Invest., 62,* 1061-1068

Bickers, D.R., Wroblewski, D., Choudhury, T.D. & Mukhtar, H. (1982) Induction of neonatal rat skin and liver aryl hydrocarbon hydroxylase by coal tar and its constituents. *J. invest. Dermatol., 78,* 227-229

Birdwood, G.T. (1938) Keratitis from working with creosote. *Br. med. J., iii,* 18

Black, J.J. (1982) Movement and identification of a creosote-derived PAH complex below a river pollution point source. *Arch. environ. Contam. Toxicol., 11,* 161-166

Bloch, B. & Dreifuss, W. (1921) Experimental induction of carcinomas with lymph nodes and lung metastases with tar constituents (Ger.). *Schw. med. Wochenschr., 51,* 1033-1037

Blome, H. (1981) Measurement of polycyclic aromatic hydrocarbons in the work place - Assessment of results (Ger.). *Staub-Reinhalt Luft, 41,* 225-229

Blome, H. (1983) *Polyzyklische Aromatische Kohlenwasserstoffe (PAH) am Arbeitsplatz* (Polycyclic aromatic hydrocarbons in the workplace) (*BIA-Report 3/83*), Sankt Augustin, Federal Republic of Germany, Berufsgenossenschaftliches Institut für Arbeitssicherheit

Bonser, G.M. (1932) Tumours of the skin produced by blast-furnace tar. *Lancet, i,* 775-776

Borwitzky, H. & Schomburg, G. (1979) Separation and identificaton of polynuclear aromatic compounds in coal tar by using glass capillary chromatography including combined gas chromatography-mass spectrometry. *J. Chromatogr., 170,* 99-124

Bos, R.P., Hulshof, C.T.J., Theuws, J.L.G. & Henderson, P.Th. (1983) Mutagenicity of creosote in the *Salmonella*/microsome assay. *Mutat. Res., 119,* 21-25

Bos, R.P., Theuws, J.L.G., Leijdekkers, C.-M. & Henderson, P.T. (1984a) The presence of the mutagenic polycyclic aromatic hydrocarbons benzo[a]pyrene and benzo[a]anthracene in creosote P1. *Mutat. Res., 130,* 153-158

Bos, R.P., Hulshof, C.T.J., Theuws, J.L.G. & Henderson, P.Th. (1984b) Genotoxic exposure of workers creosoting wood *Br. J. ind. Med., 41,* 260-262

Boutwell, R.K. & Bosch, D.K. (1958) The carcinogenicity of creosote oil: Its role in the induction of skin tumors in mice. *Cancer Res., 18,* 1171-1175

Bridbord, K., Finklea, J.F., Wagner, J.K., Moran, J.B. & Caplan, P. (1976) Human exposure to polynuclear aromatic hydrocarbons. In: Freudenthal, R.I. & Jones, P.W., eds, *Carcinogenesis - A Comprehensive Survey,* Vol. 1, *Polynuclear Aromatic Hydrocarbons, Chemistry, Metabolism and Carcinogenesis,* New York, Raven Press, pp. 319-324

Burchill, P., Herod, A.A. & Pritchard, E. (1983a) Investigation of nitrogen compounds in coal tar products. 2. Basic fractions. *Fuel, 62,* 20-29

Burchill, P., Herod, A.A. & Pritchard, E. (1983b) Investigation of nitrogen compounds in coal tar products. 1. Unfractionated materials. *Fuel, 62*, 11-19

Cabot, S., Shear, N. & Shear, M.J. (1940) Studies in carcinogenesis. XI. Development of skin tumors in mice painted with 3:4-benzpyrene and creosote oil fractions. *Am. J. Pathol., 16*, 301-312

Carli, G. (1958) Spinocellular epithelioma of the left labium major following the application of a tar pomade (Ital.). *Minerva dermatol., 33*, 280-281

Carlton, W.W. (1966) Experimental coal tar poisoning in the white Pekin duck. *Avian Dis., 10*, 484-502

Casto, B.C., Hatch, G.G., Huang, S.L., Lewtas, J., Nesnow, S. & Waters, M.D. (1981) Mutagenic and carcinogenic potency of extracts of diesel and related environmental emissions: *In vitro* mutagenesis and oncogenic transformation. *Environ. int., 5*, 403-409

Černiková, M., Horáček, J. & Dubský, H. (1979) Resorption of some of the components of black coal tar by the human skin (Czech.) *Cesk. Dermatol., 54*, 321-325

Chrostek, W.J. (1980) *Health Hazard Evaluation Determination Report No. HE 79-43-663, Harbison-Walker Refractories, Clearfield, Pennsylvania*, Cincinnati, OH, National Institute for Occupational Safety and Health

Claxton, L. & Huisingh, J.L. (1980) *Comparative mutagenic activity of organics from combustion sources.* In: Saunders, G.L., Cross, F.T., Dagle, G.E. & Mahaffey, J.A., eds, *Pulmonary Toxicology of Respirable Particles (DOE Symposium Series No. 53)*, Springfield, VA, US Department of Energy, Technical Information Center, pp. 453-465

Coalite and Chemical Products Ltd (1984) *Product Data Sheets*, Chesterfield, UK

Codd, L.W., Dijkhoff, K., Fearon, J.H., von Oss, C.J., Roebersen, H.G. & Stanford, E.G., eds (1971) *Chemical Technology: An Encyclopedic Treatment*, Vol. II, New York, Barnes & Noble, pp. 593-600, 606-618, 624-627, 636-644, 654-678, 694-699

Collin, G. & Zander, M. (1982) Tar and pitch (Ger.). In: *Ullmanns Encyklopädie der technischen Chemie* (Ullmann's encyclopedia of technical chemistry), Vol. 22, Weinheim, Verlag Chemie, pp. 411-446

Colucci, J.M. & Begeman, C.R. (1965) The automotive contribution to air-borne polynuclear aromatic hydrocarbons in Detroit. *J. Air Pollut. Control Assoc., 15*, 113-122

Considine, D.M., ed. (1974) *Chemical and Process Technology Encyclopedia*, New York, McGraw-Hill, pp. 297-302

Cookson, H.A. (1924) Epithelioma of the skin after prolonged exposure to creosote. *Br. med. J., i*, 368

Coulter, P.D. & Petro, B.A. (1972) *Coal Tar Pitch Sampling Analysis (Contract No. HSM-099-71-0007)*, Cincinnati, OH, National Institute for Occupational Safety and Health

Crow, K.D. (1970) Chloracne. A critical review including a comparison of two series of cases of acne from chlornaphthalene and pitch fumes. *Trans. St John's Hosp. dermatol. Soc.*, *56*, 79-99

Crow, K.D., Alexander, E., Buck, W.H.L., Johnson, B.E., Magnus, I.A. & Porter, A.D. (1961) Photosensitivity due to pitch. *Br. J. Dermatol.*, *73*, 220-232

Curren, R.D., Kouri, R.E., Kim, C.M. & Schechtman, L.M. (1981) Mutagenic and carcinogenic potency of extracts from diesel related environmental emissions: Simultaneous morphological transformation and mutagenesis in BALB/c 3T3 cells. *Environ. int.*, *5*, 411-415

Decouflé, P., Stanislawczyk, K., Houten, L., Bross, I.D.J. & Viadana, E. (1977) *A Retrospective Survey of Cancer in Relation to Occupation*, Prepared under Contract No. HSM 99-73-5 for the US Department of Health, Education and Welfare, Cincinnati, OH, National Institute for Occupational Safety and Health

Doig, A.T. (1970) Epithelioma of the scrotum in Scotland in 1967. *Health Bull.*, *28*, 45-51

Dumber, F.G. (1962) Hygienic characteristics of creosote use at Karaganda coal dressing plants (Russ.). *Gig. Tr. prof. Zabol.*, *6*, 50-52

Durkin, W., Sun, N., Link, J., Paroly, W. & Schweitzer, R. (1978) Melanoma in a patient treated for psoriasis *South. med. J.*, *71*, 732-733

Elgjo, K. & Larsen, T.E. (1973) Alterations in epidermal growth kinetics induced by coal tar ointment and methotrexate. *J. invest. Dermatol.*, *61*, 22-24

Emmett, E.A., Stetzer, L. & Taphorn, B. (1977) Phototoxic keratoconjunctivitis from coal-tar pitch volatiles. *Science*, *198*, 841-842

Emmett, E.A., Bingham, E.M. & Barkley, W. (1981) A carcinogenic bioassay of certain roofing materials. *Am. J. ind. Med.*, *2*, 59-64

Federal Office for Worker's Safety (1982) *Verordnung über gefährliche Arbeitsstoffe* (Regulation for hazardous substances in the workplace) (*Bundesgesetzblatt I, 11 Februar*), Dortmund, Federal Republic of Germany, p. 140

Fibiger, J. & Bang, F. (1920) Experimental induction of tar cancer in white mice (Fr.). *C.R. Soc. Biol.*, *83*, 1157-1160

The Finnish Wood Preserving Association (1982) *Annual Report*, Helsinki

Fisher, R.E.W. (1953) Occupational skin cancer in a group of tar workers. *Arch. ind. Hyg.*, *7*, 12-18

Fisher, R.E.W. (1954) Skin cancer in tar workers. *Trans. Assoc. ind. Med. Off.*, *3*, 315-318

Foreman, M.I., Picton, W., Lukowiecki, G.A. & Clark, C. (1979) The effect of topical crude coal tar treatment on unstimulated hairless hamster skin. *Br. J. Dermatol.*, *100*, 707-715

Fritz, W. & Engst, R. (1971) About the environmental contamination of foodstuffs with carcinogenic hydrocarbons (Ger.). *Z. ges. Hyg.*, *17*, 271-275

Fysh, J.M., Andrews, L.S., Pohl, L.R. & Nebert, D.W. (1980) Differing degrees of coal-tar shampoo-induced mutagenesis in the *Salmonella*/liver test system *in vitro*. *Pharmacology*, *20*, 1-8

Gilbertson, W.E. (1982) *Handbook of Nonprescription Drugs*, 7th ed., Washington DC, American Pharmaceutical Association, The National Professional Society of Pharmacists, pp. 1-7, 579, 585-587, 590-592

Gloor, M., Dressel, M. & Schnyder, U.W. (1978) The effect of coal tar distillate, cadmium sulfide, ichthyol sodium and omadine MDS on the epidermis of the guinea pig. *Dermatologica*, *156*, 238-243

Gorman, R. & Liss, G.M. (1984) *Occupational exposure to pitch in a non-heated process: Phytotoxicity and pre-malignant warts*. In: *Polynuclear Aromatic Hydrocarbons, 8th Int. Symposium*, Columbus, OH, Battelle Press (in press)

Górski, T. (1959) Experimental investigations on the carcinogenic properties of some pitches and tars produced from Silesian pit coal (Pol.). *Med. Pract.*, *10*, 309-317

Götz, H. (1976) The relationship between ultraviolet light sensitization and tar exposure of the skin. *Aust. J. Dermatol.*, *17*, 57-60

Graham, R., Hester, H.R. & Henderson, J.A. (1940) Coal-tar-pitch poisoning in pigs. *J. Am. vet. med. Assoc.*, *96*, 135-140

Gunter, B.J. & Ligo, R. (1976) *Health Hazard Evaluation Determination Report No. 75-13-265, Protecto Wrap Company, Denver, Colorado* (*NIOSH-75-13-265*), Cincinnati, OH, National Institute for Occupational Safety and Health

Haldin-Davis, H. (1935) Multiple warts in a creosote worker. *Proc. roy. Soc. Med.*, *29*, 89-90

Halprin, K.M., Comerford, M. & Taylor, J.R. (1982) Cancer in patients with psoriasis. *J. Am. Acad. Dermatol.*, *7*, 633-638

Hammond, E.C., Selikoff, I.J., Lawther, P.L. & Seidman, H. (1976) Inhalation of benzpyrene and cancer in man. *Ann. N.Y. Acad. Sci.*, *271*, 116-124

Hawley, G.G., ed. (1981) *The Condensed Chemical Dictionary*, 10th ed., New York, Van Nostrand Reinhold, pp. 257-258, 285

Henry, S.A. (1946) *Cancer of the Scrotum in Relation to Occupation*, New York, Oxford University Press

Henry, S.A. (1947) Occupational cutaneous cancer attributable to certain chemicals in industry. *Br. med. Bull.*, *4*, 389-401

Henry, S.A., Kennaway, N.M. & Kennaway, E.L. (1931) The incidence of cancer of the bladder and prostate in certain occupations. *J. Hyg.*, *31*, 125-137

Hervin, R.L. & Emmett, E.A. (1976) *Health Hazard Evaluation/Toxicity Determination. Western Roofing Co., Sellers & Marquis Roofing Co., A.J. Shirk Roofing Co., Quality Roofing Co., Kansas City, Missouri 64130* (*NIOSH-TR-HHE-75-194-324*), Cincinnati, OH, National Institute for Occupational Safety and Health, pp. 1-5, 8-11, 15-26

Heyl, T. & Mellett, W.A. (1982) Creosote dermatitis in an ammunition depot (Dutch). *S. Afr. med. J.*, *62*, 66-67

Hilfrich, J. & Mohr, U. (1972) Comparison of carcinogenic effects of coal tar and a new synthetic tar (Ger.). *Arch. derm. Forsch.*, *242*, 176-178

Hirohata, T., Masuda, Y., Horie, A. & Kuratsune, M. (1973) Carcinogenicity of tar-containing skin drugs: Animal experiment and chemical analysis. *Gann*, *64*, 323-330

Hittle, D.C. & Stukel, J.J. (1976) Particle size distribution and chemical composition of coal-tar fumes. *Am. ind. Hyg. Assoc. J.*, *37*, 199-204

Hjorth, N. & Trolle-Lassen, C. (1963) Skin reactions to ointment bases. *Trans. St John's Hosp. dermatol. Soc.*, *49*, 127-140

Hodgson, G. (1948) Epithelioma following the local treatment of *pruritus ani* with *liquor picis carbonis*. *Br. J. Dermatol.*, *60*, 282

Hodgson, G.A. & Whiteley, H.J. (1970) Personal susceptibility to pitch. *Br. J. ind. Med.*, *27*, 160-166

Hoiberg, A.J. (1965) *Bituminous materials*. In: Bikales, N.M., ed., *Encyclopedia of Polymer Science and Technology*, Vol. 2, New York, John Wiley & Sons, pp. 402-437

Hoover, J.E., ed. (1975) *Remington's Pharmaceutical Sciences*, 15th ed., Easton, PA, Mack Publishing Company, p. 721

Horton, A.W., Tye, R. & Stemmer, K.L. (1963) Experimental carcinogenesis of the lung. Inhalation of gaseous formaldehyde or an aerosol of coal tar by C3H mice. *J. natl Cancer Inst.*, *30*, 31-43

Hueper, W.C. & Payne, W.W. (1960) Carcinogenic studies on petroleum asphalt, cooling oil, and coal tar. *Arch. Pathol.*, *70*, 372-384

IARC (1979a) *IARC Monographs on the Evaluation of the Carcinogenic Risk of Chemicals to Humans*, Vol. 20, *Some Halogenated Hydrocarbons*, Lyon, pp. 303-325

IARC (1979b) *Environmental Carcinogens. Selected Methods of Analysis*, Vol. 3, *Analysis of Polycyclic Aromatic Hydrocarbons in Environmental Samples* (*IARC Scientific Publications No. 29*), Lyon

IARC (1980) *IARC Monographs on the Evaluation of the Carcinogenic Risk of Chemicals to Humans*, Vol. 23, *Some Metals and Metallic Compounds*, Lyon, pp. 39-141

IARC (1982a) *IARC Monographs on the Evaluation of the Carcinogenic Risk of Chemicals to Humans*, Vol. 29, *Some Industrial Chemicals and Dyestuffs*, Lyon, pp. 93-148, 391-398

IARC (1982b) *IARC Monographs on the Evaluation of the Carcinogenic Risk of Chemicals to Humans*, Vol. 28, *The Rubber Industry*, Lyon

IARC (1984a) *IARC Monographs on the Evaluation of the Carcinogenic Risk of Chemicals to Humans*, Vol. 34, *Polynuclear Aromatic Compounds, Part 3, Industrial Exposures in Aluminium Production, Coal Gasification, Coke Production, and Iron and Steel Founding*, Lyon

IARC (1984b) *IARC Monographs on the Evaluation of the Carcinogenic Risk of Chemicals to Humans*, Vol. 33, *Polynuclear Aromatic Compounds, Part 4, Carbon Blacks, Mineral Oils and Some Nitroarenes*, Lyon, pp. 35-85

International Labour Office (1980) *Occupational Exposure Limits for Airborne Toxic Substances*, 2nd (rev.) ed. (*Occupational Safety and Health Series No. 37*), Geneva, pp. 76-77, 242, 287

Jach, Z. & Mašek, V. (1973) Elimination of benzo[a]pyrene from the underwear worn by workers in a pitch manufacturing plant (Ger.). *Zbl. Arbeitsmed.*, 23, 143-148

Jonas, A.D. (1943) Creosote burns. *J. ind. Hyg. Toxicol.*, 25, 418-420

Kaidbey, K.H. & Kligman, A.M. (1977) Clinical and histological study of coal tar phototoxicity in humans. *Arch. Dermatol.*, 113, 592-595

Kandus, J., Mašek, V. & Jach, Z. (1972) Determination of the content of benzo[a]pyrene of working clothes and underclothing of workers in a pitch manufacturing plant (Ger.). *Zbl. Arbeitsmed.*, 22, 138-141

Kapitul'skii, V.B. (1966) Physiologic changes encountered in workers employed in the pitch-coke industry. *Hyg. Sanit.*, 31, 408-412

Kennaway, E.L. (1924) On cancer producing tars and tar fractions. *J. ind. Hyg.*, 5, 462-488

Kennaway, N.M. & Kennaway, E.L. (1936) A study of the incidence of cancer of the lung and larynx. *J. Hyg.*, 36, 236-267

Kireeva, I.S. (1968) Carcinogenic properties of coal-tar pitch and petroleum asphalts used as binders for coal briquettes. *Hyg. Sanit.*, 33, 180-186

Kleffner, H.W., Talbiersky, J. & Zander, M. (1981) Simple scheme for the estimation of concentrations of polycyclic aromatic hydrocarbons in tars. *Fuel*, 60, 361-362

Korhonen, K. (1980) *Airborne Pollutants in Wood Impregnation* (Finn.), Helsinki, Institute of Occupational Health

Korhonen, K. & Mulari (1983) *Airborne Pollutants in Ties Workers* (Finn.), Helsinki, Institute of Occupational Health

Kosuge, T., Zenda, H., Nukaya, H., Terada, A., Okamoto, T., Shudo, K., Yamaguchi, K., Iitaka, Y., Sugimura, T., Nagao, M., Wakabayashi, K., Kosugi, A. & Saito H. (1982) Isolation and structural determination of mutagenic substances in coal tar. *Chem. pharm. Bull.*, 30, 1535-1538

Kruber, O. & Oberkobusch, R. (1952) On new components of coal-tar pitch (Ger.). *Ber. dtsch. chem. Ges.*, *85*, 433-436

Lang, K.F., Buffleb, H. & Schweym, E. (1959) On the production of new hydrocarbons from the highest boiling fractions of coal tar (Ger.). *Brennstoff-Chem.*, *40*, 369-370

Larson, B.A. (1978) Occupational exposure to coal tar pitch volatiles at pipeline protective coating operations. *Am. ind. Hyg. Assoc. J.*, *39*, 250-255

Lavker, R.M., Grove, G.L. & Kligman, A.M. (1981) The atrophogenic effect of crude coal tar on human epidermis. *Br. J. Dermatol.*, *105*, 77-82

Layman, P.L. (1982) New biocides find acceptance difficult. *Chem. Eng. News*, 12 April, 10-13

Lehmann, E., Auffarth, J. & Häger, J. (1984) Determination of selected polycyclic aromatic hydrocarbons in coal-tar, coal-tar pitch and tar products (Ger.). *Staub-Reinhalt. Luft*, *44*, 452-455

Lewis, F.A. (1979) *Hazard Evaluation and Technical Assistance Report No. TA 79-6, Fiber Materials, Inc., Biddeford, Maine* (*NIOSH Report No. TA-79-6*), Cincinnati, OH, National Institute for Occupational Safety and Health

Lewtas, L., Bradow, R.L., Jungers, R.H., Harris, B.D., Zweidinger, R.B., Cushing, K.M., Gill, B.E. & Albert, R.E. (1981) Mutagenic and carcinogenic potency of extracts of diesel and related environmental emissions: Study design, sample generation, collection, and preparation. *Environ. int.*, *5*, 383-387

Lijinsky, W., Saffiotti, U. & Shubik, P. (1957) A study of the chemical constitution and carcinogenic action of creosote oil. *J. natl Cancer Inst.*, *18*, 687-692

Lijinsky, W., Domsky, I., Mason, G., Ramahi, H.Y. & Safavi, T. (1963) The chromatographic determination of trace amounts of polynuclear hydrocarbons in petrolatum, mineral oil, and coal tar. *Anal. Chem.*, *35*, 952-956

Linnik, A.B. (1970) On the carcinogenic effect of tar ointment 'Locacorten-tar' (Russ.). *Vestn. Dermatol. Venerol.*, *44*, 32-36

Lorenz, L.F. & Gjovik, L.R. (1972) *Analyzing creosote by gas chromatography: Relationship to creosote specifications*. In: *Proceedings of the Annual Meeting of the American Wood-Preservers' Association 1972*, Stevensville, MD, pp. 32-42

Luke, D. (1954) Liver dystrophy associated with coal tar pitch poisoning in the pig. *Vet. Rec.*, *66*, 643-645

Mackenzie, S. (1898) A case of tar eruption. *Br. J. Dermatol.*, *10*, 417

MacLean, C.W. (1969) Observations on coal tar poisoning in pigs. *Vet. Rec.*, *84*, 594-598

Malaiyandi, M., Benedek, A., Halko, A.P. & Bancsi, J.J. (1982) *Measurement of potentially hazardous polynuclear aromatic hydrocarbons from occupational exposure during roofing and paving operations*. In: Cooke, M., Dennis, A.J. & Fisher, G.L., eds, *Polynuclear*

Aromatic Hydrocarbons: Physical and Biological Chemistry, 6th International Symposium, Columbus, OH, Battelle Press, pp. 471-489

Mariich, L.I. & Lenkevich, Z.K. (1973) Capillary chromatography as a method for the rapid determination of the main components in coal tar and coal tar fractions. *Zh. anal. Khim.*, *28*, 1193-1198

Margolis, E.M. (1964) Eye lesions from coal tar pitch (Russ.). *Gig. Tr. prof. Zabol.*, *8*, 48-49

Markel, H.L., Jr, Ligo, R.N. & Lucas, J.B. (1977) *Health Hazard Evaluation Determination Report No. 75-117-372, Koppers Company, Inc., North Little Rock, Arkansas*, Cincinnati, OH, National Institute for Occupational Safety and Health

Mašek, V. (1966) On the 3,4-benzpyrene contents in black-coal tars. *Prac. Lék.*, *18*, 65-68

Mašek, V. (1973) Thin-layer chromatographic dermination of 3,4-benzpyrene in high-temperature coal tars and their α-, β- and γ-fractions (Ger.). *Gesundheits-Ingenieur*, *94*, 340-343

Maughan, W.Z., Muller, S.A., Perry, H.O., Pittelkow, M.R. & O'Brien, P.C. (1980) Incidence of skin cancers in patients with atopic dermatitis treated with coal tar. *J. Am. Acad. Dermatol.*, *3*, 612-615

McLaughlin, J.K., Blot, W.J., Mandel, J.S., Schuman, L.M., Mehl, E.S. & Fraumeni, J.F., Jr (1983) Etiology of cancer of the renal pelvis. *J. natl Cancer Inst.*, *71*, 287-291

McNeil, D. (1983) Tar and pitch. In: Kirk, R.E. & Othmer, D.F., eds, *Encyclopedia of Chemical Technology*, 3rd ed., Vol. 22, New York, John Wiley & Sons, pp. 564-600

McQuilkin, S.D. (1980) *Health Hazard Evaluation Determination Report No. 78-102-677, Continental Columbus Corporation, Columbus, Wisconsin (NIOSH-HE-78-102-677)*, Cincinnati, OH, National Institute for Occupational Safety and Health

Menck, H.R. & Henderson, B.E. (1976) Occupational differences in rates of lung cancer. *J. occup. Med.*, *18*, 797-801

Milham, S. (1982) *Occupational Mortality in Washington State 1950-1979*, Prepared under Contract No. 210-80-0088, Cincinnati, OH, National Institute for Occupational Safety and Health

Mitchell, A.D. & Tajiri, D.T. (1978) *In Vitro Mammalian Mutagenicity Assays of Creosote P1 and P2*. SRI International report for EPA, Contract no. 68-01-2458. Cited in *Fed. Regist.* (1978) *43*, no. 202, 48154-48214

Mitchell, A.D., Evans, E.L., Jotz, M.M., Riccio, E.S., Mortelmans, K.E. & Simmon, V.F. (1981) Mutagenic and carcinogenic potency of extracts of diesel and related environmental emissions: In vitro mutagenesis and DNA damage. *Environ. int.*, *5*, 393-401

Möller, P. (1925) Primary pulmonary carcinoma in pie rats painted with coal tar (Fr.). *Acta pathol. microbiol. scand.*, *1*, 412-437

Morando, A. (1960) Observations on some particular cutaneous and hepatic manifestations due to processing of tar (Ital.). *Med. Lav.*, *51*, 144-149

Murphy, J.B. & Sturm, E. (1925) Primary lung tumors in mice following the cutaneous application of coal tar. *J. exp. Med.*, *42*, 693-700

National Institute for Occupational Safety and Health (1974) *Report on Analytical Methods Used in a Coke Oven Effluent Study, The Five Oven Study* (HEW Publication No. (NIOSH) 74-105), Cincinnati, OH, pp. iii-v, 1-3

National Institute for Occupational Safety and Health (1977) *Criteria for a Recommended Standard... Occupational Exposure to Coal Tar Products* (DHEW (NIOSH) Publication No. 78-107), Washington DC, US Department of Health, Education and Welfare

National Institute for Occupational Safety and Health (1983a) *NIOSH Tradename Ingredient Data Base - NOHS* (Computer printout for coal tar and coal tar pitch, 11/28/83), Cincinnati, OH

National Institute for Occupational Safety and Health (1983b) *Industrial Hygiene Surveys of Occupational Exposure to Wood Preservative Chemicals* (DHHS Publication (NIOSH) No. 83-106), Washington DC, US Government Printing Office, pp. 30, 33-36, 70, 71, 75

Nesnow, S. & Lewtas, J. (1981) Mutagenic and carcinogenic potency of extracts of diesel and related environmental emissions: Summary and discussion of the results. *Environ. int.*, *5*, 425-429

Nesnow, S., Triplett, L.L. & Slaga, T.J. (1982a) Comparative tumor-initiating activity of complex mixtures from environmental particulate emissions on SENCAR mouse skin. *J. natl Cancer Inst.*, *68*, 829-834

Nesnow, S., Triplett, L.L. & Slaga, T.J. (1982b) *Comparison of the skin tumor initiating activities of emission extracts in the SENCAR mouse.* In: Cooke, M., Dennis, A.J. & Fisher, G.L., eds, *Polynuclear Aromatic Hydrocarbons: Physical and Biological Chemistry, 6th International Symposium*, Columbus, OH, Battelle Press, pp. 585-595

Nesnow, S., Triplett, L.L. & Slaga, T.J. (1983) Mouse skin tumor initiation-promotion and complete carcinogenesis bioassays: Mechanisms and biological activities of emission samples. *Environ. Health Perspect.*, *47*, 255-268

Nestler, F.H.M. (1974) *The Characterization of Wood-Preserving Creosote by Physical and Chemical Methods of Analysis* (USDA Forest Service Research Paper FPL 195), Madison, WI, US Department of Agriculture

Novotny, M., Strand, J.W., Smith, S.L., Wiesler, D. & Schwende, F.J. (1981) Compositional studies of coal tar by capillary gas chromatography/mass spectrometry. *Fuel*, *60*, 213-220.

O'Donovan, W.J. (1920) Epitheliomatous ulceration among tar workers. *Br. J. Dermatol. Syphilis*, *32*, 215-228

Pekker, R.Y. (1967) The state of the oral cavity in workers having contact with coal tar and pitch (Russ.). *Stomatologiya*, *46*, 35-38

Perch, M. & Muder, R.E. (1974) *Coal carbonization and recovery of coal chemicals*. In: Kent, J.A., ed., *Riegel's Handbook of Industrial Chemistry*, 7th ed., New York, Van Nostrand Reinhold, pp. 193-206

Pittelkow, M.R., Perry, H.O., Muller, S.A., Maughan, W.Z. & O'Brien, P.C. (1981) Skin cancer in patients with psoriasis treated with coal tar. A 25-year follow-up study. *Arch. Dermatol.*, *117*, 465-468

Poel, W.E. & Kammer, A.G. (1957) Experimental carcinogenicity of coal-tar fractions: The carcinogenicity of creosote oils. *J. natl Cancer Inst.*, *18*, 41-55

Rao, T.K., Epler, J.L. & Eatherly, W.P. (1979) Mutagenicity testing of extracts from petroleum and coal tar pitches. *Ext. Abstr. Program bienn. Conf. Carbon*, *14*, 63-64

Reed, L.D. (1983) *Health Hazard Evaluation Report HETA 82-067-1253, Anchor Hocking Glass Company, Roofing Site, Lancaster, Ohio*, Cincinnati, OH, National Institute for Occupational Safety and Health

Ritschel, W.A., Siegel, E.G. & Ring, P.E. (1975) Biopharmaceutical evaluation of topical tar preparations. *Sci. Pharm.*, *43*, 11-21

Roe, F.J.C., Bosch, D. & Boutwell, R.K. (1958) The carcinogenicity of creosote oil: The induction of lung tumors in mice. *Cancer Res.*, *18*, 1176-1178

Romanowski, T., Funcke, W., Grossmann, I., König, J. & Balfanz, E. (1983) Gas chromatographic/mass spectrometric determination of high-molecular-weight polycyclic aromatic hydrocarbons in coal tar. *Anal. Chem.*, *55*, 1030-1033

Rook, A.J., Gresham, G.A. & Davis, R.A. (1956) Squamous epithelioma possibly induced by the therapeutic application of tar. *Br. J. Cancer*, *10*, 17-23

Saffiotti, U. & Shubik, P. (1956) The effects of low concentrations of carcinogen in epidermal carcinogenesis. A comparison with promoting agents. *J. natl Cancer Inst.*, *16*, 961-969

Salfeld, K., Rupec, M. & Hoos, I. (1966) About the effect of some antieczemetics on the normal skin of guinea-pigs. III. Tar and shale oil at low concentration (Ger.). *Arch. klin. exp. Dermatol.*, *224*, 392-401

Sall, R.D. & Shear, M.J. (1940) Studies in carcinogenesis. XII. Effect of the basic fraction of creosote oil on the production of tumors in mice by chemical carcinogens. *J. natl Cancer Inst.*, *1*, 45-55

Saperstein, M.D. & Wheeler, L.A. (1979) Mutagenicity of coal tar preparations used in the treatment of psoriasis. *Toxicol. Lett.*, *3*, 325-329

Sarkany, I. & Gaylarde, P.M. (1976) Effect of coal tar fractions on guinea-pig and human skin. *Clin. exp. Dermatol.*, *1*, 51-58

Sawicki, E., Fox, F.T., Elbert, W., Hauser, T.R. & Meeker, J. (1962) Polynuclear aromatic hydrocarbon composition of air polluted by coal-tar pitch fumes. *Am. ind. Hyg. Assoc. J.*, *23*, 482-486

Schecker, H.-G. (1984) *Systematische Untersuchung der Verbreitung und Verwendung von Pech, Teer, Teeröl in Bitumen, Bitumen und der dabei auftretenden Arbeitsplatzbelastungen, Report no. 50577* (Systematic study of the distribution and use of pitch, tar, tar oil in bitumen, bitumen and their occurrence in the work place), Dortmund, Federal Republic of Germany, Bundesanstalt fur Arbeitsschutz

Schimberg, R.W. (1981) Polycyclic aromatic hydrocarbons in the work environment (Finn.). *Kemia-Kemi, 9*, 537-541

Schoental, R. (1957) Isolation of 3:4-9:10-dibenzopyrene from coal-tar. *Nature, 180*, 606

Schulz, G. (1983) The importance of creosote for wood conservation and the necessity for harmonizing technical specifications (Ger.). *Holz Roh- Werkst., 41*, 387-391

Schunk, W. (1979) On the relationship between exposure to polycyclic hydrocarbons and the frequency of tumours in two chemical factories at Thüringen (Ger.). *Z. ärztl. Fortbild, 73*, 84-88

Schwartz, L. (1942) Dermatitis from creosote treated wooden floors. *Ind. Med., 11*, 387

Shabad, L.M., Khesina, A.Ja., Linnik, A.B. & Serkovskaya, G.S. (1970) Possible carcinogenic hazards of several tars and of locacorten-tar ointment (spectro-fluorescent investigations and experiments in animals). *Int. J. Cancer, 6*, 314-318

Shabad, L.M., Linnik, A.B., Tumanov, V.P. & Rubetskoy, L.S. (1971) On the possibility of blastomogenic properties of ointments containing tar (Russ.). *Eksp. Naia Khir. Anesteziol., 16*, 6-9

Shustova, M.N. & Samoilovich, L.N. (1971) Blastomogenicity of neutralized soots from the sulfate shop of a coke plant. *Gig. Sanit., 36*, 103-104

Simmon, V.F. & Poole, D.C. (1978) In vitro *Microbiological Mutagenicity Assays of Creosote P1 and Creosote P2*. SRI International report for EPA, Contract no. 68-01-2458. Cited in *Fed. Reg.* (1978) *43*, no. 202, 48154-48214.

Slaga, T.J., Triplett, L.L. & Nesnow, S. (1981) Mutagenic and carcinogenic potency of extracts of diesel and related environmental emissions: Two-stage carcinogenesis in skin tumor sensitive mice (SENCAR). *Environ. int., 5*, 417-423

Söderkvist, P. & Toftgård, R. (1982) *Effect of creosote on xenobiotic metabolizing enzymes in liver and lungs of rats. A preliminary report* (Swed.). In: Proceedings of the 31st Scandinavian Meeting on Occupational Health, Reykjavik, Iceland, Vinnueftirlit Mikisins

Starke, J.C. & Jillson, O.F. (1961) Photosensitization to coal tar. *Arch. Dermatol., 84*, 935-936

Stasse, H.L. (1954) Fractional distillation of creosote and composition of preservatives used in the cooperative creosote program. *Proc. Am. Wood-Preservers' Assoc., 50*, 13-40

Stern, R.S., Zierler, S. & Parrish, J.A. (1980) Skin carcinoma in patients with psoriasis treated with topical tar and artificial ultraviolet radiation. *Lancet, i*, 732-735

Stoughton, R.B., DeQuoy, P. & Walter, J.F. (1978) Crude coal tar plus near ultraviolet light suppresses DNA synthesis in epidermis. *Arch. Dermatol., 114*, 43-45

Superstein, M.D. & Wheeler, L.A. (1979) Mutagenicity of coal tar preparations used in the treatment of psoriasis. *Toxicol. Lett., 3*, 325-329

Susorov, N.A. (1970) Group eye lesions by coal-tar pitch (Russ.). *Voenno-med. Zh., 3*, 326

Swallow, W.H. & Curtis, J.F. (1980) Levels of polycyclic aromatic hydrocarbons in some coal tar skin preparations. *Aust. J. Dermatol., 21*, 154-157

Tanenbaum, L. & Parrish, J.A. (1975) Tar phototoxicity and phototherapy for psoriasis. *Arch. Dermatol., 111*, 467-470

Tharr, D.J. (1982) *Health Hazard Evaluation Report HETA 81-432-1105, Roofing Sites, Rochester and Buffalo, New York*, Cincinnati, OH, National Institute for Occupational Safety and Health

Thomas Ness Ltd (1984) *Product Data Sheets*, Nottingham, UK

Todd, A.S. & Timbie, C.Y. (1981) *Industrial Hygiene Report No. PB82-174160, Comprehensive Survey of Wood Preservative Treatment Facility at Cascade Pole Company, McFarland Cascade, Tacoma, Washington*, Cincinnati, OH, National Institute for Occupational Safety and Health

Todd, A.S. & Timbie, C.Y. (1983) *Industrial Hygiene Surveys of Occupational Exposure to Wood Preservative Chemicals*, Washington DC, US Government Printing Office

Tomkins, B.A., Kubota, H., Griest, W.H., Caton, J.E., Clark, B.R. & Guerin, M.R. (1980) Determination of benzo[a]pyrene in petroleum substitutes. *Anal. Chem., 52*, 1331-1334

Trosset, R.P., Warshawsky, D., Menefee, C.L. & Bingham, E. (1978) *Investigation of Selected Potential Environmental Contaminants: Asphalt and Coal Tar Pitch*. Prepared under Contract No. 68-01-4188 for the Office of Toxic Substances, US Environmental Protection Agency, Springfield, VA, Technical Information Service

Tsutsui, H. (1918) On the artificial cancer produced in mice (Ger.). *Gann, 12*, 17-21

US Department of Agriculture (1980a) *The Biologic and Economic Assessment of Pentachlorophenol, Inorganic Arsenicals, Creosote*, Vol. 1, *Wood Preservatives (Technical Bulletin Number 1658-I)*, Washington DC, pp. 193, 201-209, 214-218, 220-221, 223-224

US Department of Agriculture (1980b) *The Biologic and Economic Assessment of Pentachlorophenol, Inorganic Arsenicals, Creosote*, Vol. 2, *Non-Wood Preservatives (Technical Bulletin Number 1658-II)*, Washington DC, pp. 214, 220-221, 234-235, 239

US Department of Energy (1981) *Coke and Coal Chemicals in 1980 (DOE/EIA-0120(80))*, Washington DC, Energy Information Administration, p. 19

US Department of Energy (1983) *Quarterly Coal Report (DOE/EIA-0121 (83/1Q))*, *January-March 1983*, Washington DC, Energy Information Administration, p. 62

US Department of the Treasury (1983) Formulas for denatured alcohol and rum. *Code Fed. Regul., Title 27*, Part 21; *Fed. Regist., 48(No. 107)*, 24672-24675, 24680, 24681

US Environmental Protection Agency (1978) Notice of rebuttable presumption against registration and continued registration of pesticide products containing coal-tar, creosote, and coal tar neutral oil. *Fed. Regist., 43(No. 202)*, 48154-48156

US Environmental Protection Agency (1979) *Toxic Substances Control Act (TSCA) Chemical Substance Inventory, Initial Inventory*, Vol. 1, Appendix A, Washington DC, Office of Toxic Substances, pp. 3, 16

US Environmental Protection Agency (1981) Creosote, pentachlorophenol and the inorganic arsenicals; preliminary notice of determination concluding the rebuttable presumption against registration of the wood preservative uses of pesticide products; notice of availability of position document 2/3. *Fed. Regist., 46(No. 33)*, 13020-13036

US Environmental Protection Agency (1982) Hazardous waste management system: Identification and listing of hazardous waste. *Code Fed. Regul., Title 40*, Part 261

US Environmental Protection Agency (1983) Notification requirements; Reportable quantity adjustments. *Fed. Regist., 48(No. 102)*, 23552-23601

US Food and Drug Administration (1982a) Drugs for human use: Oligosaccharide antibiotic drugs. *US Code Fed. Regul., Title 21*, Part 444.542a

US Food and Drug Administration (1982b) OTC drug products for the control of dandruff, seborrheic dermatitis, and psoriasis; Establishment of a monograph. *Fed. Regist., 47(No. 233)*, 54646-54684

US International Trade Commission (1982) *Synthetic Organic Chemicals, US Production and Sales, 1981* (*USITC Publication 1292*), Washington DC, US Government Printing Office, pp. 9, 11

US International Trade Commission (1983) *Synthetic Organic Chemicals, US Production and Sales, 1982* (*USITC Publication 1422*), Washington DC, US Government Printing Office, pp. 9-14

US Occupational Safety and Health Administration (1983a) Air contaminants. *US Code Fed. Regul., Title 29*, Parts 1910.1000, 1910.1002, pp. 48, 66

US Occupational Safety and Health Administration (1983b) Occupational exposure to coal tar pitch volatiles; modification of interpretation. *US Code Fed. Regul., Title 29*, Part 1910; *Fed. Regist., 48(15)*, 2764-2769

US Pharmacopeial Convention (1980) *The United States Pharmacopeia*, 20th rev., Rockville, MD, p. 161

US Tariff Commission (1919) *Report on Dyes and Related Coal-Tar Chemicals, 1918* (*Tariff Information Series No. 20*), Washington DC, US Government Printing Office, pp. 14-18

US Tariff Commission (1973) *Synthetic Organic Chemicals, United States Production and Sales, 1971* (*TC Publication 614*), Washington DC, US Government Printing Office, p. 9

Vo-Dinh, T. & Gammage, R.B. (1980) *Use of a Fiber-Optics Skin Contamination Monitor in the Worplace*, Oak Ridge, TN, Oak Ridge National Laboratory

Wallcave, L., Garcia, H., Feldman, R., Lijinsky, W. & Shubik, P. (1971) Skin tumorigenesis in mice by petroleum asphalts and coal-tar pitches of known polynuclear aromatic hydrocarbon content. *Toxicol. appl. Pharmacol.*, *18*, 41-52

Walter, J.F., Stoughton, R.B. & DeQuoy, P.R. (1978) Suppression of epidermal proliferation by ultraviolet light, coal tar and anthralin. *Br. J. Dermatol.*, *99*, 89-96

Wang, J.-D., Wegman, D.H. & Smith, T.J. (1983) Cancer risks in the optical manufacturing industry. *Br. J. ind. Med.*, *40*, 177-181

Wheeler, L.A., Saperstein, M.D. & Lowe, N.J. (1981) Mutagenicity of urine from psoriatic patients undergoing treatment with coal tar and ultraviolet light. *J. invest. Dermatol.*, *77*, 181-185

Wilson, P.J., Jr & Wells, J.H. (1950) *Coal, Coke and Coal Chemicals*, Chemical Engineering Series, New York, McGraw-Hill, pp. 1-5, 372-399

Windholz, M., ed. (1983) *The Merck Index*, 10th ed., Rahway, NJ, Merck & Co., p. 345

Wiskemann, A. & Hoyer, H. (1971) Phototoxicity of tar preparation (Ger.). *Hautarzt*, *22*, 257-258

Wolff, M.S., Taffe, B., Boesch, R.R. & Selikoff, I.J. (1982) Detection of polycyclic aromatic hydrocarbons in skin oil obtained from roofing workers. *Chemosphere*, *11*, 595-599

Woodhouse, D.L. (1950) The carcinogenic activity of some petroleum fractions and extracts. Comparative results in tests on mice repeated after an interval of eighteen months. *J. Hyg.*, *48*, 121-134

Wulf, K., Unna, P.J. & Willers, M. (1963) Experimental studies on the photodynamic effect of coal tar constituents (Ger.). *Hautarzt*, *14*, 292-297

Yamagiwa, K. & Ichikawa, K. (1915) On the experimental induction of papillomas (Ger.). *Verh. Jpn. Pathol. Ges.*, *5*, 142-148

Zorn, H. (1978) Occupational and environmental risk of polycyclic aromatic hydrocarbons in petroleum and coal distillates due to their technological use in industry and road construction (Ger.). *Arbeitsmed. Sozialmed. Präventivmed.*, *1*, 6-9

GLOSSARY

ANTHRACENE PASTE. An anthracene-rich fraction isolated from high-temperature LIGHT ANTHRACENE OIL

BASE OIL. See HEAVY ANTHRACENE OIL

BLACK VARNISH. A varnish produced by fluxing soft PITCH with HEAVY ANTHRACENE OIL and used as a protective coating for industrial steelwork and timber buildings and as an antifouling paint for marine application

BLAST-FURNACE TAR. A by-product tar from an unusual blast furnace used previously to make iron in Scotland. In this process, splint coal was used in addition to coke.

BRIQUETTING PITCH. A PITCH softening at 80°C and used for the briquetting of SOLID SMOKELESS FUEL. A high toluene content and a high coking value are required.

CARBOLIC OIL. A primary fraction (boiling-point, 180-205°C) from the distillation of high-temperature COAL-TARS

CARBON-BLACK OIL. Feedstock for the manufacture of carbon black from CREOSOTE and anthracene oils; can also be made directly from WAX OIL

CARBONIZATION. The destructive distillation of coal to produce coke and by-product COAL-TARS

COAL OIL. A synonym for high-temperature COAL-TARS

COAL-TARS. A by-product of the destructive distillation of coal to produce coke

COAL-TAR ENAMELS. Protective coatings made by fluxing a coke-oven pitch with anthracene oil. PIPE-COATING ENAMELS are one example.

COAL-TAR FUELS. Fuels made by blending various higher-boiling distillation fractions and PITCH from the distillation of COAL-TARS

COAL-TAR PITCH VOLATILES (CTPVs). Determination of CTPVs is a common method for measuring the level of organic fumes in the air. It involves air-sample collection with filters followed by solvent extraction and gravimetric analysis. The extracts are also called 'benzene solubles' and 'cyclohexane solubles', depending on the solvent used.

COKE-OVEN TARS. COAL-TARS produced as by-products of the manufacture of coke for blast furnaces

CONTINUOUS VERTICAL-RETORT (CVR) TARS. Tars from coal carbonization in continuous vertical retorts for the manufacture of coke and gas for domestic heating

CREOSOTE. Fractions or blends of COAL-TAR distillation fractions, sometimes including coal-tar pitch, which are used for timber preservation

COAL-TARS AND DERIVED PRODUCTS

CREOSOTE OIL/WASH OIL. Primary fractions (boiling-point, 230-290°C) from the distillation of high-temperature COAL-TARS

ELECTRODE PITCH. See HARD PITCH

FIBRE-PIPE PITCH. A pitch of 65-70°C softening-point and low quinoline-insoluble content used in the manufacture of fibre pipes

FLUXING OIL. A blend of NAPHTHALENE OILS, WASH OILS and LIGHT ANTHRACENE OILS made by distillation of high-temperature COAL-TARS (Note: This term is also used for petroleum-derived products; see monograph on bitumens in this volume.)

HARD PITCH. Residue from the distillation of COAL-TARS made by HIGH-TEMPERATURE COAL-TAR PROCESSES when the depth of distillation is controlled to leave a residue with a high softening-point

HEAVY ANTHRACENE OIL/BASE OIL. Primary fractions (boiling-point, >310°C) from the distillation of high-temperature COAL-TARS

HEAVY/WAX OIL. Primary fractions (boiling-point, 240-360/235-450°C) from the distillation of low-temperature COAL-TARS

HIGH-TEMPERATURE COAL-TAR PROCESSES. Processes using a temperature of >700°C in the carbonization of coal. The main processes employ coke ovens and continuous vertical retorts.

IMPREGNATION PITCH. A PITCH with a softening-point of 85°C and a low quinoline-insoluble content used by the graphite industry for impregnation of electrodes

LIGHT ANTHRACENE OIL. A primary fraction (boiling-point, 260-360°C) from the distillation of high-temperature COAL-TARS

LIGHT OIL. A primary fraction (boiling-point, 90-160°C) from the distillation of low-temperature COAL-TARS

LIGHT OIL/OVERHEADS. A primary fraction (boiling-point, <180°C) from the distillation of high-temperature COAL-TARS

LIGNITE TAR. Tar from carbonization of lignite (brown coal)

LIQUOR PICIS CARBONIS. A product used for pharmaceutical preparations in the UK, made by macerating prepared COAL-TARS (20%) and quillaia (10%) in 90% ethanol for seven days and filtering

LOW-TEMPERATURE COAL-TAR PROCESSES. Processes using a temperature of <700°C in the destructive distillation of coal to produce either SOLID SMOKELESS FUELS for industrial and home heating (the Coalite and Rexco processes) or synthetic natural gas (the LURGI GASIFICATION process)

LURGI GASIFICATION. A low-temperature process to produce synthetic natural gas from coal

MEDIUM-SOFT PITCH. Residue from the distillation of COAL-TARS made by HIGH-TEMPERATURE COAL-TAR PROCESSES when the depth of distillation is controlled to leave a residue with a low softening-point (see also BRIQUETTING PITCH)

MIDDLE OIL. A primary fraction (boiling-point, 90-160°C) from the distillation of low-temperature COAL-TARS

NAPHTHALENE OIL. A primary fraction (boiling-point, 200-230°C) from the distillation of high-temperature COAL-TARS

PHARMACEUTICAL COAL-TARS. Crude COAL-TARS, as such, or after further processing, used as components of pharmaceutical products

PIPE-COATING ENAMEL. An enamel made by fluxing a coke-oven pitch with anthracene oil and used to protect buried pipelines from corrosion

PITCH. Residues from the distillation of COAL-TARS (Note: This term is also used for residues from distillations of other materials, e.g., petroleum pitch.)

PITCH COKE. A coke made from coal-tar PITCH

PITCH/POLYMER COATINGS. Coal-tar PITCH used as an extender for epoxy, methane and other resins, which are used as damp-proofing and protective coatings in corrosive environments

PITCH REFRACTORIES. Coal-tar PITCH is used to impregnate brick refractories: a soft pitch or refined tar seeps into the brick and is fired to produce a PITCH COKE which prolongs the life of the oven.

PIX CARBONIS PRAEPARATA. Commercial COAL-TAR heated at 50°C for one hour to remove some of the lighter oil fractions and used for pharmaceutical preparations in the UK

PIX CARBONIS. Crude COAL-TAR used for pharmaceutical preparations in the UK

PIXALBOL. Purified colourless COAL-TAR

PRIMARY DISTILLATION FRACTIONS. Distillate fractions with particular boiling ranges which are typically collected during commercial-scale distillation of COAL-TARS

REFINED TAR. A product made by fluxing a high-temperature PITCH to a low softening-point PITCH using strained HEAVY ANTHRACENE OILS or HEAVY OILS

REDUCED CRUDE. Crude COAL-TAR after the removal of water, LIGHT OIL and MIDDLE OIL

ROAD TAR. A paving material produced by fluxing MEDIUM-SOFT PITCH with high-boiling TAR OILS

ROOFING PITCH. A high-temperature coal-tar PITCH used for roofing

COAL-TARS AND DERIVED PRODUCTS

SOLID SMOKELESS FUEL. Coal partly combusted under controlled conditions to remove volatile components. The residual material is chiselled and pelletted using coal-tar PITCH as a binder and is used as a household fuel. Volatiles removed in the processing are used as components in other coal-tar products.

STRAINED ANTHRACENE OIL. The product obtained from LIGHT ANTHRACENE OIL after centrifugation to recover an anthracene-rich portion

TAR ACIDS. Acidic compounds of crude COAL-TARS, including phenol, cresols and xylenols

TAR BASES. Basic compounds of COAL-TARS (e.g., quinoline, isoquinoline, methylquinolines, pyridine, picolines, ethylpyridine)

TAR DECANTER SLUDGE. A synonym for high-temperature COAL-TAR describing the method by which the tar is recovered

TARGET PITCH. A very hard PITCH used with a clay or limestone filler to produce brittle clay pigeons used for target practice

TAR OILS. A collective term for the primary distillation fractions of crude COAL-TAR

WASH OIL. See CREOSOTE OIL.

WAX OIL. See HEAVY OIL.

SHALE-OILS

1. Chemical and Physical Data

1.1 Synonyms and trade names

Chem. Abstr. Services Reg. No.: 68308-34-9

Chem. Abstr. Name: Shale oils

1.2 Description

Crude shale-oil is the product of thermal processing of raw oil shale. Oil shale is a sedimentary rock containing mainly mineral components and organic matter called kerogen, which has a low solubility in organic solvents with a low boiling-point but produces liquid organic products (oils) on thermal decomposition. Crude shale-oils differ principally from crude petroleum in that they contain higher concentrations of organic nitrogen compounds and arsenic.

The terminology used in this monograph and its relationship to processing operations are illustrated by Figure 1. Important terms are defined in the glossary, p. 215. Materials of concern to oil-shale processing include raw oil shale, crude shale-oil, spent shale, oil-shale ash, synthetic crude oil (or 'syncrude') and refined products. Operations include retorting, upgrading and refining.

Crude shale-oils are viscous, waxy liquids made up of hydrocarbons (alkanes, alkenes and aromatic compounds) and polar components (organic nitrogen, oxygen and sulphur compounds). Crude shale-oils are very complex mixtures, and only a small percentage of the compounds have been identified. One crude shale-oil, produced from raw oil shale from the Green River Formation of the western USA (in the states of Colorado, Utah and Wyoming), was reported to contain C_{10}-C_{29} alkanes and alkenes, aromatic hydrocarbons (mainly with 1-4 rings), and a number of heterocyclic nitrogen compounds (Crowley *et al.*, 1980). In contrast, 'Arabian light' crude petroleum, for example, is a material with a wider boiling range than crude shale-oil, is lower in all hetero atoms except sulphur, and contains few aromatics and no olefins (Fenton *et al.*, 1981).

Fig. 1. Processing of raw oil shale

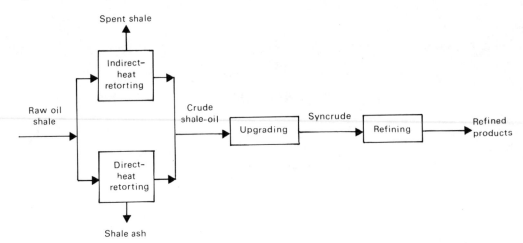

Another source reported that a typical Green River crude shale-oil contained 40 wt% hydrocarbons and 60 wt% organic nitrogen, oxygen and sulphur compounds. About 60 wt% of the non-hydrocarbon organic components were nitrogen compounds (nitriles and ring compounds such as pyridines and pyrroles), 30 wt% were oxygen compounds (phenols and carboxylic acids), and 10 wt% were sulphur compounds (thiophenes, sulphides and disulphides). Most crude shale-oils (especially those from the western USA) also contain traces of metals, such as arsenic and iron (Dickson, 1981).

In the crude shale-oil produced from USSR Baltic raw oil shale, oxygen compounds make up 65-70% of the total crude shale-oil (Aarna, 1978).

The proportions of the different components in a given crude shale-oil depend on the source of the raw oil shale as well as the method used to recover the oil (Dickson, 1981). For instance, raw oil shales that provide a high yield of crude shale-oil tend to generate a higher proportion of paraffin and olefin products and lesser amounts of aromatics than raw oil shales that provide lower oil yields (Futrell, 1980). The reported ranges of components for crude shale-oils (excluding the crude shale-oils with a high oxygen content from the USSR) are: saturates, 12-54 wt%; olefins, 2-44 wt%; and aromatics, 19-64 wt% (Dickson, 1981; Qian, 1982).

1.3 Chemical and physical properties

The properties of several crude shale-oils produced from Scottish, US and USSR oil shales, using a variety of shale-retorting techniques, are given in Table 1.

The properties of crude shale-oils produced from USSR Baltic oil shales by three retorting methods in current use in the USSR are presented in Table 2.

The levels of selected polynuclear aromatic compounds (PACs) in several crude shale-oils and crude petroleum oils have been analysed. These data are presented in Table 3.

Table 1. Properties of crude shale oils from various sources

Property	Source									
	USA								UK	USSR
	Union	Paraho	Paraho	Tosco II	Occidental	US Bureau of Mines	US Bureau of Mines	Scottish		Estonian
Type of retorting[a]	AGI	AG	AGD	AGI	MIS	AGD	TIS	AGI		AGI
Reference	Union Oil Company of California (1982)	Winward & Burdett (1979)	Bartick et al. (1975)	Dickson (1981)	Office of Technology Assessment (1980)	Sell (1951)	Winward & Burdett (1979)	Smith et al. (1938)		Kogerman (1938)
Specific gravity	NA[c]	0.929	0.938	0.927	0.904	0.952	0.912	0.878		0.943
Distillation[b], °C (°F)										
Initial BP[d]	40 (105)	NA	114 (238)	NA	NA	101 (225)	NA	161 (327)		83 (181)
10% recovered	175 (347)	232 (500)	NA	NA	227 (440)	264 (507)	183 (361)	225 (437)		140 (285)
50% recovered	428 (802)	388 (730)	NA	NA	371 (700)	338 (641)	287 (513)	316 (600)		271 (520)
90% recovered	575 (1067)	527 (980)	NA	NA	493 (920)	NA	411 (772)	NA		NA
Final BP	595 (1104)	NA	NA	NA	NA	NA	NA	NA		NA
Carbon, wt%	NA	84.9	84.9	NA	84.86	NA	84.7	85.03		82.9
Hydrogen, wt%	NA	11.5	11.5	NA	11.8	NA	11.8	12.64		10.6
Sulphur, wt%	0.80	0.70	0.61	0.7	0.71	0.7	1.09	0.46		0.7
Nitrogen, wt%	1.70	2.20	2.19	1.9	1.5	2.2	1.44	0.14		5.8[e]
Oxygen, wt%	1.52	1.05	1.40	0.8	1.13	NA	0.80	NA		
Metals, µg/g										
Arsenic	34	29	19.6	NA	NA	NA	5.4	NA		NA
Nickel	1.6	3.2	2.5	NA	NA	NA	4.1	NA		NA
Iron	9.1	57	71.2	NA	NA	NA	164	NA		NA
Vanadium	1.0	0.28	0.37	NA	NA	NA	0.5	NA		NA

[a]AGI, above-ground retort, indirectly heated; AGD, above-ground retort, directly heated; MIS, modified in situ
[b]Distillation data by gas chromatographic method
[c]NA, not available
[d]BP, boiling-point
[e]Nitrogen + oxygen, wt%

Table 2. Properties of USSR Baltic crude shale-oils[a]

Property	Type of retort[b]	
	Gas generator (proptotype to Kiviter)	Solid heat carrier (prototype to Galoter)
Specific gravity	1.000-1.010	0.960-0.975
Viscosity at 75°C	3-4	1.3-1.6
Phenols, wt%	25-30	10-15
Calorific value, mj/kg	39.5-40	39.25-40.25
Flash-point, °C	75-110	~10
Initial boiling-point, °C	180-200	70-90
Fractions, vol%		
distilled at 100°C	-	2-4
200°C	1-3	20-25
200°C	18-24	55-69
360°C	45-55	~70

[a]From Rooks (1980)
[b]These retorting processes are described on pp. 167-172.

Table 3. Measurements of polynuclear aromatic compounds (PACs) in some crude shale-oils and crude petroleum oils[a]

PAC	Crude shale-oils						Crude petroleum oils (range in 4 oils[b,d])
	No. 1[b]	No. 2[b]	No. 3[b]	No. 4[b]	No. 5[c]	No. 6[d]	
Naphthalene	632	714	397	203	672	1390	402-900
Acenaphthalene	542	903	693	271	147	260	147-348
Anthracene	502	598	480	231	986	620[e]	204-231
Fluorene	262	381	203	104	114	940	106-220
1-Methylfluorene	-[f]	-	-	-	-	980	110-140
9-Methylfluorene	-	-	-	-	-	410	IR[g]
Phenanthrene	526	572	526	221	842	620[e]	301-322
Benzo[a]fluorene	-	-	-	-	-	53	IR-22
Benzo[b]fluorene	-	-	-	-	-	140	IR-13
Chrysene	43	49	33	28	52	-	19-26
7,12-Dimethylbenz[a]anthracene	1.9	12.6	4.7	0.8	2.7	-	0.9-1.4
Fluoranthene	110	73	57	41	43	400	35-326
1-Methylpyrene	-	-	-	-	-	70	IR-36
Pyrene	139	421	200	63	177	170	193-216
Benzo[a]pyrene	3.3	29.1	13.6	17.7	192	-	2.8-3.7
Benzo[e]pyrene	1.3	19.2	3.1	5.6	61	-	1.9-2.0
Dibenz[a,h]anthracene	4.2	5.2	1.1	2.1	1.4	-	0.4-0.7
3-Methylcholanthrene	1.4	5.7	1.3	0.8	1.1	-	0.8
Perylene	4.7	19.8	5.7	6.2	68	22	6.2-31
Benzo[ghi]perylene	1.7	4.8	2.0	1.1	3.6	ND[h]	ND-0.7

[a]Crude shale-oils designated as Nos 1, 2, 3, 4 and 6 are from US sources; crude shale-oil No. 5 was obtained from Fischer assay (see glossary) of Estonian oil shale; concentrations reported for petroleum crude oils represent ranges of values for four crude oils or blends of crude oils reported by Spall (1984) and Griest et al. (1979).

[b]From Spall (1984)
[c]From Peterson and Spall (1983)
[d]From Griest et al. (1979)
[e]Anthracene and phenanthrene not separated in analysis
[f]-, analysis not performed
[g]IR, incomplete chromatographic resolution
[h]ND, not detected

1.4 Technical products, specifications and impurities

Crude shale-oils are refinery or chemical feedstocks rather than end-products, so that no specification (or test method) exists specifically for shale-oil products. Products derived from crude shale-oils will meet the specifications established for equivalent petroleum-derived products. With today's technology, crude shale-oils are used primarily as materials for manufacturing liquid fuels similar to those refined from petroleum. In the USSR, crude shale-oils are also used as feedstocks for the chemical industry, e.g., for the production of phenols. In western USA, crude shale-oils must be pretreated to remove suspended solids and trace metals (particularly arsenic) and to reduce nitrogen compounds. As most crude shale-oils have a high viscosity, their waxy components are hydrocracked or thermally cracked to decrease the pour-point in order to facilitate pipelining. The upgraded shale-oil then can be used as a feedstock for oil refineries (Williams, 1982).

2. Production, Use, Occurrence and Analysis

2.1 Historical review

The basic property of oil shale, that is, its capacity to burn, has been known for many centuries. Stories exist in both the western USA and Estonia (USSR) of rocks being burned in open fires. The Estonian name for oil shale is still 'põlevkivi', meaning 'stone that burns'.

The industrial use of oil shale first developed in France and the UK. The first patent for shale-oil production from oil shale was granted in the UK in 1694. Commercial production began in France in 1838, and in Scotland in 1850. Since that time, shale-oil production industries of various sizes have existed throughout the world (Russell, 1980). Table 4 contains a summary of shale-oil production throughout the world, giving countries, principal production areas or regions, operating periods and production rates, and estimated current production. Recent reviews of past developments in oil-shale mining and shale-oil production have been published (Baker & Hook, 1979; Office of Technology Assessment, 1980).

Table 4. World shale-oil industries

Country	Principal area or region[a]	Operating period	Estimated peak shale throughput (tonnes/year)	Estimated current shale-oil production (tonnes/year)
Australia	New South Wales	1862-1952[a]	320 000 (1947)	0
Brazil	Irati formation	1862-1946 and 1970-[a]	660 000	0
China	Kuantung (Manchuria)	1940-	10 000 000 (1980)	320 000
France		1840-1957[a]	450 000 (1950)	0
Germany, Federal Republic of		1857-[a]		0
South Africa		1935-1962[a]	225 000	0
Spain		1922-1966[a]	900 000 (late 1950s)	0
Sweden	Kvarntorp	1920s-1966[a]	3 000 000	0
United Kingdom	Scotland	1859-1962[a]	3 280 000 (1913)	0
USA	Colorado, Utah	~1858[a], 1984-	3 000 000 (1984)	0
USSR	Estonia	~1920-[a]	36 900 000 (1981)	1-1 500 000

[a]From Office of Technology Assessment (1980)

The Scottish shale-oil industry operated for more than a century before closing in 1964 (Russell, 1980). Deposits of oil shale in Scotland were dwindling at the time of the industry's closure (Stewart, 1970).

A small shale-oil industry existed in Sweden from the 1920s up to 1966, when it ceased operation. Two types of above-ground retorts were used, in addition to an in-situ technique using electric heaters for heat input (Office of Technology Assessment, 1980).

At the present time, commercial-scale shale-oil production occurs only in China and the USSR. A commercial facility has been constructed in the USA that is expected to begin operation in early 1984. Brazil is also in the process of developing commercial-scale operations (Allred, 1982); a prototype with a capacity of 2000 tonnes per day has been built and operated since 1972 in that country.

Raw oil shale was used for centuries as a solid fuel in Fushun (China) and Manchuria in the ceramic and porcelain industries. Crude shale-oil production was begun in 1926 by the Japanese, who occupied Manchuria at that time. Currently, three shale-oil production plants are in operation (two in Fushun and one in Maoming, near Canton), producing about 320 thousand tonnes of oil per year (Baker & Hook, 1979).

Shale-oil production began in Estonia around 1860, but was irregular until about 1930. Production has steadily increased since that time, except during the Second World War. Approximately 36 million tonnes of raw oil shale were mined in the USSR in 1980, with production in both Estonia and the Leningrad region. About 70% of the raw oil shale is burned directly in boilers for the generation of electricity, and the remainder is processed to produce crude shale-oil.

The most extensive world oil-shale deposits occur in the USA, the most important of which lie in the Green River Formation in the states of Colorado, Utah and Wyoming. A number of oil-shale processing technologies have been developed, but no industrial operation has yet been implemented. Several demonstration-scale shale-oil production facilities have been constructed and operated over the past seven to eight years. The first commercial facility, with a production capacity of 500 000 tonnes per year of crude shale-oil, is expected to begin operation in 1984 in Colorado.

2.2 Production and use

(a) *Production*

The manufacture of products from raw oil shale can entail a large number of combinations of process steps. Basically, however, the oil shale must be retorted (heated) to break the kerogen down into a crude shale-oil[1], and, in turn, the crude shale-oil must be refined to produce the desired products. Refining can consist of several steps employed in parallel or succession, and may be done in a refinery designed for petroleum crude oil. Most of the crude shale-oil will be 'upgraded' to make its characteristics more nearly like those of petroleum crude oil. Such upgrading may be done at a site close to the retort. In the USSR, some of the crude shale-oil is processed to recover chemical products (e.g., phenols) or for the manufacture of special products.

---

[1]Translations of USSR reports on shale-oil production often use the term 'tars' for liquid products of oil-shale retorting. In this monograph, products of all raw shale retorting processes are termed 'crude shale-oil'.

SHALE-OILS

(i) Retorting

A number of retorting processes for the production of shale-oils have been developed. These processes may be classified as (1) above-ground retorting, in which the oil shale is mined and transported to retorting vessels located on the surface; and (2) in-situ retorting, in which the oil shale is retorted in the ground. The in-situ processes are less well developed, and it is expected that most oil-shale processing in the near future will continue to be in above-ground retorts.

(1) Above-ground retorting

The major operations in above-ground processing consist of mining (either underground or surface mining), preparation and handling of oil shale, retorting, disposal of retorted or 'spent' shale, cleaning and upgrading of crude shale-oil, and treatment of retort offgases. Auxiliary operations include shale-oil storage and loading, process-water treatment, recovery of ammonia and sulphur as by-products, and plant utilities. A short description of each of the major operations follows.

Mining. Both underground and surface techniques are used in oil-shale mining, the type used being dependent upon the characteristics of the oil-shale deposit. Both techniques are conducted in the USSR, with about 40-45% of the oil shale being obtained by surface mining (Parakhonskiy, 1980). Oil shale in the Green River Formation of the western USA is expected to be mined mostly underground (Hargis & Jackson, 1983). Underground mining will also be required in support of modified in-situ retorting (as explained on p. 172).

Shale preparation. The principal preparation steps are crushing and screening of the raw oil shale to the appropriate size for feed to the retort. In the USSR, the oil shale is beneficiated by sorting to remove low quality material. Some retorting processes may also require that the raw oil shale be dried prior to being fed to the retort.

Retorting. A number of above-ground oil-shale retorting processes have been developed. An oil-shale processing facility may use a combination of two or more different retorting processes in order to better utilize the various sizes of material produced in mining and crushing.

Comprehensive descriptions of the numerous retorting processes are given by Shih *et al.* (1979), Opik and Kagayavich (1981), Allred (1982), and Opik (1983).

In general, retorting methods can be classified as follows:

- *direct heating*: where the heat is supplied by combustion within the retort;
- *indirect heating*: where the heat is supplied by combustion external to the retort and transferred to the oil shale through the vessel walls, or by a heated gas or solid material that circulates between the retort and the combustion zone.

Retorting can be done at 'low temperature' (450-600°C) or at 'high temperature' (800-1000°C) (Bogovski, 1961). Crude shale-oils produced at high temperatures generally have a higher content of aromatics, including polynuclear aromatic compounds (Dickson, 1981). This is an important difference with respect to the biological effects of the materials, as discussed in section 3. The retort offgases produced by indirectly-heated retorting processes contain higher concentrations of both organic and inorganic constituents, except nitrogen, than do the offgases from directly-heated retorts. Indirectly-heated retorts also produce a spent shale that is higher in levels of carbonaceous residue.

Table 5 categorizes the commonly-considered above-ground retorting technologies into the above subdivisions. Some retorting processes that are presently in commercial use, or expected to be in commercial use in the near future, are described in more detail below.

Table 5. Classification of some above-ground retorting technologies with respect to retorting temperature and heating method

Retorting process	Retorting temperature		Heating method	
	Low (450-650 °C)	High (850-950 °C)	Direct	Indirect
Paraho (USA)[a]	x		x	x
Petrosix (Brazil)[a]	x			x
Superior (circular grate) (USA)[a]	x			x
Lurgi-Ruhrgas (FRG)[a,b]	x			x
NTU (USA)[c]	x		x	
Tosco II (USA)[a]	x			x
Union Retort-B (USA)[a]	x			x
LETC (USA)[a]	x		x	x
Kiviter (USSR)[a]	x		x	
Gas generator (USSR)	x		x	
Galoter (USSR)[a]	x			x
Chamber oven (USSR)		x		x
Fushun (China)[d]	x		x	

[a]From Piper (1981)
[b]FRG, Federal Republic of Germany
[c]See Glossary and Dickson (1981)

In most retorts the design is such that the shale flows down inside a vertical vessel and the hot gas flows up countercurrently.

A US company has built the only commercial upflow shale retort (Fig. 2). A piston forces crushed shale up into a conical retort bed. Hot, indirectly-heated gas flows downward through the bed, heating the bed to the retorting temperature. The fresh, cold shale at the bottom of the bed cools the recycle gas and the crude shale-oil. The gas leaves the bed via slots at the bottom of the retort vessel and the oil leaves through the shale-feeder case. A retort with a capacity of 12 000 tonnes per day has been constructed near Parachute Creek, Colorado, USA, and will be in operation in 1984 (Duir et al., 1983).

The USSR has developed a retort to handle kukersite oil shale, in which the shale passes through a softening stage before retorting is complete. This process is called the Kiviter retorting process (Fig. 3). Raw shale is fed downward through a narrow bed, and hot combustion gases pass through slots in the side of the bed in crosscurrent flow. This geometry eliminates the agglomeration problem that occurred in other types of retorts with kukersite oil shale. The carbonaceous residue on the spent shale is gasified at the bottom of the bed by injecting a steam-air mixture through a tuyère. Similar principles are used in retorts of the gas generator type, which are in operation at Kohtla-Järve, Estonia (Yefimov, 1982) and at Slantso, Leningrad district (Öpik & Kaganovich, 1981).

The Galoter retort (USSR) (Fig. 4) operates using spent shale as the recirculating heat carrier (Baker & Hook, 1979).

In the last three years, the first Kiviter retort and two Galoter retorts have been started up in the USSR. The Kiviter retort, using 1000 tonnes per day of beneficiated coarse shale, has a daily production of 180 tonnes of shale-oil (Yefimov et al., 1982). The Galoters produce 388 tonnes of shale-oil per day from 3000 tonnes of shale fines (Tyagunov et al., 1980).

'Chamber-oven batteries' are also operated in the USSR, in which heat is transferred through chamber walls of refractory brick for high-temperature (850-950°C) oil-shale decomposition. This retort produces gas of a high calorific value with chamber-oven crude shale-oil as a by-product (Serebryannikov, 1970).

Disposal of spent shale: Spent shale and shale ash from processes employing direct heating or combustion of the shale are discharged from the retorting vessel at temperatures of 100°C or higher. The material may be discharged dry, quenched with water, or dumped into a water bath. The material is then conveyed or hauled by trucks to the disposal site.

Treatment of retort-product gases: Methods for treating retort-product gases or 'retort offgases' vary. However, the gases can be treated to remove organic, ammonia and sulphur compounds of high molecular weight.

Fig. 2. Diagram of the Union-B oil shale retorting process (from Duir et al., 1983)

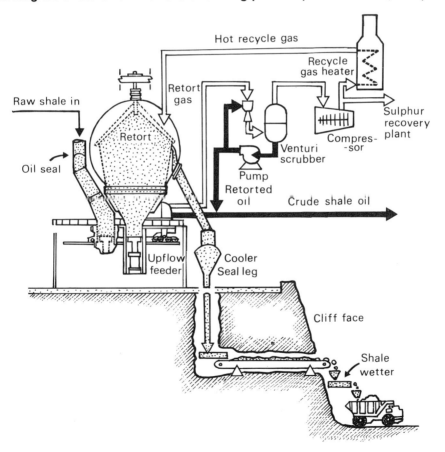

Fig. 3. Diagram of the Kiviter oil shale retorting process (from Piper, 1981)[a]

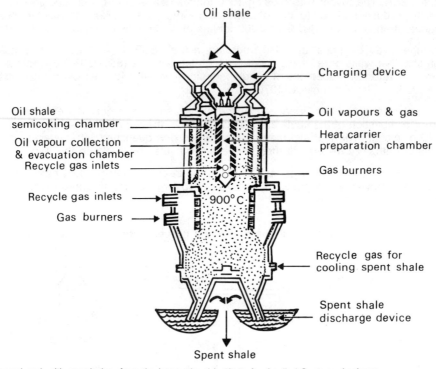

[a]Reproduced, with permission, from the International Institute for Applied Systems Analyses

Fig. 4. Diagram of the Galoter oil shale retorting process (from Piper, 1981)[a]

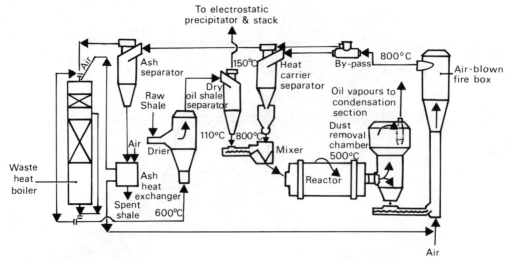

[a]Reproduced, with permission, from the International Institute for Applied Systems Analysis

Handling of liquid products from the retort: The handling system can include a collection system for condensed liquid products from the retort, a separator for the crude shale-oil and retort-water, and a system for transporting the crude shale-oil to the upgrading facility. Water from the retort and other processes is sent to a waste-water treatment facility prior to disposal and/or used for wetting of the spent shale.

(2) *In-situ retorting*

In-situ oil-shale retorting (Allred, 1982) is attractive because it offers the prospect of greatly reducing or eliminating underground mining as well as the disposal of spent shale after retorting. Much of the above-ground retort hardware is also eliminated.

The in-situ processes may be classified as 'true in-situ' processes in which no mining is performed before the creation of a retort in the ground, and 'modified in-situ' (MIS) retorting processes, in which 20-40% of the raw oil shale is mined before the creation of a retort in the ground.

True in-situ retorting: A number of true in-situ oil-shale retorting processes are under development. These include the injection of superheated steam into the deposit, the use of radio-frequency radiation, and the use of explosives to create retorts within the ground. A true in-situ technique using electric heaters to provide the heat input was also used in Sweden (Office of Technology Assessment, 1980). The Geokinetics horizontal in-situ process (USA) is the only current true in-situ process in which commercial-sized retorts have been demonstrated. More than 8 million litres of shale-oil have been produced using this process. The oil shale is fragmented with explosives, with a portion of the overburden uplifted to provide void space needed to break the shale into the desired size and make it permeable to gas flow. The zone of fragmented oil shale constitutes the in-situ retort. Wells are drilled for injection of air, removal of offgases, and removal of liquid products (Hargis et al., 1983a). A diagram of a Geokinetics retort is presented in Figure 5.

Fig. 5. Diagram of a Geokinetics true in-situ oil-shale retort (from Lekas, 1982)[a]

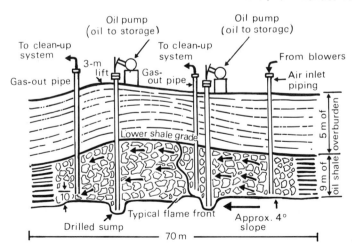

[a] Reproduced with permission from the Institute of Gas Technology

The oil shale in the retort is ignited at the end at which air is injected, and a flame front is established that moves horizontally across the retort toward the wells from which retort offgases are removed. Oil shale is retorted ahead of the flame front by the hot combustion gases as they flow through the retort. Vapours condense as they contact cooler oil shale; the liquids flow to sumps at the bottom of the retort, and are pumped to the surface (Hargis et al., 1983a).

Modified in-situ (MIS) retorting: Examples of MIS retorting processes are the Occidental (USA) and Rio Blanco (USA) MIS processes. Both of these depend upon mining out a portion of the oil shale underground to create void space, so that the remaining oil shale can be fragmented by explosives to produce the proper-sized pieces and make it permeable to gas flow. The retorts are large vertical columns of fragmented oil shale. Wells for the injection of air and steam are drilled from the surface to the top of the retorts. Large ducts for the removal of retort offgases are connected to the bottom of the retorts. The retorts are ignited at the top, and a flame front is established that moves down the retort in the same direction as the gas flow. The raw oil shale is retorted ahead of the flame front by the hot gases. Crude shale-oil is condensed as it contacts cooler shale, flows to sumps at the bottom of the retort, and is pumped to the surface.

Downstream operations: Facilities are similar to these for above-ground processing, in that operations for cleaning and upgrading of crude shale-oil, treatment of retort offgases, and treatment of process-water are required. Facilities that use the MIS process are also likely to use above-ground retorting to process the raw oil shale removed in the creation of the MIS retorts.

(ii) *Refining*

Crude shale-oil must be refined to give finished products, unless it is to be used as boiler fuel at or near the retorting site. Crude shale-oil is olefinic (owing to the thermal decomposition process), which causes the oil to be gum-forming and unstable for storage. Crude shale-oil also contains high concentrations of impurities, such as organic nitrogen and sulphur compounds, which must be removed before it can be used by existing shale-oil refineries. Some US crude shale-oil also has a high pour-point, making shipping by unheated pipeline impracticable. Crude shale-oil produced in Estonia contains relatively high concentrations of oxygenated compounds, such as phenols, which are recovered for chemical usage. Special chemical products are also produced from selected fractions of the crude shale-oil.

Crude shale-oil can be converted or upgraded to synthetic crude oil (frequently called 'syncrude') by either severe hydrotreating of the whole oil or by coking followed by hydrotreating of the lighter coke products (Chevron Oil Company, 1978). Severe hydrotreating of crude shale-oil must be done in several steps, as depicted in Figure 6. The hydrogen requirement may be satisfied by reforming the gaseous by-products, by partial oxidation of shale-oil, or by reforming natural gas (if available). The process results in a synthetic crude shale-oil very similar to high-quality petroleum crude oil, except that the synthetic crude shale-oil has virtually no impurities such as metals, nitrogen and sulphur, has no asphaltic residue content, and is low in pour-point.

Coking and hydrotreating will produce an even lower-boiling syncrude, but disposal of the coke by-product may present a problem and coking reduces the yield of liquids compared to severe hydrotreating (Fig. 6). The syncrude produced in this manner is likely to contain somewhat higher levels of organic nitrogen and sulphur compounds than that yielded by hydrotreatment alone.

Fig. 6. Shale-oil upgrading methods

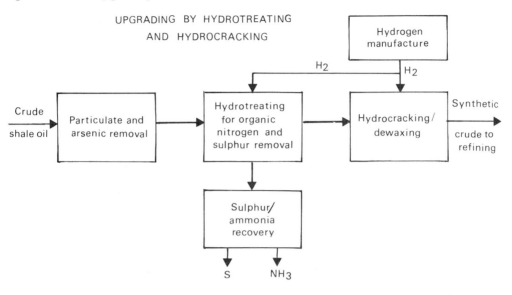

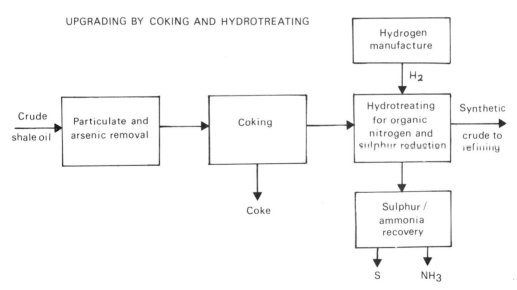

(b) Use

Early shale-oil processing operations in Scotland produced paraffin waxes and burning oils for lamps. Later, with the development of internal combustion engines, the light distillates were used to produce gasoline, the intermediate distillates for diesel oils, and the heavy distillates for lubricants.

Table 6. Shale-oil products from the Scottish industry in the early 1900s

Description	Use	Percentage of total products
Gasoline and naphthas	Motor spirits, cleaners and solvent naphthas	10
Burning oils	Lamp oils, light-house oils, railway-signal lamp oils, engine distillates	25
Intermediate oils	Gas and fuel oils	25
Lubricating oils	Lubrication	7
Paraffin wax	Wax, candles	10
Still coke and sludge	Fuel	3
Gas and loss (including tar)	Fuel	20

Shale-oil products made and marketed in the early 1900s at the height of the Scottish shale-oil industry are presented in Table 6.

One of the earliest uses of shale-oil was for medicinal purposes. Raw oil shale found near the village of Seefeld in the Austrian Tyrol was retorted and the oil was sulphonated and then neutralized with ammonia. The product is an ointment, trade-name Ichthyol, used in the treatment of skin disorders in human and veterinary medicine. Ichthyol is also produced in the USSR in the Syzran (Volga basin). In 1934, 123 tonnes of this product were marketed (Kozhevnikov, 1947). A similar product, Tolichthtol, is also produced in Estonia.

The profitability of today's oil-shale processing industry in the USSR is based on the availability of raw materials of specific character. In the early stages of its development, fuel production was the basis for the utilization of crude shale-oil. Shale-oil was used as a feedstock for producing fuel oil for boilers and automobile gasoline. After the Second World War, production of oil-shale fuel (town) gas was started in chamber-oven batteries (850-950°C) producing chamber-oven crude shale-oil as a by-product. In 1967, 550 million m^3 of gas were produced in 12 chamber-oven batteries at Kohtla-Järve (Serebryannikov, 1968).

In the USSR, internal combustion (gas-generator type) retorts (using 190-200 tonnes of beneficiated coarse oil shale per retort per day) produced approximately 35 tonnes of shale-oil per retort per day (Öpik & Kaganovich, 1981). Table 7 shows the specific products obtained from shale-oil operations of the gas-generator type.

Table 7. Shale-oil products from internal combustion retorts (gas-generator type, prototypes of the Kiviter retort) (USSR)[a]

Product	Annual production at Kohtla-Järve plants (tonnes) (1980)
Fuel shale-oil	215 000
Wood preservatives	165 000
'Nerozin' (preparation for fixing soil and drift sands)	130 800
Bitumen-mastic 'Kukersol'	71 300
Electrode coke	42 600
Grouting mastics	23 800
Synthetic tanning agent	4 600
Alkyl resorcinols	2 200
Glue resins (epoxy-glue and others)	6 300
Rubber modifier (for the tyre industry)	15 100
Cast resin	2 600

[a]From Rebane & Öpik (1982)

The fuel oils, representing the major part of liquid products from the internal combustion (gas-generator type) retorts, are used in the manufacture of gas-turbine fuel oil, automobile gasoline, additives for high-sulphur petroleum fuel oils, etc. (Aarna, 1978; Öpik & Kaganovich, 1981).

In China, shale-oil is used to generate electric-power (this use accounted for about 33% of 1975 production) and as a refinery feedstock (about 67% of 1975 production) (Dickson, 1981). Products include gasoline, kerosene, diesel fuel and coke (Qian, 1982).

The Brazilian plant (PETROBRAS) produces 140 tonnes per day of a synthetic oil with a specific gravity of 0.893 (Decora & Freitas, 1980).

Figure 7 is a diagram to show the products that may be produced in a shale-oil production facility. A typical commercial facility is unlikely to include all of these operations.

Fig. 7. Products that may be produced by above-ground retorting of oil shale

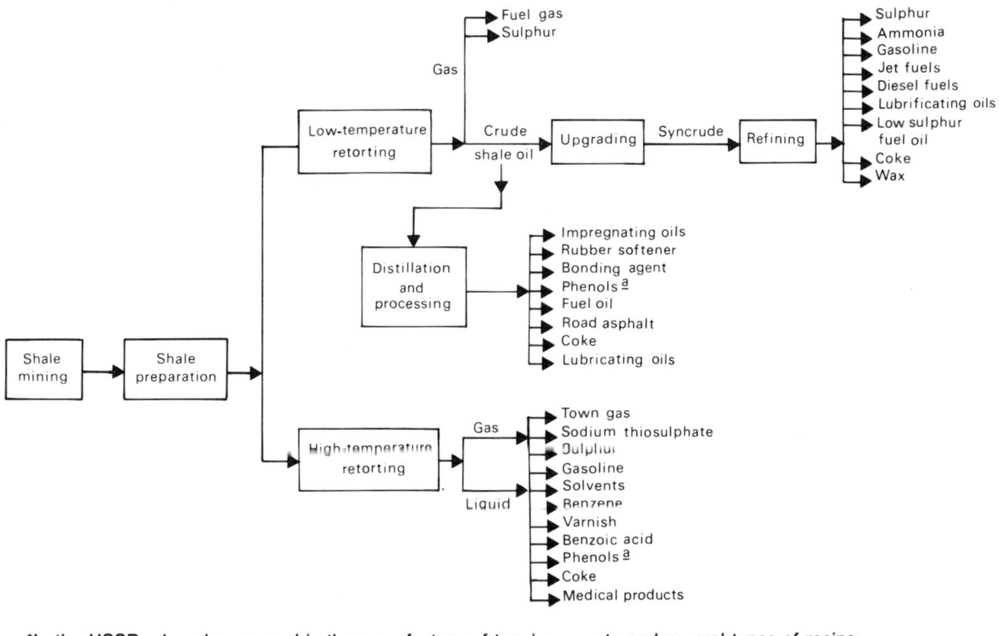

*In the USSR, phenols are used in the manufacture of tanning agents and several types of resins.

2.3 Occupational exposure

Reviews of general health hazards in oil-shale processing have been published by Akkerberg et al. (1980) and Weaver and Gibson (1979).

Oil shale contains significant levels of free (crystalline) silica in the form of quartz. Quartz levels in raw oil-shale dust are reported to range from 3-4% in Estonian facilities (Shmidt et al., 1980) and from 3-9% in US demonstration-scale oil-shale facilities (Hargis et al., 1983b).

A dust analysis in a Scottish oil-shale mine was reported to show <3% free silica, but quartz levels in Scottish oil-shale rock are reported at 6-12% (Seaton et al., 1981).

Trace elements of potential concern have been reported for an oil shale from the Green River Formation, USA. They include: arsenic, 44.3 µg/g; cadmium, 0.64 µg/g; lead, 26.5 µg/g; mercury, 0.089 µg/g; and nickel, 27.5 µg/g (Fruchter et al., 1980).

Exposures to hydrogen sulphide occurred in the Swedish shale-oil industry during the 1940s and early 1950s. Concentrations of hydrogen sulphide were reported to range from <20 ppm to >600 ppm in air (Ahlborg, 1951). Uranium levels in Swedish oil shale (known as 'alum shale') are also reported to be high, resulting in high radon and radon-daughter products in Swedish houses built of concrete containing the spent shale (Axelson & Edling, 1980).

Carbon monoxide and phenol are present in the air at oil-shale processing facilities in the USSR, usually in low concentrations (Akkerberg, 1980). Benzo[a]pyrene is also present in the air of such facilities. Benzo[a]pyrene concentrations ranged from 0.05-0.60 µg/m^3 (average 0.4 µg/m^3) on the loading platform and 0.02-0.32 µg/m^3 (mean 0.15 µg/m^3) on the extraction platform of retorts of the chamber-oven type. On the unloading platform of the electrode coke shop, benzo[a]pyrene levels ranged from 0.59-48.5 µg/m^3 (mean 29.5 µg/m^3), and during the unloading of coke from vats benzo[a]pyrene levels ranged from 216.1-800 µg/m^3 (mean 467.4 µg/m^3) (Akkerberg et al., 1976).

Carbon monoxide was detected in concentrations of up to 30 mg/m^3 (mean 10 mg/m^3) in the air around the shale-feeding platforms of gas-generating stations. Phenol was present in the air of the tray and pan premises. The concentration of phenol in air is a direct function of the concentration of phenols in the water fed to the gas-generator pans for quenching the ash on the tray. When the concentration of phenol in the water is no higher than 1 g/litre, the concentration of phenol in the air will not exceed 5 mg/m^3, which was considered the sanitary standard until 1975. Higher concentrations of carbon monoxide (62-625 mg/m^3), sulphur dioxide (30-50 mg/m^3) and spent shale dust (23-30 mg/m^3) were found during periods of cleaning of the gas-overflow pipes (during 30 minutes) and gas generators (4 hours). Concentrations decreased at the end of the cleaning periods (Akkerberg, 1980).

The results of industrial hygiene air-sampling studies at experimental US oil-shale retorting facilities are summarized in Table 8. These results indicate that high dust levels can exist near operations involved with mining, crushing and above-ground retorting of oil shale. Measurable levels of a number of polynuclear aromatic compounds (PACs) were observed at the facilities. Diesel exhaust from mining equipment may contribute a substantial portion of the total airborne particles and PACs within US oil-shale mines.

The highest levels of PACs were measured at the true in-situ facility in the area of a fan leaking retort offgases and shale-oil mist. Fugitive emissions of dusts, gases and vapours may also occur in charging oil shale to an above-ground retort, and in removing spent shale.

In modern facilities, skin contact with oil-shale liquid products (shale-oil and retort process-water) is more likely to occur for maintenance personnel, who work in proximity to fugitive emissions or leaks in repairing systems and also work with contaminated equipment and materials.

2.4 Occurrence

Shale-oils do not exist in Nature, but are produced during pyrolysis (retorting) of oil shales.

Table 8. Results of industrial hygiene surveys at experimental US oil-shale retorting facilities

Air contaminant (units)	Above-ground retorting facility[a]	True in-situ facility[b]	Modified in-situ facility A[c]	Modified in-situ facility B[d]
Total dust (mg/m^3)	0.8-90.8	0.1-0.54	0.27-3.94	0.5-1.0
Respirable dust (mg/m^3)	0.1-20.0	-	0.09-1.20	0.4-1.2
Cyclohexane extractables (µg/m^3)	-	1-85.2	-	0.25-0.28
Particulate-phase PACs[e] (ng/m^3)	-	<10-2820[f]	26-2648[g]	0.5[h]
Vapour-phase PACs (ng/m^3)	-	230-6800[f]	<5-200[g]	-
Total hydrocarbons (ppm)	<1-400	<1-800	-	-
Formaldehyde (ppm)	<1-2	<1-2	<0.3	-
Benzene (ppm)	-	<0.1-0.2	<0.001-<5	-
Toluene (ppm)	-	<0.1-0.7	<0.001-<0.02	-
Xylenes (ppm)	-	<0.1-88.6	<0.007	-
Phenols (ppm)	-	<0.1	<0.002	-
Amines (ppm)	-	<0.4	<0.4-1.3	-
Pyridine (ppm)	-	<0.1	-	-
Carbon monoxide (ppm)	<1-700	5-200	<5	-
Hydrogen sulphide (ppm)	<1-20	1-10	<0.5-10	-
Arsine (ppm)	-	<0.05	<0.05	-
Silica	-	3-7% (of total dust)	-	-
α-Quartz (µg)	-	-	<10-346	-

[a]From Garcia et al. (1981)

[b]From Hargis et al. (1983a)

[c]From Gonzales et al. (1982)

[d]From Hargis et al. (1983b)

[e]PACs, polynuclear aromatic compounds

[f]Sum of anthracene, benzo[a]pyrene, chrysene, fluoranthene, fluorene, phenanthrene and pyrene; analysis of material extracted from filters and sorbent vapour trap by high-performance liquid chromatography

[g]Sum of anthracene, benzo[a]pyrene, chrysene, fluoranthene, naphthacene, phenanthrene and pyrene; analysis of material extracted from filter and sorbent vapour trap by high-performance liquid chromatography

[h]Sum of acenaphthene, anthracene, benzo[a]pyrene, benzo[e]pyrene, benzo[ghi]perylene, chrysene, dibenz[a,h]anthracene, 7,12-dimethylbenz[a]anthracene, fluoranthene, fluorene, 3-methylcholanthrene, naphthacene, naphthalene, perylene, phenanthrene, pyrene, acridine, 1-azapyrene, benzo[f]quinoline, benzo[h]quinoline and carbazole; analysis of material extracted from filters by gas chromatography/mass spectrometry

2.5 Analysis

No data on methods for the determination of shale-oil were available to the Working Group.

3. Biological Data Relevant to the Evaluation of Carcinogenic Risk to Humans

3.1 Carcinogenicity studies in animals

Raw and spent oil shale

(a) *Skin application*

Mouse: Berenblum and Schoental (1944) investigated a chloroform extract of raw, solid Scottish oil shale. After removal of the chloroform, a dark-brown product of the consistency of soft wax remained (1% of the powdered shale). The material was diluted with benzene [concentration not indicated] and applied to the skin of 20 white mice in the interscapular region twice a week for life (up to 51.5 weeks). No tumour was found at the site of application.

Hueper (1953) tested a benzene extract of a Green River raw oil shale by weekly skin application in 100 strain A mice and in 50 hairless mice for a year. Only one strain A mouse out of 65 survivors at eight months developed a papilloma at the application site. In treated hairless mice, no tumour considered by the authors to be related to contact with the material studied was observed. No tumour was observed in 42 untreated strain A controls.

Rowland et al. (1980) tested benzene extracts of raw oil shale (mined at the Colony Oil Shale Development at Parachute Creek, Colorado) and spent oil shale from a TOSCO II retort in two groups of 50 Swiss mice. One drop (1/60 ml) of the extract containing 2.5 mg of the oil-shale solids was applied twice weekly for life on the dorsal skin between the flanks. The raw oil-shale extract (benzo[a]pyrene content, 0.66 µg/ml, i.e., 0.66 ppm) had no carcinogenic effect, whereas the extract of spent oil shale (benzo[a]pyrene content, 1.4 µg/ml [1.4 ppm]) induced skin tumours in six mice (three papillomas and three squamous-cell carcinomas).

(b) *Subcutaneous and/or intramuscular administration*

Mouse: A benzene extract of Green River raw oil shale was tested in 30 strain A mice by i.m. injection of 0.1 ml of the extract in 0.2 ml of lanolin. No tumour at the administration site was observed in 20 mice surviving 10 months (Hueper, 1953). [The Working Group noted that a single injection was given.]

(c) *Intratracheal administration*

Rat: Another pair of similar samples (raw oil shale from Anvil Points, Colorado and the spent shale from an unspecified run from the 10-ton experimental retort at Laramie Energy Technology Center (LETC), Laramie, Wyoming) was tested by Renne et al. (1980). Suspensions of these materials, ground into particles with a size of 0.5-1.5 µm (average 0.9 µm) in diameter, in physiological saline (5% w/v) were administered intratracheally to young adult female Wistar rats. Two dose levels were used, 5 mg (0.1 ml) and 20 mg (0.4 ml), repeated at two-month intervals until five doses (25 mg and 100 mg) had been given. Vehicle-control animals received 0.4 ml of saline only. A positive-control group of 40 rats was given five doses of 3 mg benzo[a]pyrene and 3 mg iron oxide at two-week intervals. The numbers of animals killed in each group were: after 12 months, 10 animals; after 18 months, 10; and after 24 months, 40. The only primary respiratory neoplasm observed in this study was a mucinous adenocarcinoma in the positive-control group; the test substances gave negative results.

Hamster: Rowland *et al.* (1980) tested suspensions of Green River raw oil shale and spent shale (specified above). The suspensions were instilled intratracheally into 50 female and 50 male Syrian golden hamsters aged 6-7 weeks in doses of 3 and 0.5 mg of solid material in 0.2 ml of saline once a week for 15 and 30 weeks, respectively. A positive-control group (receiving 3 mg of benzo[a]pyrene and 3 mg of iron oxide in 0.2 ml of saline once a week for 15 weeks) and a vehicle-control group (0.2 ml of saline once a week for 30 weeks) were used. In contrast with the positive-control material, neither the raw oil shale nor the spent shale caused any tumour in the larynx, trachea or lung.

(d) Inhalation

Rat: One raw-shale sample (PRS) from Anvil Points, Colorado, and one spent-shale sample (PSS) from a direct-heated retort were packed in Wright Dust Feed (mass median aerodynamic diameter 2.45 μm ± 1.68 μm) from which aerosols were generated. Groups of 62 Fischer 344 rats [sex not specified] were exposed to each sample at a level of 90 mg/m^3 (respirable fraction) for five hours per day, on four days per week for 24 months. Of 50 PRS-treated rats examined, one had a lung adenoma and 11 had lung carcinomas. Of 57 PSS-treated rats examined, three had lung adenomas and 10 had lung carcinomas. In a control group of 57 rats treated similarly with quartz (10 mg/m^3 respirable fraction), four had lung adenomas, and 13 had lung carcinomas. No lung tumour was observed in 15 sham controls examined nor in 17 untreated controls (Holland *et al.*, 1983). [The Working Group noted that no data on mortality or on the latency period of tumour detection were reported.]

Hamster: In the same study as reported above (Holland *et al.*, 1983), the raw-shale (PRS) and spent-shale (PSS) samples together with a second sample of spent shale designated TSS (from an indirectly-heated retort) were tested in Syrian golden hamsters. Groups of 64 animals [sex not specified] were exposed to each sample at a level of 60 mg/m^3, for four hours per day, on four days per week for 16 months. No lung tumour was observed among 55, 58 and 57 hamsters examined and treated with the three samples, respectively, nor in 21 sham-treated controls. [The Working Group noted that the results were reported as preliminary.]

Crude shale-oils from low-temperature retorting

(a) Skin application

Mouse: An oil processed in a Fischer retort from Jurassic Chuvash oil shale (USSR) was applied on the intrascapular area of the skin of two groups of random-bred mice. In the first group, 35 mice were painted twice weekly (52 times), and in the second group 14 mice were painted three times weekly (85 times). Of 31 mice surviving six months in the first group, six had squamous-cell papillomas and two, squamous-cell carcinomas. Of 13 survivors in the second group, four mice developed papillomas and two, epidermal carcinomas (Larionov *et al.*, 1934).

Soboleva (1936) investigated the carcinogenicity of a similar oil processed from siluric shales of the Gdov region (Leningrad district, USSR) at low temperature in a rotating Fischer retort (elementary composition: carbon, 82.52%; hydrogen, 10.13%; nitrogen, 0.17%; sulphur, 0.81%; oxygen, 6.37%), and the 275-350°C-fraction of the oil. Of 40 random-bred mice, 30 received skin applications of the whole oil once every five days, and 10 mice twice weekly (60 times). There were 14 survivors at the end of the application period. One mouse developed a papilloma after nine months, which became a malignant exulcerating tumour in 12 months. Another mouse had a *cornu cutaneum* after 12 months and a small squamous-cell papilloma after 14.5 months.

Estonian shale-oils were studied first by Larionov (1947). Crude shale-oil retorted industrially in Kohtla-Järve at 450-500°C, called 'generator oil' [a low-temperature oil], was tested by skin application in 54 random-bred white mice. One drop of oil was applied to the interscapular area of the skin twice weekly for six months. Of 22 mice surviving this period, only one developed a papilloma at the application site, which later regressed. No skin tumour was induced in this study by similar application of the heavy fraction (boiling at over 270°C) of the same generator oil (70 mice, 18 surviving six months), of the heavy fraction of the tunnel-oven oil from Kohtla-Järve, liquefied by addition of benzene (52 mice, 15 survivors), or of the generator oil of the Kashpir oil shales (69 mice, 15 survivors).

As the Estonian shale-oil industry expanded rapidly after the Second World War, with technical innovations and the construction of new retorting facilities, especially the so-called gas generators, the generator oil was studied repeatedly. Gortalum (1955) applied 20-40 mg of the generator oil from Kohtla-Järve (generator station No. 5) to the skin in the interscapular area of 120 random-bred white mice 52 times over seven months; 20 untreated animals served as controls. During the study, 23 experimental animals and eight controls were killed at various times to investigate early changes morphologically. Over a period of 12 months after cessation of skin applications, no malignant skin tumour was observed.

Taking into account that in the new semi-coking retorts (gas generator No. 5) in Kohtla-Järve the processing temperatures may at times be higher than 500-600°C and that commercial oils are often blended, Bogovski (1958) investigated a blend of shale-oil from various generators in Kohtla-Järve, with a benzo[a]pyrene content of 0.01% (100 ppm) (characteristics: density (g/cm^3) at 20°C, 0.956; viscosity at 75°C (Engler), 2.5°; flash-point, 110°C; moisture, 0.8%). In 105 random-bred white mice, one drop (30-40 mg) of oil was applied to the skin twice weekly 50 times. Over a period of 271 days, 21 mice developed squamous-cell papillomas (28.8% of the 73 effective mice surviving 72 days, the latency period of the first tumours). No malignant tumour was observed.

An industrial intermediate fraction of the generator oil from the oil-shale processing plant in Kiviyli (Estonia), retorting temperature, 450-600°C (density (g/cm^3) at 20°C, 0.966; moisture, 1.2%; flash-point, 94°C; ash, 0.1%; viscosity at 80°C (Engler), 1.97°; boiling-point, 206°C; evaporation at 360°C, 60%) was tested by Vinkmann (1972a). A group of 68 male CC57Br mice was treated twice weekly 50 times on the interscapular area of the skin with 35-40 mg of undiluted oil. During the 18 months of the experiment, 17 mice developed skin tumours, six of which were squamous-cell carcinomas. There was no tumour in 36 untreated control animals. The incidence of tumours of the internal organs was identical in experimental and control animals.

A crude shale-oil retorted in Fushun (China) at 550°C in generators (and containing 22.4-24.6% of unsaturated compounds; on boiling at up to 200°C, 2.5% of the oil was evaporated; up to 250°C, 10%; up to 375°C, 45%) was applied by Sun *et al.* (1961) in doses of 30-50 mg two to three times a week for 69 times on the interscapular skin of 120 random-bred mice. All but four mice, which died during the first two months, developed topical skin tumours. Histologically, of 73 mice dead or killed before 195 days, 12 mice had squamous-cell carcinomas and 61 mice epidermal papillomas; 23 mice with skin papillomas were still alive after 195 days. In 20 untreated control mice no skin tumour was observed.

Oil shale was retorted industrially in horizontal tunnel ovens in Kohtla-Järve over several decades. Larionov (1947) tested an oil from such an oven in a small number of mice and did not observe skin tumours. Bogovski (1961) investigated another heavy tunnel-oven oil retorted in Kohtla-Järve at 460-550°C (density (g/cm^3) at 20°C, 0.982; viscosity at 75°C (Engler), 2.3°; flash-point, 90°C; phenol content, 26.2%; boiling-point, 180°C; fractionation: up to 200°C, 2.0% comes over, up to 250°C, 14%, up to 300°C, 30%, and up to 330°C, 44%). A benzo[a]pyrene level of 0.001% (10 ppm) was determined in the oil. A group of 70 random-bred white mice was painted with undiluted oil twice a week, 50 times. During the first two months, 40 mice died owing to the toxicity of the oil. Of 34 mice surviving at the time of appearance of the first tumour (59th day), 12 had developed skin tumours, two of which were squamous-cell carcinomas.

A sample of tunnel-oven oil from the Kiviyli plant (density (g/cm^3) at 20°C, 0.977; flash-point, 73°C; viscosity at 80°C (Engler), 1.3°; boiling-point, 178°C; evaporation at 360°C, 92%) was tested by Vinkmann (1972a) in 68 male CC57Br mice. Undiluted oil in a dose of 35-40 mg was applied twice weekly (50 times) to the skin. During the 18 months of the experiment, 16 of 46 mice developed topical skin tumours, two of which were squamous-cell carcinomas. No skin tumour was observed in 36 untreated control animals.

Turu (1961) tested the solid heat-transfer (SHT) oil produced at 500-530°C in a pilot plant in Kiviyli (density (g/cm^3) at 20°C, 1.005; boiling begins at 100°C, 10% boils over up to 215°C, 20% up to 253°C, 50% up to 310°C, 65% up to 325°C and 75% at 328°C; viscosity at 75°C (Engler), 1.95° in 122 random-bred white mice. Mice received twice-weekly applications of 40-50 mg of the oil on the interscapular skin 50 times. Of 68 mice surviving 90 days, when the first skin tumours appeared, skin tumours (including four squamous-cell carcinomas) developed in 46 (Bogovski, 1961). In related experiments in 68 male CC57Br mice, Vinkmann (1972) investigated an industrial SHT oil (medium fraction) (characteristics: density (g/cm^3) at 20°C, 0.989; flash-point, 66°C; viscosity at 80°C (Engler), 1.38°; boiling begins at 165°C, 96% comes over up to 360°C). Of 47 effective mice receiving 50 twice-weekly skin applications of 35-40 mg oil, 17 developed skin tumours during the 18 months of the experiment; in six mice the tumours were squamous-cell carcinomas.

As part of a large study, Smith et al. (1951) tested a shale-oil from Colorado, USA, processed in the Nevada-Texas-Utah (NTU) retort, in a group of 30 white male albino mice which received skin applications of 15 mg of material three times a week for life. In 330 days, eight benign skin tumours were induced by the whole oil. [The Working Group noted that a fraction of this oil distilling at 550-700°F (288-371°C) was reported to be more active, producing five carcinomas and nine papillomas.]

Hueper (1953) investigated the carcinogenicity of two Green River crude shale-oils, one obtained in the NTU retort at a temperature of 1000-1500°F (538-816°C) and the other in the modified Fischer-assay retort at a temperature of 700-1000°F (371-538°C). Weekly skin applications were made for a year. During the first six months, the oils were diluted 1:10 in xylene, for the next six months they were diluted in ethyl ether (40%:60% w/w). The NTU retort oil was applied to 100 strain A mice and 25 C57Bl mice. Of the mice surviving at eight months, one of the 38 strain A mice had a skin papilloma, whereas four of the 19 C57Bl mice developed squamous-cell carcinomas. The Fischer-retort oil was applied to 50 strain

A mice and 30 hairless mice. Of the mice surviving eight months, two of 45 strain A mice developed squamous-cell carcinomas of the skin, as did one of 10 hairless mice. No skin tumour was observed in 42 untreated strain A control mice surviving more than eight months.

In the experiments of Bingham and Barkley (1979), three crude Green River shale-oils, two samples (Nos 1 and 2) from a heat-transfer process and one (No. 3) from a retort-combustion process, were applied to the interscapular region of the backs of C3H male mice with a microlitre pipette, usually in doses of 50 µl two or three times per week. [Number of applications and duration of experiment not specified. It was not indicated how the usually viscous crude oils were solubilized to be applied by a micropipette.] Of 20 mice [effective number not specified] treated with shale-oil sample No. 1 (benzo[a]pyrene content, <0.00001% [<100 ppb]), one developed a papilloma and 17 developed carcinomas of the skin within an average latent period of 43 $\pm$ 4 weeks. Of 30 mice treated with shale-oil sample No. 2, one developed a papilloma and 18 developed skin carcinomas within an average latent period of 36 $\pm$ 2 weeks. Of 30 mice treated with shale-oil sample No. 3 (benzo[a]pyrene content, <0.00001%), 22 mice developed skin carcinomas and three developed papillomas within a latent period of 43 $\pm$ 5 weeks. [The Working Group noted the limited reporting of the data.]

In a dose-response study (Holland et al., 1979), a crude shale-oil produced in a simulated in-situ, above-ground retorting of Green River oil shale (LETC) [retorting temperature not specified] containing benzo[a]pyrene at a level of 20 µg/g [20 ppm] was investigated in SPF male and female C3H/fBd mice. The crude shale-oil was dissolved in a mixed solvent consisting of acetone and cyclohexane (30%/70% v/v). In the first experiment, a single concentration (25 mg) was applied on the skin of mice three times weekly for 22 weeks and the mice were observed for a further 22 weeks. In the second experiment, doses of 25, 12, 6 and 3 mg were applied for 30 weeks twice weekly with a 20-week follow-up. In the third experiment, doses of 2.5, 0.5, 0.3 and 0.1 mg of shale-oil were applied three times a week for 24 months. Of the 30 mice in the first experiment, 11 died within 44 weeks and 14 developed skin carcinomas (average latency, 154 $\pm$ 9 days). No papilloma was recorded. In the dose-response series, application of 25 mg of shale-oil twice weekly for 30 weeks induced skin carcinomas in seven of 20 mice (average latency, 208 $\pm$ 19 days) and skin carcinomas were induced in one of 20 mice by 12 mg of shale-oil (average latency, 213 days). The 6-mg and 3-mg dose levels produced no skin carcinoma. In the low-dose prolonged study (applications three times weekly for 24 months), 2.5 mg per application induced skin carcinomas in 45/50 mice (average latency, 483 $\pm$ 15 days); 0.5 mg per application resulted in skin carcinomas in one of 50 mice (average latency, 315 days); and a skin carcinoma was found in one of 50 mice who received 0.3 mg per application. No papilloma was observed. [The Working Group noted that the criteria for diagnosis of skin carcinoma were not presented.]

In a study of the effect of application frequency on epidermal carcinogenicity assays, Wilson and Holland (1982) used two Green River crude shale-oils: PCSO II from the Paraho retort (Anvil Points, Colorado, USA) and OCSO No. 6 produced at Logan Wash, Colorado, USA (in a modified in-situ retort) [no characteristic given]. It was reported that the more irritating OCSO No. 6 oil masked the tumour response owing to early toxic changes in the epidermis (progressive necrotizing dermatitis and exulceration) which the authors called 'epidermal degeneration'. Both crude shale-oils induced skin tumours in C3Hf/Bd mice with a higher incidence when the oil (dissolved in cyclohexane-acetone) was applied at 40 mg once per week, whereas 10 mg four times per week produced the highest frequency of epidermal degeneration and a low incidence of tumours. With OCSO No. 6, for example, the corresponding tumour incidences were 13/20 compared with 2/20. [No distinction was made between benign and malignant tumours.]

Two crude shale-oils (OCSO 6 and PCSO II) (see above) and a hydrotreated product of the second crude shale-oil (PCSO-UP) were tested by application to the skin of mice. Doses of 50 μl of dilutions in cyclohexane and acetone (30%/70%) were applied thrice weekly to the interscapular area of groups of 20 male and 20 female C3Hf/He mice. Although the experiments were still in progress at the time of first reporting, both crude shale-oils produced grossly visible skin tumours after 238 days at dose levels of 5, 20 and 40 mg per dose. The hydrotreated sample produced fewer skin tumours than the parent crude shale-oil at all three dose levels (1/40 versus 10/40 at the 5-mg dose level; 0/40 versus 14/40 at 20 mg; and 2/40 versus 12/40 at 40 mg) (Holland et al., 1981).

Rabbit: Vahter (1959) tested the heavy fraction of the generator (semi-coking) oil obtained from the Estonian oil shale in gas generators at Kohtla-Järve (boiling-point, 153°C; viscosity (Engler), 6.5° at 75°C; phenols, 15-20 wt %; sulphur, 0.8-1.0 wt %). A group of 47 male and female rabbits [unidentified breed] (bw, 1800-2400 g) was treated with oil at about 5 mg/cm^2 applied twice weekly during one year (about 100 paintings) to the internal skin surface of both ears. The rabbits were observed for 2.5 years. Multiple papillomas and skin horns started to appear after 3 months of painting; some papillomas regressed. Finally, in all 47 rabbits permanent skin tumours developed, in the majority of cases squamous-cell papillomas and keratoacanthomas. In four rabbits (8.5%) squamous-cell carcinomas were diagnosed, in one of which (surviving two years and three months) metastases of the carcinoma were found in regional lymph nodes, in the liver and in the lungs.

(b) Subcutaneous and/or intramuscular administration

Mouse: Hueper (1953) investigated the carcinogenicity of two crude shale-oils (NTU-retort oil and Fischer-assay-retort oil; see reference to Hueper (1953) in section (a) above) by injecting 0.1 ml of each oil dissolved in 0.2 ml lanolin i.m. into strain A mice. Of the 30 mice in each group, 27 and 17 mice, respectively, survived 10 months. No tumour was found at the injection site. [The Working Group noted that a single injection was given.]

(c) Intratracheal administration

Mouse: In the framework of an extensive study carried out to validate the mouse-lung tumour assay (Smith & Witschi, 1983), a sample of Paraho crude shale-oil was investigated at three dose levels: 2500 mg/kg, which was the maximum tolerated dose (MTD), one half of this dose (1250 mg/kg) and one fifth of the MTD (500 mg/kg). The oils were administered in corn oil by 24 intratracheal injections (three times weekly for eight consecutive weeks) to groups of 30 strain A/Jax male mice for each dose. Three lung tumours were observed in three of five surviving mice who received 2500 mg/kg of shale-oil, in six of 12 survivors who received 1250 mg/kg and in seven of 16 survivors with the lowest dose. The incidence of lung tumours was significantly higher ($p < 0.05$) than in the vehicle-control group, as was the multiplicity of lung tumours in two groups receiving the smaller doses of shale-oil.

Crude shale-oils from high-temperature retorting

(a) Skin application

Mouse: A sample of crude oil obtained from a pilot plant in Kohtla-Järve (chamber-oven type II) at 1000°C [no other characteristic reported] was tested by Larionov (1947) in 66 random-bred white mice aged about three months. One drop of tar was applied to the skin of the interscapular region twice a week for six months. Of 22 mice surviving this period, seven developed squamous-cell carcinomas; one mouse had metastases in regional lymph nodes and in the lung.

The carcinogenic action of a sample of chamber-oven oil obtained at about 900°C in an industrial installation (chamber-oven type III) in Kohtla-Järve (density (g/cm^3) at 20°C, 1.060; viscosity at 75°C (Engler), 1.50°; flash-point, 80°C, benzo[a]pyrene content 0.1% (1000 ppm)) was investigated. A group of 111 random-bred white mice received one drop (30-40 mg) of tar twice weekly on the skin of the interscapular region. The mice did not survive the intended skin application period (50 times): all mice died before the 48th application. The first papillomas appeared after 68 days. Of 88 mice surviving to this point, 49 had skin tumours. In 27 mice the tumours were malignant (cornifying, non-cornifying and anaplastic spindle-cell squamous-cell carcinomas). In seven mice the carcinomas metastasized (three in lymph nodes, four in lungs). By 128 days from the beginning of the experiment, 54/59 mice had skin neoplasms and by the end of the experiment at 188 days all mice at risk had developed skin neoplasms (Bogovski, 1958, 1961; Bogovski & Vinkmann, 1979).

Another sample of chamber-oven oil [density (g/cm^3), 1.0813; flash-point, 112°C; viscosity at 50°C (Engler), 2.58°: boiling begins at 171°C, 10% comes over up to 230°C, 20% up to 251°C, 30% up to 275°C, 40% up to 302°C and 50% up to 325°C; benzo[a]pyrene content, 0.17% (1700 ppm)] was tested as a positive control in a study of solid-heat-transfer oil (see Turu, 1961 in section (a)). A group of 122 random-bred white mice received 50 twice-weekly applications of one drop (40-50 mg) of the tar on the skin of the interscapular region. The first papillomas appeared after 56 days, when 74 mice were alive. Of these mice, 15 had skin tumours and seven developed squamous-cell carcinomas. By 140 days all mice had skin neoplasms (Bogovski, 1961; Turu, 1961).

Rabbit: A sample of chamber-oven oil [boiling-point, 61°C, viscosity at 75°C (Engler), 1.2°; phenols, 6-8 wt%; sulphur, 0.6-0.9 wt%] was tested by Vahter (1959) in 60 rabbits. The oil was applied twice weekly for one year to the skin of the inner surface of both ears of the rabbits at about 5 mg/cm^2. The rabbits were observed for 2.5 years. In all rabbits, multiple squamous-cell papillomas and keratoacanthomas were induced, and in 13 (21.7%) cornifying and non-cornifying squamous-cell carcinomas developed. In one rabbit metastases in the lung and liver were found.

Shale-oil fractions

Fractionation of shale-oil and carcinogenicity testing of the fractions has been undertaken repeatedly to find the more active fractions and to correlate the carcinogenic activity with contents of known carcinogens, primarily benzo[a]pyrene.

(a) *Skin application*

Mouse: Blue Scottish shale-oil and its chromatographic fractions, prepared by adsorption on aluminium oxide and elution with various solvents, were tested by Berenblum and Schoental (1943) in white mice. Test fractions were applied to the skin twice weekly using about 10 mice for each fraction and continuing each experiment for 18-20 weeks. It was found that the whole oil was very active; in 7.5 weeks, the first skin tumour appeared and six out of 10 mice developed skin tumours (one tumour became malignant, Berenblum & Schoental, 1944). One fraction (weakly adsorbed) had about the same activity; the latency of the first skin tumour was 6.5 weeks, and five of nine survivors developed skin tumours. Benzo[a]pyrene was found in this fraction as in the whole oil (0.01%) by fluorescence spectroscopy [the first report on benzo[a]pyrene in oil-shale products]. Another fraction, weakly adsorbed, which was of about the same carcinogenicity (first tumour in 10 weeks, five animals with skin tumours among eight survivors) did not contain benzo[a]pyrene. The authors concluded that the carcinogenicity of shale-oils might be due (as in coal-tars), in part, if not entirely, to the effects of destructive distillation.

Chromatographic fractions of the high-temperature (800-1000°C, chamber-oven) oil from Kohtla-Järve (see experiments of Turu, 1961, above) were investigated by Bogovski (1961, 1962). About 2 kg of the tar were fractionated by vacuum distillation and repeated chromatographic fractionation on silica gel and aluminium oxide. The aromatic part of the tar (87.6 g) was obtained on silica gel and was further fractionated into five fractions on aluminium oxide eluted with various solvent combinations. In comparative tests these fractions were used as 3% solutions in benzene. Each fraction was tested in a group of 50 random-bred white mice of both sexes. Two drops of the benzene solution (31 mg) were applied with a calibrated pipette twice weekly 70 times. Two fractions (identified as Nos 4 and 5) contained no measurable amount of benzo[a]pyrene; a third fraction (identified as No. 6) contained 70 ppm; a fourth fraction (identified as No. 7), 2000 ppm; a fifth fraction (identified as No. 8), 50 ppm; and a sixth (a 1:1 combination of fractions, identified as Nos 4 and 6), 35 ppm. In a positive-control group, mice received a 2000 ppm solution of pure benzo[a]pyrene in benzene by the same schedule. The results showed that fraction No. 7 was the most carcinogenic, inducing skin tumours in 34 mice, which were malignant in 24 mice (four with metastases); however the latency was considerably longer (150 days) than in the group receiving skin application of the 2000 ppm solution benzo[a]pyrene (83 days). Fraction No. 5 (without benzo[a]pyrene) induced skin tumours in 15 mice, which were malignant in seven mice (one with metastases). The 1:1 mixture of fractions No. 4 (without benzo[a]pyrene) and No. 6, resulting in a benzo[a]pyrene concentration of 35 ppm, had a considerably greater carcinogenic effect - 22 tumours (12 carcinomas, five with metastases) - than fraction No. 6 alone - 15 tumours (11 sarcomas, two with metastases).

Rabbit: The fractions of the chamber-oven tar described above for experiments in the mouse were applied to the dorsal shaved skin of 48 rabbits, eight animals for each fraction. The 3% benzene solutions were applied twice weekly 70 times and the rabbits were observed for 12-13 months. Four fractions (except No. 8 and Nos 4+6) each induced squamous-cell papillomas in three rabbits out of five or six surviving 10 months. No malignant tumour was found (Bogovski, 1961, 1962).

(b) Subcutaneous and/or intramuscular administration

Mouse: Hueper and Cahnmann (1958) investigated various thermo-distillation and multiple chromatography fractions of crude shale-oil, processed in the NTU retort, by injecting 0.25-0.3 ml of each fraction into the muscle tissue of the right thigh of 30 C57Bl mice once a week for three successive weeks. These series of injections were repeated up to three times after treatment-free intervals of three weeks with some of the fractions. Five fractions boiling below the temperature range at which a benzo[a]pyrene-containing fraction is obtained, and in which no benzo[a]pyrene was identified, possessed definite carcinogenic properties, inducing sarcomas at the site of i.m. injection. Analysing the behaviour of the fraction containing 'oxy' compounds, which is very toxic, and of which a 25% dilution induces cancer, the authors drew attention to the fact that excessive toxicity of an agent, which may manifest itself in necrotizing effects or in an undue shortening of the lifespan, interferes with its carcinogenic action.

Bogovski (1961, 1962) injected chromatographic fractions No. 5, No. 6 and No. 7 of the chamber-oven tar (see above), in doses of 10 mg per injection dissolved in 0.5 ml peach oil, twice with an interval of five months i.m. into the left thigh of groups of 20 random-bred white mice for each fraction. The total dose of benzo[a]pyrene per mouse was 0.046 mg for fraction No. 6 and 1.32 mg for fraction No. 7; eight of 17 and 13 of 14 autopsied mice out of 20, respectively, developed sarcomas at the injection site; four and five mice developed pulmonary adenomas with these two fractions, respectively. Fraction No. 5, not containing benzo[a]pyrene, induced lung tumours in five mice but failed to induce topical sarcomas. [The Working Group noted the absence of vehicle controls and that the author reported a low incidence of spontaneous lung tumours in historical controls.]

Shale-oil distillates, blends and other commercial products

The first shale-oil products obtained on a commercial scale by the industry in Scotland were mainly distillates of crude shale-oil, and it is these that were used in the first historical carcinogenicity studies published. A number of commercial products of the Estonian oil-shale industry representing distillates, residues, blends and chemical products have been tested in animal experiments.

(a) Skin application

Mouse: The first reported carcinogenicity study in which West Lothian (Scotland) oil-shale processing products were tested was published by Leitch (1922), who applied four shale-oils - 'green' oil, 'blue' oil, unfinished gas oil and unfinished lubricating oil - to the skin of mice [strain not specified]. He described the 'green oil' as a dark, thick, viscid fluid which contains a considerable amount of semi-solid paraffin in suspension. The crude distillate that comes from the retorts is treated with sulphuric acid and soda and again distilled, and the product (green oil) is the first with which workers come in contact. According to Smith (1952), the oil in Scotland was retorted at 1600-1900°F (871-1038°C). 'Blue oil', according to Leitch, differs from green oil only in that it contains less solid and semi-solid paraffin, which has been removed by refrigeration. [According to Hueper (1942), blue oil is the heavy fraction of distillation of green oil, from which the paraffin has been separated (by cooling and passing through filter-presses).] Leitch (1922) applied green oil by means of a camel-hair brush to the skin of 100 mice [strain not specified] between the shoulders three times a week for several months. Of 29 mice surviving 100 days, eight developed papillomas. One mouse developed a malignant tumour (spindle-cell sarcoma), and no tumour was found in 16. Of 100 mice receiving similar skin applications of blue oil, 22 were alive after 100 days; of these, 11 mice had papillomas, one of which regressed; two developed malignant tumours, an epithelioma and a spindle-cell sarcoma; and no tumour was found in 9. The unfinished gas oil (a golden-yellow, transparent fluid) was tested similarly in 50 mice, of which 13 survived 100 days; only one papilloma, which regressed, was observed. The unfinished lubricating oil (a dark-green oil still containing solid paraffin in solution) was administered similarly to 50 mice. Of 10 mice surviving 100 days, two developed papillomas and one an 'epithelioma' (squamous-cell carcinoma).

Twort and Ing (1928), in the framework of an extensive investigation of many petroleum products, tested a refined shale-oil, which was used to lubricate the mule machines in the cotton industry. The oil was applied twice weekly to the skin of 100 white mice. The first tumours were observed after 17 weeks; during 50 weeks, 13 mice developed papillomas and 18 mice had 'epitheliomas' (squamous-cell carcinomas). It was also found that fractions boiling at lower temperatures (140-222°C) were more carcinogenic (44% of mice with tumours) than higher-boiling fractions (240-260°C, 8%; 260-280°C, 12%; 220-240°C, 23%) and that treatment with sulphuric acid inhibited the carcinogenicity.

Hueper (1953) investigated a range of Green River shale-oil distillation products referring to the study of Smith et al. (1951), who claimed that fractions of NTU shale-oil boiling below 550°F (288°C) and above 700°F (371°C) had no carcinogenic activity. The heavy shale-oil obtained at temperatures of up to 900°F (482°C) was tested by Hueper in 50 strain A mice by skin application once a week in a 1:10 dilution in xylene for six months and in a diethyl ether solution (40%/60% w/w) for the following six months. In two of 41 mice alive after eight months, papillomas were found at the application site, and in one mouse a carcinoma occurred. The light distillate of shale-oil, boiling over at up to 580°F (305°C), induced only one papilloma in one strain A mouse of 54 that survived eight months (initial number, 100).

The fact that heavy fractions of shale-oils are more carcinogenic than light fractions or than the total oil is also illustrated by an experiment (Bogovski & Vinkmann, 1979) in which the heavy residue of an Estonian generator oil, diluted 1:1 in olive oil and with a benzo[a]pyrene content of only 10 ppm, was administered twice weekly to 50 random-bred mice (total number of applications, 50). All 50 mice survived the latency of the first tumour (10 weeks); 22 mice developed skin tumours, which were malignant in 15 mice. The initial number of mice was equal to the effective number owing to the lowered toxicity of the sample (50% in olive oil) and to the short latent period (10 weeks). A group of 32 CC57Br mice was treated twice weekly (50 times) on the interscapular region with an intermediate fraction of the solid-heat-transfer (SHT) oven tar, benzo[a]pyrene content, 20 ppm. Of the 15 mice alive after 22 weeks (latency of first tumour), six had benign skin neoplasms. A further group of 31 CC57Br mice was similarly treated with the heavy fraction, benzo[a]pyrene content, 30 ppm. Of the 29 mice alive after 10 weeks (latency of first tumour), 11 had benign tumours and four had carcinomas. In spite of the almost equal concentration of benzo[a]pyrene, the heavy fraction showed a much shorter latency and a stronger carcinogenic action.

Two types of shale-oil bitumen were tested (Bogovski, 1961) in a comparative skin application experiment - a sample of blown (oxidized) bitumen obtained industrially from heavy fractions of shale-oils (benzo[a]pyrene content, 0.003% (30 ppm)), and a laboratory sample representing a vacuum-distillation residue corresponding to the residue boiling above 400°C at atmospheric pressure (benzo[a]pyrene content, 0.1% (1000 ppm)). Each of two groups of 70 random-bred white mice received one drop of one of the test materials, liquefied with a small amount of benzene, twice weekly, 50 times. The blown bitumen (density (g/cm^3), 1.012 of the initial tar; softening temperature (Ring and Ball test), 42°C; blowing time, 8.5 h, blowing temperature (max), 180°C) induced skin tumours in 12 of 48 mice alive at the time (104 days) when the first tumour appeared. Squamous-cell carcinomas were diagnosed in two mice and papillomas in 10. The residual bitumen induced the first tumours much earlier, at the 73rd day, when 44 mice were alive. In 24 mice, skin tumours appeared; in seven mice the tumours were squamous-cell carcinomas. Comparison of the tumour incidence curves shows that the tumours appeared much earlier in more mice and more malignant tumours were found in the group given the residual bitumen than in that given the blown bitumen.

In the Estonian shale industry, various blends of oils and their intermediate and heavy fractions have been and are being produced and commercialized in large amounts as fuel oils and impregnating oils. Bogovski (1954) showed that fuel oil blended from generator oil and chamber-oven oil (about 10-15%) [of which 2% comes over up to 200°C, 20% up to 250°C, 36% up to 300°C, and 60% up to 350°C] was highly carcinogenic and toxic. Doses of 30-40 mg of the undiluted fuel oil were applied to the interscapular region of 170 random-bred mice (14-18 g bw) three times a week for one month (because of toxicity, mice then had no application for two weeks) and then twice a week. Of 50 mice surviving five months, 23 developed skin tumours. In 12 mice, the tumours were squamous-cell carcinomas, and in five of these highly anaplastic stages of epidermal carcinoma, resembling spindle-cell sarcomas, developed. In two mice, metastases were found in various organs.

Another blend, containing 40% of the high-temperature chamber-oven tar and commercialized as wood-impregnating oil (density (g/cm³), 1.000; viscosity at 80°C (Engler), 1.6-1.8°, flash-point, 87-90°C; sulphur, 1 wt%; moisture, up to 2 wt%; ash, 0.3 wt%), was applied to 100 random-bred mice twice weekly 50 times to the interscapular region. Of the 13 mice that survived five months, nine developed skin tumours (two of them malignant). In one mouse, metastases were found in the regional lymph nodes (Bogovski, 1955).

A study (Bogovski, 1961) designed to compare the carcinogenicity of solutions containing 5, 10, 15 and 20% by weight of chamber-oven tar showed that in all four groups, each consisting of 105 (63 male and 42 female) random-bred mice, receiving twice-weekly skin applications (total, 50) of 50 mg of the mixture, the number of benign and malignant skin tumours was similar, but the latent period of the tumours induced by the 5% solution was about twice as long as in the other three groups.

Two commercial products obtained from the Estonian low-temperature generator oil, not containing the carcinogenic high-temperature tar, were tested in random-bred mice with negative results (Bogovski, 1961). One was a rubber-softener from Kohtla-Järve, representing a heavy fraction of the generator oil neutralized with alkali [flash-point, not less than 130°C; viscosity at 75°C (Engler), not more than 3.6°]. Twice-weekly (50) applications of the material (30-40 mg) in 90 mice for 175 days caused the death of 72 mice in five months. The majority of mice had large ulcerations at the application site. No tumour was observed. Another product, a typographic ink, contained a shale-oil varnish, representing a fraction of the low-temperature generator oil (40%), 20% of tall oil, 30% of Blue 5 T and 10% of spindle oil. The composition was applied to the skin of 75 random-bred mice twice weekly (50 times) during 172 days. No skin tumour was found in 64 mice surviving 172 days or in 38 animals surviving 272 days, at which time the experiment was terminated. Skin irritation was moderate, and no ulcer was observed.

Various chemical products of the Estonian oil-shale industry have been tested. Vinkmann (1972b) reported experiments in CC57Br mice with the lacquer LSP-1, the main constituent of which is the residue of pyrolysis of the oil-shale gasoline, containing 0.05% (500 ppm) of benzo[a]pyrene. A group of 52 mice (28 females and 24 males) received twice-weekly applications of 40-50 mg of the lacquer (50 applications). Skin tumours developed in 45 mice; of these, 27 had squamous-cell carcinomas. It appeared that, compared with males, female mice showed a shorter latency, had more malignant tumours earlier and had a considerably shorter lifespan.

Tolichthtol, a product obtained from the acidic residue of rectification of shale-oil aromatic fractions containing up to 22% by weight of sulphur compounds, which is used in veterinary medicine (and is comparable to ichthyol), was tested in 50 CC57Br mice. To the skin of the interscapular region, 50 mg of the ointment was applied twice weekly 50 times. A group of 50 controls were untreated, and most treated and control animals survived 18 months. No skin tumour was observed during 24 months (Vinkmann & Mirme, 1975).

Bogovski et al. (1963) demonstrated that the liquid coking distillate was less active in inducing skin tumours than the initial chamber-oven tar (1/50 compared to 23/50 white mice with skin tumours under similar experimental conditions).

Rowland et al. (1980) tested a sample of shale-oil coke [a raw shale-distillation residue] from the Colony Development operation, Denver, Colorado. A 15% (w/v) benzene solution of the sample, benzo[a]pyrene content, 9.2 µg/ml (9.2 ppm), was applied to the skin of 50 Swiss mice at a dose of one drop (1/60 ml) twice weekly for their lifespan. Of 50 animals, 48 developed skin tumours; among them, 44 mice had 55 squamous-cell carcinomas. In addition, local fibrosarcomas were seen in two animals.

(b) Intratracheal administration

Hamster: The same shale-oil coke described above [a raw shale-distillation residue] was tested intratracheally in 50 male and 50 female Syrian golden hamsters aged six to seven weeks. Shale-oil coke suspensions in doses of 3 and 0.5 mg in 0.2 ml physiological saline were administered once a week for 15 and 30 weeks, respectively. A positive-control group (3 mg of benzo[a]pyrene + 3 mg of haematite in 0.2 ml of saline once a week for 15 weeks) and a vehicle-control group (0.2 ml of saline once a week for 30 weeks) served for comparison. No tumour was found in the lungs of 36 and 56 animals surviving 70 weeks at the two dose levels, respectively (Rowland *et al.*, 1980).

No relevant data on oil-shale retort process-waters were available to the Working Group.

3.2 Other relevant biological data

(a) Experimental systems

Toxic effects

The oral LD_{50} of crude shale-oil in male and female BALB/c mice was 11.3 g/kg bw. The i.p. LD_{50}s were 6.1 and 4.3 g/kg bw, respectively, and those of two jet fuels derived from shale-oils were 8.0 and 6.4 g/kg bw in female mice. A hydrotreated shale-oil, a hydrotreated residue, and two jet fuels derived from shale-oil did not cause acute oral lethality at doses of up to 16 g/kg bw (Smith *et al.*, 1981). In rats given various jet fuels derived from shale-oils by gavage, the LD_{50} values ranged from 26-60 ml/kg bw (Parker *et al.*, 1981). The oral LD_{50} of undiluted shale-oil retort-water was 3-4 ml/kg bw in mice and 33 ml/kg bw in rats (Gregg *et al.*, 1981).

After a single dermal application of 2 ml/kg bw of crude shale-oil or its derivatives to rats, no skin lesion, manifestation of central nervous toxicity or death was observed (Smith *et al.*, 1981). Repeated long-term application of crude shale-oil or shale-oil distillates to the skin of mice induced local inflammatory, degenerative and hyperplastic skin changes, accompanied by hair loss, and skin thickening and pigmentation; males were more affected than females (Holland *et al.*, 1981a,b). The application of crude shale-oil and subsequent exposure to ultraviolet light elicited much more severe skin damage in hairless mice than either agent alone (Gomer & Smith, 1980). Middle distillates of shale-oil applied to the backs of C3Hf/Bd mice three times weekly for 60 weeks caused renal papillary necrosis and atrophy of nephrons (Easley *et al.*, 1982).

Rats and monkeys exposed to 10 or 30 mg/m^3 of respirable dusts of both raw (Anvil Points) and spent (above-ground, direct-fired retort) shale six hours per day, five days per week, for two years, showed accumulations of pigment- and particle-laden macrophages in and around end airways, with inflammatory reactions in the lungs (MacFarland *et al.*, 1982).

Effects on reproduction and prenatal toxicity

Two studies have used shale-oil and derivatives from Anvil Points, Colorado, USA.

Groups of 25 female rats (Charles River CRL: COBS-CO (SD) BR) were exposed by inhalation for six hours per day from days 6-15 of gestation to either raw or spent oil-shale dust at concentrations of 25 and 100 mg/m^3 or to shale-oil aerosol at concentrations of 5 and 100 mg/m^3. On day 20 of gestation, foetuses were removed and examined for skeletal and soft-tissue malformations and resorptions. No significant difference from control groups was observed (Conaway et al., 1984). [Doses used were below maternally toxic concentrations.]

Groups of about 25 ICR/DUB female mice were given 0, 0.1, 0.3 and 1% filtered Paraho oil-shale retort-water (pH 8.4) as the only drinking-water and mated with untreated males. There was no reproductive effect except for a possible increase in preimplantation losses in mice consuming 1.0% retort-water. No maternal toxicity was observed at this concentration (the oral LD_{50} of the undiluted water in mice is given as 3-4 ml/kg bw). An increase in 14th (additional) ribs was noted, and abnormal palates (clefts or partial absence of the palates) were observed in 6.5, 10 and 16.4% of the low-, middle- and high-dose groups, respectively (significant at the 1% level, Mann-Whitney test for statistical significance, for the high-dose group) (Gregg et al., 1981).

Preimplantation losses were demonstrated by a decrease in the total number of implants in female mice and were most pronounced in females bred the third week after treatment of males with raw or upgraded shale-oil (Barkley et al., 1979).

Pregnant New Zealand white rabbits were administered spent shale suspended in distilled water by oral intubation at 250 or 500 mg/kg bw on days 8 and 12 of gestation (Barkley et al., 1979). [At 500 mg/kg bw there seemed to be an increase in skull malformations and soft-tissue anomalies of the head and brain.]

Absorption, distribution, excretion and metabolism

Oral administration of shale-oil retort-water from an in-situ process to male Holtzman rats resulted in an increase in levels of liver microsomal cytochrome P-450 and some associated drug metabolism activities (Nelson et al., 1978).

Mutagenicity and other short-term tests

A large number and variety of short-term test systems have been applied to the study of shale-derived oils and/or precursors or by-products of shale-oil processing (Tables 10-15). The samples derive from US sources and from low-temperature retorting processes. The resulting picture is quite complex, the experimental results being affected by a number of factors, including: (1) the nature of the material under study (raw shale, spent shale, crude, hydrotreated or refined shale-oil, retort process-water, or oil-shale ash); (2) the mineral composition of the original shale and the type of retort process; (3) the activation system used in the in-vitro assays - either metabolic (various rodent-liver homogenates) or photoactivation (near ultraviolet radiation, artificial visible light or sunlight); (4) the solubilizing or dispersing agent used for sample preparation and, in some cases, the fractionation procedure used in order to separate and/or characterize the active constituents of the complex mixture.

Raw and spent oil shale (Tables 9 and 10)

In a comparative study, dimethyl sulphoxide (DMSO) extracts of samples of raw shale and of spent shale (Paraho retorting process) were tested in three in-vitro short-term test systems. Neither of the extracts induced reverse mutations in *Salmonella typhimurium* TA1535, TA1537, TA1538, TA98 or TA100, mitotic gene conversion in yeast (strain D4 of *Saccharomyces cerevisiae*) or changes at the thymidine kinase (TK) locus in Fischer L5178Y mouse lymphoma cells. (A slight, but not significant increase in mutation frequency was seen with raw shale). Contrary to spent shale, the raw-shale sample (presumably the kerogen component) induced a significant increase in chromosomal aberrations (chromatid breaks and fragments) in bone-marrow cells of male Sprague-Dawley rats when administered daily by gavage on five consecutive days (Conaway et al., 1984). A Tosco II spent oil shale was mutagenic to *S. typhimurium* TA1537 and TA98 in the presence of Aroclor-induced rat liver homogenate (S9). In contrast, a sample of Paraho spent shale yielded negative results. The corresponding raw-shale sample was also negative in *S. typhimurium* (Hinchee et al., 1983).

Table 9. Summary of results from short-term tests: Raw oil shale

Test	Organism/assay	Reported result	Reference
PROKARYOTES			
Mutation	*Salmonella typhimurium* (his$^{-/+}$)	Negative	Hinchee et al. (1983); Conaway et al. (1984)
FUNGI			
Mutation	*Saccharomyces cerevisiae* D4 (mitotic gene conversion)	Negative	Conaway et al. (1984)
MAMMALIAN CELLS *IN VITRO*			
Mutation	Mouse lymphoma L5178Y cells (TK$^{+/-}$)	Weakly positive	Conaway et al. (1984)
MAMMALIAN CELLS *IN VIVO*			
Chromosomal effects	Rat bone-marrow cells (gavage) (aberrations)	Positive	Conaway et al. (1984)

Table 10. Summary of results from short-term tests: Spent oil shale

Test	Organism/assay	Reported result	Reference
PROKARYOTES			
Mutation	*Salmonella typhimurium* (his$^{-/+}$)	Positive	Hinchee et al. (1983)
	Salmonella typhimurium (his$^{-/+}$)	Negative	Conaway et al. (1984)
FUNGI			
Mutation	*Saccharomyces cerevisiae* D4 (mitotic gene conversion)	Negative	Conaway et al. (1984)
MAMMALIAN CELLS *IN VITRO*			
Mutation	Mouse lymphoma L5178Y cells (TK$^{+/-}$)	Negative	Conaway et al. (1984)
MAMMALIAN CELLS *IN VIVO*			
Chromosomal effects	Rat bone-marrow cells (gavage) (aberrations)	Negative	Conaway et al. (1984)

Crude shale-oil (Table 11)

Many studies (Table 11) showed that various samples of shale-derived crude oils, irrespective of the origin of the shale and of the retorting process, were mutagenic in *S. typhimurium* TA100, TA1537, TA1538 and/or TA98, but only in the presence of S9. Mutagenicity was also observed following photoactivation with artificial visible light (Selby *et al.*, 1983).

When shale-derived oils were fractionated, mutagenic activity was found in neutral and basic fractions. Neutral components included known carcinogenic/mutagenic polynuclear aromatic hydrocarbons (PAHs) and alkylated PAHs. High-resolution gas chromatography/mass spectrometry analyses of highly mutagenic basic subfractions showed the causative agents to be aza arenes and polynuclear aromatic primary amines (Epler *et al.*, 1978; Guerin *et al.*, 1978; Epler *et al.*, 1979a,b; Pelroy & Petersen, 1979; Guerin *et al.*, 1980, 1981; Pelroy *et al.*, 1981; Rao *et al.*, 1981a,b,c; Timourian *et al.*, 1981; Toste *et al.*, 1982).

DMSO extracts of three crude oils obtained by different retorting processes (Tosco, Paraho and Union Retort-B) did not induce mitotic gene conversion in strain D4 of *Saccharomyces cerevisiae* with or without S9 (Conaway *et al.*, 1984), whereas a sample of Paraho crude oil and its ether-soluble fraction affected both the canavanine and the histidine markers in strain XL7-10B of the same yeast in the presence and absence of S9 (Rao *et al.*, 1981c).

Crude-oil samples were negative or marginally positive in the L5178Y mouse lymphoma (TK$^{+/-}$) assay (Conaway *et al.*, 1984). A DMSO extract of Paraho crude oil induced ouabain and trifluorothymidine resistance in cultured Chinese hamster ovary (CHO) cells (only in the presence of S9) but not resistance to azaadenine or to thioguanine (Timourian *et al.*, 1981). Following exposure to near-ultraviolet light, DMSO extracts of four crude-oil samples (Paraho above-ground retorted (AGR) crude oil, LETC-simulated MIS-retorted crude oil and two oils from a MIS retort (MIS3 and MIS185)) induced 6-thioguanine resistance in CHO cells (Strniste *et al.*, 1982a; Strniste, 1984).

Emulsions of two shale-oils in Tween 80 increased sister chromatid exchange in cultured human lymphocytes in the presence of S9 (Lockard *et al.*, 1982).

In-vivo studies in mice treated with samples of crude shale-oils showed the induction of chromosomal aberrations in bone-marrow cells following i.p. administration but not after skin painting (Meyne & Deaven, 1982a). Equivocal results were reported for induction of sister chromatid exchanges in bone-marrow cells after i.p. treatment of mice (Timourian *et al.*, 1981; Lockardt *et al.*, 1982; Meyne & Deaven, 1982a); no increase in micronucleated cells in bone marrow were observed (Lockardt *et al.*, 1982). Rats treated by gavage with the crude-oil samples did not show any increase in chromosomal aberrations in bone-marrow cells (Conaway *et al.*, 1984).

Barkley *et al.* (1979) reported that the i.p. injection of crude shale-oil (in mineral oil) to male hybrid 101 x C3H mice, at various times before breeding, did not result in induction of dominant lethal mutations.

Hydrotreated and refined shale-oils (Table 12)

Hydrotreatment of shale-oil samples led to loss of activity in various short-term test systems (Guerin *et al.*, 1981; Rao *et al.*, 1981c; Timourian *et al.*, 1981; Meyne & Deaven, 1982a;

Table 11. Summary of results from short-term tests: Shale-derived crude oils

Test	Organism/assay	Reported result	Reference
PROKARYOTES			
Mutation	Salmonella typhimurium (his$^{-/+}$)	Positive	Epler et al. (1978); Guerin et al. (1978); Barkley et al. (1979); Epler et al. (1979a,b); Griest et al. (1979); Pelroy & Petersen (1979); Guerin et al. (1980); Hermann et al. (1980); Guerin et al. (1981); Pelroy et al. (1981); Rao et al. (1981a,b,c); Timourian et al. (1981); Lockard et al. (1982); Toste et al. (1982); Calkins et al. (1983); Ma et al. (1983); Selby et al. (1983); Conaway et al. (1984)
FUNGI			
Mutation	Saccharomyces cerevisiae D4 (mitotic gene conversion)	Negative	Conaway et al. (1982)
	Saccharomyces cerevisiae XL7-10B (can$^{r/s}$; his$^{-/+}$)	Positive	Rao et al. (1981c)
MAMMALIAN CELLS IN VITRO			
Mutation	Mouse lymphoma L5178Y cells (TK$^{+/-}$)	Inconclusive	Conaway et al. (1984)
	Chinese hamster ovary cells (1-4 markers)	Positive	Timourian et al. (1981); Strniste et al. (1982a); Strniste (1984)
Chromosomal effects	Human cultured lymphocytes (sister chromatid exchange)	Positive	Lockard et al. (1982)
MAMMALIAN CELLS IN VIVO			
Chromosomal effects	Mouse bone-marrow cells (i.p.) (aberrations)	Positive	Meyne & Deaven (1982a)
	Mouse bone-marrow cells (skin painting) (aberrations)	Negative	Meyne & Deaven (1982a)
	Mouse bone-marrow cells (i.p.) (sister chromatid exchange)	Positive	Meyne & Deaven (1982a)
	Mouse bone-marrow cells (i.p.) (sister chromatid exchange)	Negative	Timourian et al. (1981)
	Mouse bone-marrow cells (i.p.) (sister chromatid exchange)	Inconclusive	Lockard et al. (1982)
	Mouse bone-marrow cells (i.p.) (micronuclei)	Negative	Lockard et al. (1982)
	Rat bone-marrow cells (gavage) (aberrations)	Negative	Conaway et al. (1984)
	Mouse (i.p.) (dominant lethals)	Negative	Barkley et al. (1979)

Strniste et al., 1982a; Strniste, 1984). However, in two studies, the reversion rate in *S. typhimurium* (Timourian et al., 1981) and the induction of chromosomal aberrations in mouse bone-marrow cells (Meyne & Deaven, 1982a) were reduced but not totally eliminated by hydrotreatment. Negative findings for hydrotreated shale-oils were also reported in cases in which the corresponding crude oils were found to be inactive (Barkley et al., 1979 (dominant lethals); Timourian et al., 1981 (sister chromatid exchanges *in vivo*)).

Samples of refined shale-oils, including jet fuels and marine diesel fuels, were also negative in *S. typhimurium* (Rao et al., 1981c).

Table 12. Summary of results from short-term tests: Hydrotreated and refined shale-oils

Test	Organism/assay	Reported result	Reference
PROKARYOTES			
Mutation	*Salmonella typhimurium* (his$^{-/+}$)	Positive	Timourian et al. (1981)
	Salmonella typhimurium (his$^{-/+}$)	Negative	Guerin et al. (1981); Rao et al. (1981c)
FUNGI			
Mutation	*Saccharomyces cerevisiae* XL7-10B (can$^{r/s}$) (his$^{-/+}$)	Negative	Rao et al. (1981c)
MAMMALIAN CELLS *IN VITRO*			
Mutation	Chinese hamster ovary cells (1-4 markers)	Negative	Timourian et al. (1981); Strniste et al. (1982a); Strniste (1984)
MAMMALIAN CELLS *IN VIVO*			
Chromosomal effects	Mouse bone-marrow cells (i.p.) (aberrations)	Positive	Meyne & Deaven (1982a)
	Mouse bone-marrow cells (i.p.) (sister chromatid exchange)	Negative	Timourian et al. (1981); Meyne & Deaven (1982a)
	Mouse (i.p.) (dominant lethals)	Negative	Barkley et al. (1979)

Oil-shale retort process waters (Table 13)

All of the studies on samples of retort process waters showed positive results in *S. typhimurium* but only following metabolic or near-ultraviolet-induced activation. Combination of a process-water sample with near-ultraviolet irradiation also resulted in increased lethality in *S. typhimurium* TA98 (uvrB$^-$) compared with strain UTH8413 (uvrB$^+$) (Strniste, 1984).

Following fractionation of a process-water from an above-ground, simulated in-situ retorting process, Epler et al. (1978, 1979a,b) reported that highly active materials occurred in the basic fraction, while the neutral fraction (containing PAHs) showed only weak mutagenic activity.

Following near-ultraviolet irradiation, oil-shale process-water samples were also mutagenic in CHO cells (Barnhart & Cox, 1980; Strniste & Chen, 1981; Chen & Strniste, 1982; Strniste et al., 1982b) and in human skin fibroblasts (Strniste et al., 1982b).

Direct concentration-related increases in chromosomal aberrations (primarily chromatid breaks and exchanges) were reported in CHO cells in the absence of S9 (Chen et al., 1983).

Three retort process-water samples reported to give positive results in S. typhimurium and in the CHO/HPRT/near-ultraviolet test systems (Strniste & Chen, 1981) were also assayed for cytogenetic effects in vivo (Meyne & Deaven, 1982b). Administration of samples of above-ground retort-water (ARP) to mice in drinking-water (about 1 ml/kg bw per day for 8 weeks) induced a slight clastogenic effect in bone-marrow cells. A more pronounced clastogenic effect was obtained after i.p. treatment (but without dose-response), while no induction of sister chromatid exchanges was observed. Administration of 1% ARP in drinking-water to pregnant C3H mice (from days 1-12 of gestation) resulted in an increased frequency of chromosomal aberrations and gaps in embryo cells. However, this finding could not be confirmed in Swiss mice.

Table 13. Summary of results from short-term tests: Oil-shale retort process-waters

Test	Organism/assay	Reported result	Reference
PROKARYOTES			
DNA damage	Salmonella typhimurium (differential lethality)	Positive	Strniste (1984)
Mutation	Salmonella typhimurium (his-/+)	Positive	Epler et al. (1978, 1979a,b); Toste et al. (1982); Strniste et al. (1983); Strniste (1984)
MAMMALIAN CELLS IN VITRO			
DNA damage	Human skin fibroblasts (6-TG^r)	Positive	Strniste et al. (1982b)
	Human skin fibroblasts (differential cytotoxicity)	Positive	Strniste et al. (1982b)
Mutation	Chinese hamster ovary cells (1-5 markers)	Positive	Barnhart & Cox (1980); Strniste & Chen (1981); Chen & Strniste (1982); Strniste et al. (1982b); Chen et al. (1983); Strniste et al. (1983); Strniste (1984)
Chromosomal effects	Chinese hamster ovary cells (aberrations)	Positive	Chen et al. (1983)
MAMMALIAN CELLS IN VIVO			
Chromosomal effects	Mouse bone-marrow cells (i.p.) (aberrations)	Positive	Meyne & Deaven (1982b)
	Mouse bone-marrow cells (p.o.) (aberrations)	Positive	Meyne & Deaven (1982b)
	Mouse embryo cells (transplacental) (aberrations)	Inconclusive	Meyne & Deaven (1982b)
	Mouse bone-marrow cells (i.p.) (sister chromatid exchange)	Negative	Meyne & Deaven (1982b)

Oil-shale ash (Table 14)

A mutagenic response in *S. typhimurium* TA1538, TA98 and TA100 [metabolic conditions not specified] was induced by various extracts from baghouse fly ash produced by burning oil shale in a fluidized bed combustor (Brooks *et al.*, 1980). Three ash samples, two of which (A and B) had been collected on glass-fibre filters 50 m from a simulated oil-shale fire (burning Mahogany shale at 900°C) and one (C) obtained from accumulated ash on the wall of the burner, exhibited a high direct mutagenic activity in *S. typhimurium* TA98 and TA100 (his$^{-/+}$), which was markedly decreased by S9 fractions. A weak activity was also detected in tests with strain TA1537, while results with strain TA1535 were negative. The same extracts of sample C, tested in the absence of S9, also induced forward mutations in the *ara*^{r} strain SV-50 of *S. typhimurium*. In contrast, sample A was negative when tested for gene conversion and mitotic crossing-over in strain D7 of *Saccharomyces cerevisiae* and in the *ad-3* system (mutants N23 and N24) of *Neurospora crassa* (Whong *et al.*, 1982, 1983).

Table 14. Summary of results of short-term tests: Oil-shale ash

Test	Organism/assay	Reported result	Reference
PROKARYOTES			
Mutation	*Salmonella typhimurium* (his$^{-/+}$)	Positive	Brooks *et al.* (1980); Whong *et al.* (1982)
	Salmonella typhimurium (ara$^{r/s}$)	Positive	Whong *et al.* (1982, 1983)
FUNGI			
Mutation	*Saccharomyces cerevisiae* D7 (gene conversion; mitotic crossing over)	Negative	Whong *et al.* (1982)
	Neurospora crassa (*ad-3* reversion)	Negative	Whong *et al.* (1982)
MAMMALIAN CELLS *IN VITRO*			
Mutation	Chinese hamster ovary cells (6TG$^{s/r}$)	Negative	Whong *et al.* (1982)

(*b*) *Humans*

Toxic effects

Scott (1922a) described comedones, folliculitis, erythema, papular and chronic hyperkeratotic dermatitis and the development of hyperkeratotic lesions ('warts') in workers in the Scottish shale industry. Cutaneous cancers (further described in section 3.3) developed from hyperkeratotic lesions, indurated papules or, in some workers, on skin that was not obviously abnormal. Similar lesions have been described in workers exposed to Estonian shale-oils (Loogna & Hering, 1972).

Birmingham (quoted by Costello, 1979) reported in 1955 telangiectasia, flat warts, seborrheic keratoses, senile keratoses and pigmentation in Colorado oil-shale plant workers. Because of the small number of workers studied, he was unable to differentiate between effects of shale-oil and other possible causes.

Inflammation of the upper respiratory tract, acute and chronic bronchitis and pneumonia have been reported frequently among shale-oil workers in the USSR (Feoktistov, 1972; Luts, 1972; Maripuu, 1972). Maripuu studied the prevalence of chronic bronchitis in 1114 workers from a shale-processing plant (485 men and 629 women), in 1167 men from two mechanized mines and in 151 women from a shale-concentration plant. The prevalence of chronic bronchitis in all workers was 16.3%, and for subgroups ranged from 5% (above-ground, non-dusty work) to 32.6% (chamber ovens, high-temperature gasification). In addition to smoking habits, the main risk factors were: sex (male > female), type of activities (processing plant > mines), duration of age-adjusted exposure, and previous exposure to irritant gases. No similar observation was reported among Colorado shale-oil workers in a study designed to evaluate health status and to establish a data base for a prospective epidemiological study, but the average duration of exposure was usually less than five years (Rudnick & Voelz, 1980).

Cases of pneumoconiosis have been attributed to the inhalation of Estonian oil-shale dusts containing 3-4% of free silica and the silicates [kaolin and mica (Seaton et al., 1981)]. On X-ray examination, interstitial lung fibrosis was found to be accompanied by enlarged hilar shadows. Autopsy of workers with pneumoconiosis showed lung emphysema (Küng, 1979).

Seaton et al. (1981) described four men who worked exclusively in Scottish shale mining who developed complicated pneumoconiosis. Three had histologically confirmed massive fibrosis, while the fourth developed large pulmonary nodules.

It was reported that workers exposed to shale-oils complained more frequently of headache and fatigue than a comparison group (Kahn, 1979).

Several cases of acute and subacute hydrogen sulphide poisonings were observed in the Swedish shale-oil industry. Examination of over 800 workers indicated that the frequency of neurasthenic troubles increased with the degree of hydrogen sulphide exposure and the length of employment (Ahlborg, 1951).

No data were available to the Working Group on reproduction and prenatal toxicity, on absorption, distribution, excretion and metabolism, or on mutagenicity and chromosomal effects.

3.3 Case reports and epidemiological studies of carcinogenicity in humans

Soon after the initiation of oil-shale mining and processing, Bell (1876) reported two cases of epithelioma (skin cancer) of the scrotum among workers from the 'paraffin' area of a Scottish oil-shale works.

Scott (1922a,b) enumerated 'all' cases of epithelioma that occurred during 1900-1921 at a major shale-oil company in Scotland. The 65 cases found included three scrotal and 16 other skin cancers among paraffin workers and 28 scrotal and 18 other skin cancers among the rest of the work force. This latter group included oil workers, retortmen, labourers and stillmen, who may have come into contact with ash, coke or other gritty material. The total work force of 5000 men was fairly stable over the years, as was the work force of 200 paraffin workers. [Using these figures, the Working Group calculated the following crude incidence rates (per 10^6 person-years): 681 for scrotal cancer among paraffin workers and 265 for scrotal cancer among 'other' shale-oil workers. In comparison, Henry (1946) calculated a scrotal cancer mortality rate of $4.2/10^6$ person-years for England and Wales during the period 1911-1938.]

Of 141 scrotal cancers diagnosed between 1902 and 1922 in a single hospital in the UK, 69 were among mule-spinners (an occupation in cotton-textile mills that involves substantial exposure to lubricating oil) and 22 were among 'paraffin' workers (Southam & Wilson, 1922). The cases among mule-spinners are important because, after 1862, shale-oil in addition to petroleum and other types of oils began to be processed into lubricating oil (Henry, 1946) and gradually replaced animal and vegetable oils (Hunter, 1975).

Henry (1937) reviewed 1487 fatal cases of scrotal cancer reported to the Registrar General for England and Wales between 1911-1935; 345 of these occurred among mule-spinners.

Seaton *et al.* (1981) described two cases of localized peripheral, well-differentiated squamous-cell carcinomas of the lung in people who worked exclusively in Scottish shale mining.

Bridge and Henry (1928), Henry (1946, 1947) and Doig (1970) reviewed lists in the UK compiled from compulsory notifications of all skin cancers contracted in selected types of factories[1]. Among 3753 cases reported (Henry, 1947) from 1920-1945, 52 cases, including 15 of the scrotum, occurred among oil-shale workers; 1389, including 857 of the scrotum, occurred among cotton-textile workers; and 79, including 44 scrotal cancers, occurred among other workers exposed to mineral oil (lubricating oil). The vast majority of the cancers were squamous-cell, but 93 (2.4%) were basal-cell. The period from initial employment in an exposed job until the appearance of papilloma or tumour, for all cases associated with shale-oil or mineral oil ranged from 4-75 years and peaked at 40-59 years. [This may give some indication of the induction-latent period, although it is impossible to define when shale-derived oils were introduced at each case's work site. Therefore, the actual times since first exposure to a shale-derived mineral oil may have been less than those cited.]

Doig (1970) extended Henry's (1947) review of notifiable epitheliomas from 1944-1968. Exposures to lubricating oils accounted for 767 of the 4471 reported epitheliomas during the period. Based on a sample of cases of scrotal epitheliomas from Scotland for 1967, Doig determined that only 33% (four of 12) of eligible cases were actually reported to the government. In his sample, nine of the cases were exposed to oil, but none could specifically be related to shale-oil.

The reports of the Registrar General showed consistently high standardized mortality ratios (SMRs) for skin cancer among cotton spinners; for example, in 1911, the SMR was 343 in men aged 25-64 years (Logan, 1982).

Heller (1930) examined hospital records and all death certificates in three cities in Massachusetts, USA (a major textile area), to ascertain all cases of skin cancer. He compared the occurrence of scrotal cancer in US mule-spinners with the UK experience. He concluded that the 'occurrence of scrotal epithelioma among mule spinners is insignificant in the USA and that during the last fifteen years not a single case has occurred which can be attri-

---

[1]The UK Factory and Workshop Act 1901 required that diseases contracted in premises to which the Act applied be notified to HM Chief Inspector of Factories by the medical practitioners concerned. The list of such notifiable diseases was extended in 1919 to include 'epitheliomatous ulceration due to pitch, tar, bitumen, mineral oil or paraffin, or any compound, product or residue of these substances' (Doig, 1970).

buted to the American mule spinning occupation'. After visiting 14 cotton mills in the USA, Heller concluded that the mule-spinning equipment and work practices were very similar in the USA and the UK, resulting in the same potential for contact with mineral oil. [The Working Group noted that both shale-oil and petroleum oil were used in the UK, whereas only petroleum oils were used in the USA.]

Death certificates of all deaths due to cancer of the scrotum and penis in England and Wales for 1913, 1914 and 1921-1923 were reviewed and classed by occupation (Kennaway, 1925). Scrotal cancer accounted for 197 of 76 196 deaths from all causes during the first four of these five years. The ratio of scrotal to penile cancers was higher among chimney-sweeps (21:4), cotton textile workers (70:12), and other workers likely to have been exposed to tar, pitch and lubricating oils (32:16) compared to all other workers (126:615). [If the data are analysed as a case-control study, with controls being cases of cancer of the penis, crude odds ratios would be 24.1, 26.8 and 9.2 for chimney-sweeps, cotton-textile workers, and other workers likely to be exposed, respectively.]

A morbidity cohort study of 2003 shale-oil workers in Estonia was carried out by Purde and Etlin (1980). The cohort comprised men and women who by 1975 had been employed for at least 10 years in the industry. Cases of skin and other cancer in the cohort and in the general population were ascertained from the local cancer treatment centre. Significantly elevated rates of skin cancer (per 100 000 per annum) were found in men (147 observed, 46 expected; $p < 0.05$) and in women (193 observed, 41 expected; $p < 0.01$) (Bogovski, 1981). Non-significant excesses of stomach cancer and deficits of lung cancer were found in people of both sexes.

In a mortality study, Costello (1979, 1983) followed a cohort of 713 male shale-production workers in Colorado, USA, who had worked for more than one month (median duration of employment, nine months; average, 30 months). He found a statistically significant excess of total cancer (49 observed, 36 expected; $p < 0.05$) and of colon cancer (7 observed, 3.1 expected; $p < 0.05$), and a non-significant excess of lung cancer (16 observed, 10 expected). The excess risks of malignancy were limited to maintenance and 'miscellaneous' workers; no excess was found among miners or retortmen. [The Working Group noted that no mention was made of time since first exposure or duration of exposure.]

4. Summary of Data Reported and Evaluation

4.1 Exposure data

Commercial oil-shale processing industries presently exist only in the USSR and in China (a commercial facility is expected to begin operation in the USA in 1984), producing crude shale-oils that are used as fuels or chemical-plant feedstocks. Crude shale-oils were produced in the past in several countries; an industry existed in the UK for more than 100 years.

Workers involved in oil-shale mining and processing may be exposed to complex mixtures of dusts, gases and vapours. The dusts may contain significant levels of crystalline silica.

Inorganic gases and vapours to which workers may be exposed include carbon monoxide and hydrogen sulphide. Workers may also be exposed to gases and vapours containing organic compounds, including low levels of polynuclear aromatic compounds. The composition of spent oil shale can vary widely in respect to the remaining tar material, depending on the variation of retorting methods. Skin contact with crude shale-oil may occur, but is limited primarily to maintenance workers in modern oil-shale processing facilities.

Contact with shale-oil liquids occurred extensively in the past in the Scottish shale-oil industry, and in the British cotton-textile industry where lubricants derived from shale-oils were used.

4.2 Experimental data

Extracts of *raw oil shale* from Scotland (UK) and the Green River Formation (USA) were tested by skin application in mice, and no skin tumour was observed. A suspension of raw oil-shale dust from the Green River Formation was also tested by intratracheal instillation in rats and hamsters, and no local tumour was observed; exposure of rats by inhalation to raw oil-shale dust resulted in the induction of lung tumours.

Crude shale-oils from low-temperature retorting from different sources were tested for carcinogenicity in various experiments by skin application in different strains of mice. Samples from Jurassic Chuvash (USSR), Estonia (USSR), the Green River Formation (USA) and Fushun (China) all resulted in the induction of benign and malignant skin tumours. A sample from Estonia was also tested by skin application in rabbits and induced benign and malignant skin tumours. Lung tumours were produced in mice following intratracheal administration of a crude shale-oil from the Green River Formation.

Three samples of *crude shale-oils from high-temperature retorting* from Estonia were tested for carcinogenicity by skin application in mice, and one sample was also tested in rabbits. All the samples resulted in the induction of benign and malignant skin tumours; these samples were more carcinogenic than crude shale-oils from low-temperature retorting from the same source.

An extract of *spent oil shale* from the Green River Formation resulted in the induction of skin tumours in mice after topical application. Dusts prepared from this sample induced lung tumours in rats after inhalation exposure. No lung tumour occurred in rats or hamsters exposed by intratracheal administration to a suspension of the spent oil-shale dusts.

Various *fractions of low- and high-temperature shale-oils* were tested by skin application in mice and rabbits and by intramuscular application in mice; their carcinogenic activities did not necessarily parallel the benzo[a]pyrene content of the fractions.

Various *crude shale-oil distillation fractions* from Scotland were tested by skin application in mice; the less refined shale-oils were more highly carcinogenic to the skin than the more refined products. Heavy fractions of shale-oil from Estonia were more carcinogenic than the light fractions or the total oil when tested in mice by skin application; in contrast, in a study from Scotland, light distillation fractions of a lubricating oil induced more tumours than heavier fractions.

Comparative carcinogenicity studies in mice by skin application indicate that *residual shale-oil bitumen* (Estonia) was more active in inducing skin tumours than *blown (oxidized) bitumen*.

Commercial samples representing *blends* of shale-oils from Estonia induced skin tumours in mice after topical application; the carcinogenic effect increased with increasing content of crude shale-oil from high-temperature retorting. In similar experiments, commercial products containing low-temperature retorting oils did not induce skin tumours.

A *pot residue of shale-oil distillation* ('shale-oil coke') from the Green River Formation was carcinogenic to mouse skin after topical application in benzene; however, the same sample did not induce respiratory tumours in hamsters after intratracheal instillation.

No relevant data were available to the Working Group on the carcinogenicity in experimental animals of *oil-shale retort process-waters*.

All the shale-derived materials tested in short-term tests came from sources in the USA, and were therefore all from low-temperature processes.

Chromosomal aberrations were induced in bone-marrow cells of rats following administration by gavage of a suspension of *raw oil shale*. In-vitro tests of extracts of raw oil shale in bacteria, yeast and cultured mammalian cells gave negative results.

Preparations of *spent oil shale* yielded contradictory results in bacterial mutation assays and were negative in mutation assays with eukaryotic cells *in vitro* and in a chromosomal assay *in vivo*.

Preparations of *shale-derived crude oils* from various sources and retort processes were mutagenic in bacteria, yeast and cultured mammalian cells following metabolic or photo-induced activation. Three crude shale-oil preparations did not induce mitotic gene conversion in yeast; two others induced sister chromatid exchanges in cultured mammalian cells. Both positive and negative results were obtained in mammalian in-vivo assays for chromosomal effects.

As compared with the corresponding crude shale-oils, preparations of *hydrotreated oils* showed decreased activity or were negative in various short-term tests.

A preparation of *refined shale-oil* was not mutagenic in bacteria.

Oil-shale retort process-waters elicited DNA damage and mutations in bacteria and in cultured mammalian cells following metabolic activation or photoactivation. They induced chromosomal aberrations in cultured mammalian cells, and induced chromosomal aberrations but not sister chromatid exchanges in mouse cells *in vivo*.

Extracts of *oil-shale ash* were mutagenic in bacteria both in reversion and forward mutation assays in the absence of a metabolic system. Tests with eukaryotic systems gave negative results.

4.3 Human data

The association between shale-oils and skin cancers, particularly of the scrotum, was demonstrated by analyses of 65 cases of skin cancer, including 31 of the scrotum, from the Scottish shale-oil industry. In the UK, over 2000 cases of skin cancer ('mule-spinners' cancer') were recorded among cotton-textile workers and others exposed to lubricating oils (many of which are believed to have been shale-derived). The occupational etiology of these cases is supported by occupational mortality statistics for the UK and by an occupational comparison with fatal cases of penile cancer. In contrast, one study reported very few scrotal cancers among US cotton-textile workers employed in mills where shale-derived lubri-

Overall assessment of data from short-term tests on preparations of raw and spent shale and oil-shale ash

	Genetic activity			Cell transformation
	DNA damage	Mutation	Chromosomal effects	
Prokaryotes		–b ?c +d		
Fungi/green plants		–b,c,d		
Insects				
Mammalian cells (in vitro)		–b,c,d		
Mammals (in vivo)			+b –c	
Humans (in vivo)				
Degree of evidence in short-term tests for genetic activity: *inadequate*				Cell transformation: no data

aThe groups into which the tables are divided and the symbols +, – and ? are defined on pp. 16-18 of the Preamble; the degrees of evidence are defined on p. 18.

bRaw oil-shale preparations

cSpent oil-shale preparations

dOil-shale ash preparations

Overall assessment of data from short-term tests on preparations of low-temperature shale-derived crude oils

	Genetic activity			Cell transformation
	DNA damage	Mutation	Chromosomal effects	
Prokaryotes		+		
Fungi/green plants		?		
Insects				
Mammalian cells (in vitro)		+	+	
Mammals (in vivo)			?	
Humans (in vivo)				
Degree of evidence in short-term tests for genetic activity: *sufficient*				Cell transformation: no data

Overall assessment of data from short-term tests on preparations of hydrotreated and refined shale-oils

	Genetic activity			Cell transformation
	DNA damage	Mutation	Chromosomal effects	
Prokaryotes		?a –b		
Fungi/green plants		–a		
Insects				
Mammalian cells (*in vitro*)		–a		
Mammals (*in vivo*)			?a	
Humans (*in vivo*)				
Degree of evidence in short-term tests for genetic activity: *inadequate*				Cell transformation: no data

aHydrotreated shale-oil preparations
bRefined shale-oil preparations

Overall assessment of data from short-term tests on oil-shale retort-process waters

	Genetic activity			Cell transformation
	DNA damage	Mutation	Chromosomal effects	
Prokaryotes	+	+		
Fungi/green plants				
Insects				
Mammalian cells (*in vitro*)	ı	+	+	
Mammals (*in vivo*)			?	
Humans (*in vivo*)				
Degree of evidence in short-term tests for genetic activity: *sufficient*				Cell transformation: no data

cants were not used. A cohort study of shale-oil workers in western USA found statistically significant excesses of total cancer and of colon cancer, although data on duration and time since first exposure were not available. A cohort study of shale-oil workers in Estonia found significant excesses of skin cancer, but not of cancers at other sites.

4.4 Evaluation[1]

There is *sufficient evidence* for the carcinogenicity in experimental animals of high-temperature crude shale-oils, low-temperature crude shale-oils, fractions of high-temperature shale-oil, crude shale-oil distillation fractions, shale-oil bitumens and commercial blends of shale-oils.

There is *limited evidence* for the carcinogenicity in experimental animals of raw oil shale, spent oil shale and a residue of shale-oil distillation.

There is *sufficient evidence* that shale-oils are carcinogenic in humans.

5. References

Aarna, A. (1978) *Chemical Engineering in the Estonian SSR*, Tallinn, Perioodika, pp. 29-34

Ahlborg, G. (1951) Hydrogen sulfide poisoning in shale oil industry. *Am. med. Assoc. ind. Hyg. occup. Med.*, 3, 247-266

Akkerberg, I.I. (1980) Hygienic characteristics of working conditions in shale oil and gas production plants. In: Akkerberg, I.I., Feoktistov, G.S. & Yankes, K.Y., eds, *Industrial Hygiene and Occupational Pathology in the Estonian SSR*, Vol. 10, *Health and Safety in Oil Shale Extraction and Processing* (*NIH Library Translation NIH 80-467*), Tallinn, Valgus, pp. 127-131

Akkerberg, I.I., Komissarova, V.V. & Tshernyshova, A.I. (1976) *Benz(a)pyrene content in the air of the chamber oven shop and the electrode coke device*. In: Akkerberg, I.I., ed., *Industrial Hygiene and Occupational Pathology in the Estonian SSR*, Vol. 9, Tallinn, Valgus, pp. 18-21

Akkerberg, I.I., Feoktistov, G.S. & Yankes, K.Y., eds (1980) *Industrial Hygiene and Occupational Pathology in the Estonian SSR*, Vol. 10, *Health and Safety in Oil Shale Extraction and Processing* (*NIH Library Translation NIH 80-467*), Tallinn, Valgus

Allred, V.D., ed. (1982) *Oil Shale Processing Technology*, East Brunswick, NJ, The Center for Professional Advancement

Axelson, O. & Edling, C. (1980) Health hazards from radon daughters in dwellings in Sweden. In: Rom, W.N. & Archer, V.E., eds, *Health Implications of New Energy Technologies*, Ann Arbor, MI, Ann Arbor Science, pp. 79-87

Baker, J.D. & Hook, C.D. (1979) Chinese and Estonian oil shale. In: *Proceedings of 12th Oil Shale Symposium*, Golden, CO, Colorado School of Mines, pp. 26-31

---

[1]For definitions of the italicized terms, see Preamble pp. 15-16 and 19.

Barkley, W., Warshawsky, D. & Radike, M. (1979) *Toxicology and carcinogenicity of oil shale products*. In: White, O., Jr, ed., *Proceedings of a Symposium on Assessing the Industrial Hygiene Monitoring Needs for the Coal Conversion and Oil Shale Industries*, Upton, NY, Brookhaven National Laboratory Associated Universities, Inc., pp. 79-95

Barnhart, B.J. & Cox, S.H. (1980) Mutation of Chinese hamster cells by near-UV activation of promutagens. *Mutat. Res.*, 72, 135-142

Bartick, H., Kunchal, K., Switzer, D., Bowen, R. & Edwards, R. (1975) *Final Report: The Production and Refining of Crude Shale Oil into Military Fuels* (Contract N00014-74-C-0055), Arlington, VA, Office of Naval Research, p. 5-3

Bell, J. (1876) Paraffin epithelioma of the scrotum. *Edinb. med. J.*, 22, 135-137

Berenblum, I. & Schoental, R. (1943) Carcinogenic constituents of shale oil. *Br. J. exp. Pathol.*, 24, 232-239

Berenblum, I. & Schoental, R. (1944) The difference in carcinogenicity between shale oil and shale. *Br. J. exp. Pathol.*, 25, 95-96

Bingham, E. & Barkley, W. (1979) Bioassay of complex mixtures derived from fossil fuels. *Environ. Health Perspect.*, 30, 157-163

Bogovski, P.A. (1954) On the action of oil-shale fuel oil in chronic experiments (Russ.). *Farm. Toksikol.*, 17, 56-59

Bogovski, P.A. (1955) On the blastomogenic action of the oil shale wood impregnating oil. *Eesti NSV Tead. Akad. Toimetised*, 4, 488-494

Bogovski, P.A. (1958) *On the blastomogenic action of oil-shale generator tar* (Russ.). In: Bogovski, P.A., ed., *Problems of Industrial Hygiene in the Shale-Oil Industry in the Estonian SSR*, Vol. 3, Tallinn, Valgus, pp. 172-185

Bogovski, P.A. (1961) *Carcinogenic Action of Estonian Oil Shale Processing Products* (Russ.), Tallinn, Academy of Science Estonian SSR, pp. 108-115, 138-148, 154-161, 189, 192-204

Bogovski, P.A. (1962) On the carcinogenic effect of some 3,4-benzopyrene-free and 3,4-benzopyrene-containing fractions of Estonian shale-oil. *Acta unio int. contra cancrum*, 18, 37-39

Bogovski, P.A. (1981) Historical perspectives of occupational cancer. In: Vainio, H., Sorsa, M. & Hemminki, K., eds, *Occupational Cancer and Carcinogenesis*, Washington DC, Hemisphere Publishing Corp., pp. 921-939

Bogovski, P.A. & Vinkmann, F. (1979) Carcinogenicity of oil shale tars, some of their components, and commercial products. *Environ. Health Perspect.*, 30, 165-169

Bogovski, P.A., Gortalum, G.M. & Kozhevnikov, A.V. (1963) Decancerigenization of some shale processing products. *Acta unio int. contra cancrum*, 19, 481-482

Bridge, J.C. & Henry, S.A. (1928) Industrial cancers. In: *Report of the International Conference on Cancer*, London, Fowler Wright Ltd, pp. 258-268

Brooks, A.L., Hanson, R. & Sanchez, A. (1980) Biological and chemical characterization of baghouse fly ash from a fluidized bed combustor burning oil shale (Abstract no. Ac12). *Environ. Mutagenesis*, *2*, 243

Calkins, W.H., Deye, J.F., Hartgrove, R.W., King, C.F. & Krahn, D.F. (1983) Mutagenesis and skin tumour initiation by shale and coal-derived oils and their distillation fractions. *Fuel*, *62*, 857-864

Chen, D.J.-C. & Strniste, G.F. (1982) Cytotoxic and mutagenic properties of shale oil byproducts. II. Comparison of mutagenic effects at five genetic markers induced by retort process water plus near ultraviolet light in Chinese hamster ovary cells. *Environ. Mutagenesis*, *4*, 457-467

Chen, D.J., Deaven, L.L., Meyne, J., Okinaka, R.T. & Strniste, G.F. (1983) Determination of direct-acting mutagens and clastogens in oil shale retort process water. In: Waters, M.D., Sandhu, S.S., Lewtas, J., Claxton, L., Chernoff, N. & Nesnow, S., eds, *Short-Term Bioassays in the Analysis of Complex Environmental Mixtures III*, New York, Plenum Publishing Co., pp. 269-275

Chevron Oil Company (1978) *Refining and Upgrading of Synfuels from Coal and Oil Shales by Advanced Catalytic Processes (Report HCP/T2315-25)*. Prepared under Contract No. EF-76-C-01-2315, Washington DC, US Department of Energy

Conaway, C.C., Brusick, D.J., Mecler, F.J., Holdsworth, C.E. & Call, R.W. (1984) Mutagenesis and teratogenesis studies of shale oil. *J. Am. Coll. Toxicol.* (in press)

Costello, J. (1979) Morbidity and mortality study of shale oil workers in the United States. *Environ. Health Perspect.*, *30*, 205-208

Costello, J. (1983) *NIOSH studies of oil shale workers*. In: Wagner, W.L., Rom, W.N. & Merchant, J.A., eds, *Health Issues Related to Metal and Nonmetallic Mining*, Boston, Butterworth Publishers, pp. 497-507

Crowley, R.J., Siggia, S. & Uden, P.C. (1980) Class separation and characterization of shale oil by liquid chromatography and capillary column gas chromatography. *Anal. Chem.*, *52*, 1224-1228

Decora, A. & Freitas, I. (1980) *Report of the technical panel on oil shale and tar sands*. In: *United Nations Conference on New and Renewable Sources of Energy, New York, 7-11 January 1980*, New York, United Nations

Dickson, P.F. (1981) Oil shale. In: Kirk, R.E. & Othmer, D.F., eds, *Kirk-Othmer Encyclopedia of Chemical Technology*, 3rd ed., Vol. 16, New York, John Wiley & Sons, pp. 333-357

Doig, A.T. (1970) Epithelioma of the scrotum in Scotland in 1967. *Health Bull.*, *28*, 45-51

Duir, J.H., Griswold, C.F. & Christolini, B.A. (1983) Oil shale retorting technology. *Chem. Eng. Progr.*, February, 45-50

Easley, J.R., Holland, J.M., Gipson, L.C. & Whitaker, M.J. (1982) Renal toxicity of middle distillates of shale oil and petroleum in mice. *Toxicol. appl. Pharmacol.*, *65*, 84-91

Epler, J.L., Young, J.A., Hardigree, A.A., Rao, T.K., Guerin, M.R., Rubin, I.B., Ho, C.-H. & Clark, B.R. (1978) Analytical and biological analyses of test materials from the synthetic fuel technologies. I. Mutagenicity of crude oils determined by the *Salmonella typhimurium*/microsomal activation system. *Mutat. Res.*, *57*, 265-276

Epler, J.L., Clark, B.R., Ho, C.-H, Guerin, M.R. & Rao, T.K. (1979a) Short-term bioassay of complex organic mixtures: Part II, Mutagenicity testing. *Environ. Sci. Res.*, *15*, 269-289

Epler, J.L., Rao, T.K. & Guerin, M.R. (1979b) Evaluation of feasibility of mutagenic testing of shale oil products and effluents. *Environ. Health Perspect.*, *30*, 179-184

Fenton, D.M., Hennig, H. & Richardson, R.L. (1981) *The chemistry of shale oil and its refined products*. In: Stauffer, H.C., ed., *Oil Shale, Tar Sands, and Related Materials (ACS Symposium Series No. 163)*, Washington DC, American Chemical Society, pp. 315-325

Feoktistov, G.S. (1972) *The ailment of the workers in oil-shale mines with the disability to work* (Russ.). In: *Industrial Hygiene and Occupational Pathology in the Estonian SSR*, Vol. 2, Tallinn, Valgus, pp. 11-15

Fruchter, J.S., Wilkerson, C.L., Evans, J.C. & Sanders, R.W. (1980) Elemental partitioning in an aboveground oil shale retort pilot plant. *Environ. Sci. Technol.*, *14*, 1374-1381

Futrell, J.H. (1980) *Chemical characteristics of oil shale and shale oil*. In: Rom, W.N. & Archer, V.E., eds, *Health Implications of New Energy Technologies*, Ann Arbor, MI, Ann Arbor Science., pp. 427-443

Garcia, L.L., Schulte, H.F. & Ettinger, H.J. (1981) Industrial hygiene study at the Anvil Points oil shale facility. *Am. ind. Hyg. Assoc. J.*, *42*, 796-804

Gomer, C.J. & Smith, D.M. (1980) Acute skin phototoxicity in hairless mice following exposure to crude shale oil or natural petroleum oil. *Toxicology*, *18*, 75-85

Gonzales, M., Garcia, L.L., Vigil, F.A., Roger, G.W., Tillery, M.I. & Ettinger, H.J. (1982) *Industrial Hygiene Sampling at Rio Blanco Oil Shale Facility (Report LA-9152-MS)*, Los Alamos, NM, Los Alamos National Laboratory

Gortalum, G.M. (1955) *On the toxic properties of the Estonian oil shale generator tar* (Russ.). In: Bogovski, P.A., ed., *Problems of Industrial Hygiene in the Shale-Oil Industry in the Estonian SSR*, Tallinn, Valgus, pp. 198-208

Gregg, C.T., Tietjen, G. & Hutson, J.Y. (1981) Prenatal toxicology of shale oil retort water in mice. *J. Toxicol. environ. Health*, *8*, 795-804

Griest, W.H., Guerin, M.R., Clark, B.R., Ho, C.-H., Rubin, I.B. & Jones, A.R. (1979) *Relative chemical composition of selected synthetic crudes*. In: White, O., Jr, ed., *Proceedings of the Symposium on Assessing the Industrial Hygiene Monitoring Needs for the Coal Conversion and Oil Shale Industries*, Upton, NY, Brookhaven National Laboratory, pp. 61-78

Guerin, M.R., Epler, J.L., Griest, W.H., Clark, B.R. & Rao, T.K. (1978) *Polycyclic aromatic hydrocarbons from fossil fuel conversion processes*. In: Jones, P.W. & Freudenthal,

R.I., eds, *Carcinogenesis*, Vol. 3, *Polynuclear Aromatic Hydrocarbons*, New York, Raven Press, pp. 21-33

Guerin, M.R., Ho, C.-H., Rao, T.K., Clark, B.R. & Epler, J.L. (1980) Polycyclic aromatic primary amines as determinant chemical mutagens in petroleum substitutes. *Environ. Res., 23*, 42-53

Guerin, M.R., Rubin, I.B., Rao, T.K., Clark, B.R. & Epler, J.L. (1981) Distribution of mutagenic activity in petroleum and petroleum substitutes. *Fuel, 60*, 282-288

Hargis, K.M. & Jackson, J.O. (1983) *Industrial hygiene aspects of underground oil shale mining*. In: Wagner, W.L., Rom, W.N. & Merchant, J.A., eds, *Health Issues Related to Metal and Nonmetallic Mining*, Woburn, MA, Butterworth Publishers, pp. 463-484

Hargis, K.M., Rom, W.N., Grier, R.S., Tillery, M.I., Voelz, G.L., Ettinger, H.J. & Wheat, L.D. (1983a) *Occupational Health Study at the Geokinetics True in situ Oil Shale Retorting Facility (Report LA-9659-MS)*, Los Alamos, NM, Los Alamos National Laboratory

Hargis, K.M., Tillery, M.I., Gonzales, M. & Garcia, L.L. (1983b) *Aerosol sampling and characterization in the developing US oil shale industry*. In: Marple, V.A. & Liu, B.Y.H., eds, *Aerosols in the Mining and Industrial Work Environments*, Vol. 2, *Characterization*, Ann Arbor, MI, Ann Arbor Science, pp. 481-500

Heller, I. (1930) Occupational cancers. *J. ind. Hyg., 12*, 169-196

Henry, S.A. (1937) The study of fatal cases of cancer of the scrotum from 1911 to 1935 in relation to occupation, with special reference to chimney sweeping and cotton mule spinning. *Am. J. Cancer, 31*, 28-57

Henry, S.A. (1946) *Cancer of the Scrotum in Relation to Occupation*, London, Oxford University Press, p. 27

Henry, S.A. (1947) Occupational cutaneous cancer attributable to certain chemicals in industry. *Br. med. Bull., iv*, 389-401

Hermann, M., Chaudé, O., Weill, N., Bedouelle, H. & Hofnung, M. (1980) Adaptation of the Salmonella/mammalian microsome test to the determination of the mutagenic properties of mineral oils. *Mutat. Res., 77*, 327-339

Hinchee, R.E., Adams, V.D., Curtis, J.G. & Seierstad, A.J. (1983) *Evaluation of the Potential for Groundwater Transport of Mutagenic Compounds Released by Spent Oil Shale (Water Quality Ser. UWRL/Q-83/06)*, Logan, UT, Utah Water Research Laboratory

Holland, J.M., Rahn, R.O., Smith, L.H., Clark, B.R., Chang, S.S. & Stephens, T.J. (1979) Skin carcinogenicity of synthetic and natural petroleums. *J. occup. Med., 21*, 614-618

Holland, L.M., Wilson, J.S. & Foreman, M.E. (1981) *Comparative dermotoxicity of shale oils*. In: Griest, W.H., Guerin, M.R. & Coffin, D.L., eds, *Health Effects Investigation of Oil Shale Development*, Ann Arbor, MI, Ann Arbor Science, pp. 117-122

Holland, L.M., Gonzales, M., Wilson, J.S. & Tillery, M.I. (1983) *Pulmonary effects of shale dusts in experimental animals*. In: Wagner, W.L., Rom, W.N. & Merchant, J.A., *Health Issues Related to Metal and Nonmetallic Mining*, London, Butterworth, pp. 485-496

Hueper, W.C. (1942) *Occupational Tumors and Allied Diseases*, Springfield, IL, C.C. Thomas, p. 145

Hueper, W.C. (1953) Experimental studies on cancerigenesis of synthetic liquid fuels and petroleum substitutes. *Arch. ind. Hyg. occup. Med.*, *8*, 307-327

Hueper, W.C. & Cahnmann, H.J. (1958) Carcinogenic bioassay of benzo(a)pyrene-free fractions of American shale oils. *Arch. Pathol.*, *65*, 608-614

Hunter, D. (1975) *The Diseases of Occupations*, 5th ed., Aylesbury, UK, Hazell Watson & Viney, Ltd, pp. 783-784

Kahn, H. (1979) Toxicity of oil shale chemical products. *Scand. J. Work Environ. Health*, *5*, 1-9

Kennaway, E.L. (1925) The anatomical distribution of the occupational cancers. *J. ind. Hyg.*, *7*, 69-93

Kogerman, P.N. (1938) *Estonian Shale Oils. Science of Petroleum*, Vol. IV, London, Oxford University Press, pp. 3107-3112

Kozhevnikov, A.V. (1947) *Oil Shales. I. Origin, Occurrence and Chemical Composition*, Tartu State Publishing House 'Scientific Literature', pp. 8-11

Küng, V.A. (1979) Morphological investigations of fibrogenic action of Estonian oil shale dust. *Environ. Health Perspect.*, *30*, 153-156

Larionov, L.T. (1947) On the carcinogenic properties of Estonian shale tars. In: Lazarev, N.V., ed., *Materials on the Toxicology of Oil Shale Products. Proceedings of the Industrial Hygiene and Occupational Diseases Research Institute*, Vol. 11, Part I, Leningrad, pp. 111-118

Larionov, L.T., Soboleva, N.G. & Shabad, L.M. (1934) On the cancerogenic effect of certain shale tars (Russ.). *Vestn. Rentgenol. Radiol.*, *13*, 131-143

Leitch, A. (1922) Paraffin cancer and its experimental production. *Br. med. J.*, *2*, 1104-1106

Lekas, J.M. (1982) *The geokinetics horizontal in situ retorting process*. In: Feingold, B.W., ed., *Synthetic Fuels from Oil Shale II*, Chicago, IL, Institute of Gas Technology, pp. 251-259

Lockard, J.M., Prater, J.W., Viau, C.J., Enoch, H.G. & Sabharwal, P.S. (1982) Comparative study of the genotoxic properties of Eastern and Western U.S. shale oils, crude petroleum, and coal-derived oil. *Mutat. Res.*, *102*, 221-235

Logan, W.P.D. (1982) *Cancer Mortality by Occupation and Social Class 1851-1971 (IARC Scientific Publications No. 36/Studies on Medical Population Subjects No. 44)*, Lyon/London, International Agency for Research on Cancer/Her Majesty's Stationery Office

Loogna, N.A. & Hering, L.H. (1972) *An examination of the importance of allergy in occupation dermatoses of workers contacted with shale oil (Russ.).* In: *Industrial Hygiene and Occupational Pathology in the Estonian SSR*, Vol. 2, Tallinn, Valgus, pp. 89-95

Luts, A.E. (1972) *An analysis of the illness causing temporary disability in workers in the case of some acute diseases of the respiratory organs in the oil-shale mines of Estonian SSR during four years (1966-1969)* (Russ.). In: *Industrial Hygiene and Occupational Pathology in the Estonian SSR*, Vol. 2, Tallinn, Valgus, pp. 16-24

Ma, C.Y., Ho, C.-H., Quincy, R.B., Guerin, M.R., Rao, T.K., Allen, B.E. & Epler, J.L. (1983) Preparation of oils for bacterial mutagenicity testing. *Mutat. Res.*, 118, 15-24

MacFarland, H.N., Coate, W.B., Disbennett, D.B. & Ackerman, L.J. (1982) Long-term inhalation studies with raw and processed shale dusts. *Ann. occup. Hyg.*, 26, 213-226

Maripuu, I.P. (1972) *The prevalence of chronic bronchitis in workers of mechanized mines and of the oil-shale plant* (Russ.). In: *Industrial Hygiene and Occupational Pathology in the Estonian SSR*, Vol. 2, Tallinn, Valgus, pp. 25-32

Meyne, J. & Deaven, L.L. (1982a) Cytogenetic effects of shale-derived oils in murine bone marrow. *Environ. Mutagenesis*, 4, 639-645

Meyne, J. & Deaven, L.L. (1982b) In vivo cytogenetic effects of oil shale retort process waters. *Toxicology*, 24, 223-229

Nelson, K.F., North, D.S., Payne, G.R., Anderson, A.D., Poulson, R.E. & Farrier, D.S. (1978) The effect of an in situ-produced oil shale processing water on metabolism. *Arch. environ. Contam. Toxicol.*, 7, 273-281

Office of Technology Assessment (1980) *An Assessment of Oil Shale Technologies*, Chapter 4, Washington DC, US Government Printing Office

Öpik, I. (1983) Major industrial and demonstration retorts for thermal treatment of oil shale (Russ.). *Eesti NSV Tead. Akad. Toim. Kemia*, 32, 81-97

Öpik, I. & Kaganovich, I. (1981) *Oil shale of the Baltic basin: Power engineering and thermal processing*. In: *Proceedings of the 6th IIASA Resources Conference, World Oil-Shale Resources and Their Potential Development, Golden, CO, 1981*, Laxenburg, Austria, International Institute for Applied Systems Analysis

Parakhonskiy, E.V. (1980) *Extraction of oil shales*. In: Akkerberg, I.I., Feoktistov, G.S. & Yankes, K.Y., eds, *Industrial Hygiene and Occupational Pathology in the Estonian SSR, Vol. 10, Health and Safety in Oil Shale Extraction and Processing* (NIH Library Translation NIH-80-467), Tallin, Valgus, pp. 16-40

Parker, G.A., Bogo, V. & Young, R.W. (1981) Acute toxicity of conventional versus shale-derived JP5 jet fuel: Light microscopic, hematologic, and serum chemistry studies. *Toxicol. appl. Pharmacol.*, 57, 302-317

Pelroy, R.A. & Petersen, M.R. (1979) Use of the Ames test in evaluation of shale oil fractions. *Environ. Health Perspect.*, 30, 191-203

Pelroy, R.A., Sklarew, D.S. & Downey, S.P. (1981) Comparison of mutagenicities of fossil fuels. *Mutat. Res.*, 90, 233-245

Peterson, E.J. & Spall, W.D. (1983) *Physical and Chemical Characterization and Comparison of Solids, Liquids, and Oils from Estonian and Green River Formation Shales (Report LA-9722-MS)*, Los Alamos, NM, Los Alamos National Laboratory

Piper, E.M. (1981) *State of the art: Oil-shale process technology*. In: *Proceedings of the 6th IIASA Resources Conference, World Oil-Shale Resources and Their Potential Development, Golden, CO, 1981*, Laxenburg, Austria, International Institute for Applied Systems Analysis

Purde, M. & Etlin, S. (1980) *Cancer cases among workers in the Estonian oil shale processing industry*. In: Rom, W.N. & Archer, V.E., eds, *Health Implications of New Energy Technologies*, Ann Arbor, MI, Ann Arbor Science, pp. 527-528

Qian, J.L. (1982) *Oil shale industry of China*. In: *Synfuels' 2nd World-Wide Symposium*, New York, McGraw-Hill, pp. 1-12

Rao, T.K., Allen, B.E., Ramey, D.W., Epler, J.L., Rubin, I.B., Guerin, M.R. & Clark, B.R. (1981a) Analytical and biological analyses of test materials from the synthetic fuel technologies. III. Use of Sephadex LH-20 gel chromatography technique for the bioassay of crude synthetic fuels. *Mutat. Res.*, 85, 29-39

Rao, T.K., Epler, J.L., Guerin, M.R., Clark, B.R. & Ho, C.-H. (1981b) Mutagenicity of nitrogen compounds from synthetic crude oils: Collection, separation and biological testing. *Environ. Sci. Res.*, 22, 243-251

Rao, T.K., Epler, J.L., Guerin, M.R. & Clark, B.R. (1981c) *Short-term microbial testing of shale oil materials*. In: Griest, W.H., Guerin, M.R. & Coffin, D.L., eds, *Health Effects Investigation of Oil Shale Development*, Ann Arbor, MI, Ann Arbor Science, pp. 161-172

Rebane, K.K. & Öpik, I.P. (1982) *The Problem of Complex Utilization of Oil Shales*, Moscow, Vestnik Akademii Nauk SSSR, pp. 19-24

Renne, R.A., Lund, J.E., McDonald, K.E. & Smith, L.G. (1980) *Morphologic effects of intratracheally administered oil shale in rats*. In: Rom, W.N. & Archer, V.E., eds, *Health Implications of New Energy Technologies*, Ann Arbor, MI, Ann Arbor Science Publishers Inc., pp. 501-514

Rooks, I.K. (1980) *Production of shale tars and gas. Technology of producing shale tars and gas* (Russ.). In: Akkerberg, I.I., Feoktistov, G.S. & Yankes, K.Y., eds, *Industrial Hygiene and Occupational Pathology in the Estonian SSR*, Vol. 10, *Health and Safety in Oil Shale Extraction and Processing (NIH Translation Library NIH-80-467)*, Tallinn, Valgus, pp. 93-104

Rowland, J., Shubik, P., Wallcave, L. & Sellakumar, A. (1980) Carcinogenic bioassay of oil shale: Long-term percutaneous application in mice and intratracheal instillation in hamsters. *Toxicol. appl. Pharmacol.*, 55, 522-534

Rudnick, J. & Voelz, G.L. (1980) *An occupational health study of Paraho oil shale workers*. In: Rom, W.N. & Archer, V.E., eds, *Health Implications of New Energy Technologies*, Ann Arbor, MI, Ann Arbor Science, pp. 491-499

Russell, P.L. (1980) *History of Western Oil Shale*, East Brunswick, NJ, The Center for Professional Advancement, p. 3

Scott, A. (1922a) The occupation dermatoses of the paraffin workers of the Scottish shale oil industry. *Br. med. J.*, *ii*, 381-385

Scott, A. (1922b) On the occupation cancer of the paraffin and oil workers of the Scottish shale oil industry. *Br. med. J.*, *ii*, 1108-1109

Seaton, A., Lamb, D., Rhind-Brown, W., Sclare, G. & Middleton, W.G. (1981) Pneumoconiosis of shale miners. *Thorax*, *36*, 412-418

Selby, C., Calkins, J. & Enoch, H. (1983) Detection of photomutagens in natural and synthetic fuels. *Mutat. Res.*, *124*, 53-60

Sell, G. (1951) *Oil Shale and Cannel Coal*, Vol. 2, London, Institute of Petroleum, p. 368

Serebryannikov, N.D. (1970) *Production of domestic gas from shales*. In: *Proceedings of a United Nations Symposium on the Development and Utilization of Oil Shale Resources, 26 August - 4 September 1968*, Tallinn, Valgus, pp. 327-335

Shih, C.C., Cotter, J.E., Prien, C.H. & Nevens, T.D. (1979) *Technological Overview Reports for Eight Shale Oil Recovery Processes (EPA-600/7-79-075)*, Cincinnati, OH, US Environmental Protection Agency

Shmidt, N.A., Feoktistov, G.S., Kyung, V.A. & Karpunin, B.I. (1980) *Toxicological-hygienic characteristics of shale dust and explosive gases*. In: Akkerberg, I.I., Feoktistov, G.S. & Yankes, K.Y., eds, *Industrial Hygiene and Occupational Pathology in the Estonian SSR*, Vol. 10, *Health and Safety in Oil Shale Extraction and Processing (NIH Translation Library NIH-80-467)*, Tallinn, Estonian SSR, Valgus, pp. 53-76

Smith, G.H., Grant, G. & Allen, S. (1938) Scottish shale oil. In: *Science of Petroleum*, Vol. IV, London, Oxford University Press, pp. 3096-3106

Smith, L.H. & Witschi, H.P. (1983) *The Mouse Lung Tumor Assay: A Final Report (Oak Ridge natl Lab. 5961)*, Oak Ridge, TN

Smith, L.H., Haschek, W.M. & Witschi, H. (1981) *Acute toxicity of selected crude and refined shale oil - and petroleum-derived substances*. In: Griest, W.H., Guerin, M.R. & Coffin, D.L., eds, *Health Effects Investigation of Oil Shale Development*, Ann Arbor, MI, Ann Arbor Science, pp. 141-160

Smith, W.E. (1952) Survey of some current British and European studies of occupational tumor problems. *Arch. ind. Hyg. occup. Med.*, *5*, 242-263

Smith, W.E., Sunderland, D.A. & Sugiura, K. (1951) Experimental analysis of the carcinogenic activity of certain petroleum products. *Arch. ind. Hyg. occup. Med.*, *4*, 299-314

Soboleva, N.G. (1936) Further studies on the carcinogenicity of shale tars (oil shales of Gdov) (Russ.). *Vestn. Rentgenol. Radiol.*, *16*, 229-236

Southam, A.H. & Wilson, S.R. (1922) Cancer of the scrotum: The etiology, clinical features, and treatment of the disease. *Br. med. J.*, *ii*, 971-973

Spall, W.D. (1984) Determination of polynuclear aromatic hydrocarbons and other selected compounds in oil shale materials. *Anal. Chem.* (in press)

Stewart, D. (1970) *The history, technology and economics of shale oil in the United Kingdom*. In: *Proceedings of a United Nations Symposium on the Development and Utilization of Oil Shale Resources, 26 August - 4 September 1968*, Tallinn, Valgus, pp. 248-260

Strniste, G.F. (1984) *Light activation of genotoxic components in natural and synthetic crude oils*. In: Cowser, K.E. & Richmond, C.R., eds, *Synthetic Fossil Fuel Technologies*, Stoneham, MA, Butterworth (in press)

Strniste, G.F. & Chen, D.J. (1981) Cytotoxic and mutagenic properties of shale oil byproducts. I. Activation of retort process waters with near ultraviolet light. *Environ. Mutagenesis*, 3, 221-231

Strniste, G.F., Bingham, J.M., Okinaka, R.T. & Chen, D.J. (1982a) Genotoxicity induced in cultured Chinese hamster cells exposed to natural or synthetic crude oils and near ultraviolet light. *Toxicol. Lett.*, 13, 163-167

Strniste, G.F., Chen, D.J. & Okinaka, R.T. (1982b) Genotoxic effects of sunlight-activated waste water in cultured mammalian cells. *J. natl Cancer Inst.*, 69, 199-203

Strniste, G.F., Bingham, J.M., Spall, W.D., Nickols, J.W., Okinaka, R.T. & Chen, J.-C. (1983) *Fractionation of an oil shale retort process water: Isolation of photoactive genotoxic components*. In: Waters, M.D., Sandhu, S.S., Lewtas, J., Claxton, L., Chernoff, N. & Nesnow, S., eds, *Short-Term Bioassays in the Analysis of Complex Environmental Mixtures III*, New York, Plenum, pp. 139-151

Sun, Q.-F., Zhang, Z.-Y., Tong, L.-W., Li, S.-Z. & Liu, K. (1961) The carcinogenic action of the generator tar obtained from the oil shale of Fushun. *Eesti NSV Tead. Akad. Toim.*, 10, 296-301

Timourian, H., Carrano, A., Carver, J., Felton, J.S., Hatch, F.T., Stuermer, D.S. & Thompson, L.H. (1981) *Comparative mammalian genetic toxicology of shale oil products assayed in vitro and in vivo*. In: Griest, W.H., Guerin, M.R. & Coffin, D.L., eds, *Health Effects Investigation of Oil Shale Development*, Ann Arbor, MI, Ann Arbor Science, pp. 173-187

Toste, A.P., Sklarew, D.S. & Pelroy, R.A. (1982) Partition chromatography-high-performance liquid chromatography facilitates the organic analysis and biotesting of synfuels. *J. Chromatogr.*, 249, 267-282

Turu, H. (1961) On the blastomogenicity of tar obtained by processing of granular oil shale in a solid heat carrier installation. *Eesti NSV Tead. Akad. Toim.*, 10, 13-18

Twort, C.C. & Ing, H.R. (1928) Studies on carcinogenic agents (Ger.). *Z. Krebsforsch.*, 27, 308-351

Tyagunov, B.I., Panov, V.I. & Stelmakh, G.P. (1980) Energo-technological use of combustible shales. *Energ. stroit.*, 4, 28-30

Union Oil Company of California (1982) *Premanufacture Notice for Full Range, Dewaxed, Dearsenited Shale Oil (Syncrude) and Other Selected Oil Shale-Derived Process Streams and Products*, Vol. 1, Brea, CA

Vahter, H. (1959) *Experimental Data on the Morphogenesis of Tumours Induced by Shale Oils* (Russ.), Tartu State Publishing House 'Scientific Literature', No. 79, pp. 99-105

Vinkmann, F.Y. (1972a) *On the results of a comparative investigation on the cancerogenic action of several oil shale tars in $CC_{57}Br$ mice* (Russ.). In: Loogna, G., ed., *Experimental and Clinical Oncology*, Tallinn, Valgus, pp. 243-250

Vinkmann, F.Y. (1972b) *The carcinogenic activity of the oil-shale pyrolytic laquer LSP-1 and its components* (Russ.). In: Akkerberg, I.I., *Industrial Hygiene and Occupational Pathology in the Estonian SSR*, Vol. 2, Tallinn, Valgus, pp. 175-181

Vinkmann, F.Y. & Mirme, H.Y. (1975) *Studies on the eventual cancerogenic activities of some commercial products of the Estonian oil-shale industry (Epo-glue and Tolichthon)*. In: Loogna, G., ed., *Experimental and Clinical Oncology*, Vol. 2, Tallinn, Estonian SSR, Valgus, pp. 76-80

Weaver, N.K. & Gibson, R.L. (1979) The U.S. oil shale industry: A health perspective. *Am. ind. Hyg. Assoc. J.*, *40*, 460-467

GLOSSARY

ALUM SHALE. Swedish OIL SHALE (or shale-oil)

BENEFICIATION. Improvement in the quality of raw OIL SHALE by selective removal of contaminating material

CATALYTIC DEWAXING. A selective HYDROCRACKING process in which the catalyst is chosen to favour the specific destruction of long straight-chain paraffins (waxes), converting them to hydrocarbons of lower molecular weight. The primary purpose is to improve the flow properties of the processed oil. In addition, almost all the nitrogen, sulphur and oxygen is removed, and there is some cracking of other oil components to produce light products similar to those from other HYDROCRACKING processes.

CHAMBER-OVEN TAR. A term used for a high-temperature CRUDE SHALE-OIL produced from chamber-oven batteries in the USSR

COKING. A severe thermal process used in oil refining to crack heavy oil (usually vacuum residue) to produce distillates and gas oils useful in downstream operations. Coke is yielded as a by-product.

COMBUSTED SPENT SHALE. The crushed rock that remains when OIL SHALE is heated by a process which includes combustion of KEROGEN residues as a direct source of retort heat or when RETORTED OIL SHALE is subjected to combustion to recover residual fuel value. It usually contains virtually no carbonaceous residue.

CRUDE SHALE-OIL. The complex combination of hydrocarbons obtained by the thermal decomposition (at 400°C or higher) of KEROGEN. It consists of hydrocarbons and heterocompounds containing nitrogen, sulphur and oxygen. This term refers to the oil as obtained from the oil-shale retort and prior to any chemical or physical treatment other than water separation. Sometimes referred to as RAW SHALE-OIL. In the USSR, this material may be termed 'tar' if obtained by high-temperature RETORTING.

DIRECT HEATING. Heat is supplied to the retort from combustion within it.

FISCHER ASSAY. A laboratory procedure used to assess the recoverable oil content of an OIL SHALE and the distribution of certain products from RETORTING. The equipment used is referred to as a Fischer assay retort or FISCHER RETORT.

FISCHER RETORT. See FISCHER ASSAY.

FLUXING OILS. Obtained by mixing naphthalene oil, wash oil and light anthracene oil; used for extending the workability of road binders

GAS GENERATOR (RETORT). A directly-heated, low-temperature retort which produces gas and shale-oil as useful products. RETORTING is carried out at temperatures in the range of 450-550°C.

GENERATOR OIL. The CRUDE SHALE-OIL produced as one of the products of the GAS GENERATOR type of retort

HYDROCRACKING. A process in which an oil is reacted in the presence of hydrogen and a catalyst for the primary purpose of converting the higher-boiling portion to lower-boiling hydrocarbons, thus increasing the yield of blend stocks needed to formulate gasolines and distillate fuels. Nitrogen, sulphur and oxygen contents are also substantially reduced, and the process favours the production of saturated hydrocarbons.

HYDROTREATING. A process in which an oil is reacted with hydrogen in the presence of a catalyst for the primary purpose of reducing the nitrogen, sulphur and oxygen content of the oil. Some unsaturated aliphatic and aromatic hydrocarbons may also be converted to their saturated analogues to varying degrees, depending on process conditions and catalyst selection.

ICHTHYOL. An ointment made by sulphonating shale-oil and neutralizing with ammonia; used in the treatment of skin disorders in human and veterinary medicine. A similar material, Tolichthtol, is manufactured in Estonia, USSR.

INDIRECT HEATING. Heat is supplied by combustion external to the retort and transferred to it through the vessel walls or by a fluid or solid material which circulates between the retort and the combustion zone.

KEROGEN. An organic matrix of high molecular weight which is converted to shale-oil on heating. It is generally found in sedimentary deposits which originated from precipitation of minerals and algae.

KUKERSITE. An OIL SHALE found in the USSR which has a high content of inorganic matter of low fusion temperature. This can lead to severe agglomeration problems during RETORTING. The name is derived from the Estonian village of Kukruse.

LETC (Laramie Energy Technology Center) RETORT. A pilot plant-scale simulation of a directly-heated in-situ retort

NTU (Nevada-Texas-Utah) RETORT. A batch-type, above-ground, directly-heated retort used to produce gas of low heating value and shale-oil

OIL SHALE. A complex mineral containing organic material (KEROGEN)

OIL-SHALE ASH. See COMBUSTED SPENT SHALE.

OIL-SHALE RETORT PROCESS WATERS. Waste-waters obtained from the process of RETORTING, which have therefore been in direct contact with the materials being processed

PRE-TREATMENT (PARTICULATE AND ARSENIC REMOVAL). A process that substantially reduces the content of paticulates, arsenic and other metals in shale-oil. It may be a separate process step, or may be in sequence with a HYDROTREATING, HYDROCRACKING or catalytic cracking process, where it is essential to protect catalysts that are sensitive to metal poisoning. For example, current methods include absorption of the metals onto a solid with or without the presence of hydrogen, solvent extraction or acid extraction.

RAW OIL SHALE. A sedimentary rock containing mainly mineral components and organic matter called KEROGEN, which has a low solubility in low-boiling organic solvents, but produces liquid organic products (CRUDE SHALE-OIL) by thermal decomposition

RAW SHALE-OIL. See CRUDE SHALE-OIL.

REFINING. The processing of CRUDE or upgraded SHALE-OIL to yield finished products, such as gasoline, diesel fuel, fuel oil, etc.

REFORMING. As used in this document, refers to the treatment of hydrocarbon gases with steam gas over a solid catalyst at high temperature to manufacture hydrogen

RETORTED OIL (SPENT) SHALE. The crushed rock that remains when OIL SHALE is heated without combustion to remove substantially the organic portion. It usually contains a carbonaceous residue.

RETORTING. A thermal process applied to OIL SHALE in which the rock is heated to decompose its KEROGEN component to yield shale-oil. The retorting is carried out either *in situ*, using the surrounding earth for containment, or in above-ground vessels specifically constructed for the operation.

SHALE-OIL BITUMEN. Bitumen (see glossary in monograph on bitumens, p. 77) derived from distillation of CRUDE SHALE-OIL. Associated with shale-oil operations in the USSR

SOLID-HEAT-TRANSFER OIL. CRUDE SHALE-OIL recovered from a retort using a solid material, circulating between the retort and a heating zone to deliver heat to the retort (e.g., Galoter retort)

SPENT SHALE. See RETORTED OIL SHALE.

SYNCRUDE. The product obtained from UPGRADING of CRUDE SHALE-OIL and used in the same manner as petroleum crude oil in refining operations

THERMAL PROCESSING. A thermal cracking operation in which high temperature, in the range of 488-540°C, is used to bring about reduction in boiling range of an oil

TOLICHTHTOL. See ICHTHYOL.

TUNNEL OVEN. A type of oil-shale retort used in the USSR, introduced in about 1930 and withdrawn in the late 1960s. These were batch-type processes in which shallow hopper cars containing a shallow depth of oil shale passed through a tunnel and were heated by hot combustion gases. RETORTING was carried out in the temperature range 650-950°C.

TUNNEL-OVEN OIL. CRUDE SHALE-OIL recovered from a TUNNEL OVEN

UPGRADING. Treatment of crude shale-oil to prepare it for subsequent use as feedstock for a petroleum refinery or chemical plant. Such treatment normally involves several sequential operations.

SOOTS[1]

1. Chemical and Physical Data

1.1 Synonyms and trade names

Chemical Abstr. Name: Soot

Synonyms: Primary carbonaceous particulate emissions

1.2 Description

A glossary of terms is given on p. 241.

Soots are generally lustreless black substances which can be defined as the by-products of the incomplete combustion or pyrolysis of any kind of carbon-containing material; therefore, they exhibit a variety of compositions (often unknown) and properties. Soots comprise a family of related particulate materials consisting of variable quantities of carbonaceous and inorganic solids in conjunction with adsorbed and occluded organic tars and resins. Any carbon-containing material, such as waste oil, fuel oil, gasoline, coal, coal-tar pitch, oil shale, wood, paper, rubber, plastics and resins, or household refuse may undergo incomplete combustion, giving rise to the formation of soot as an unwanted by-product. The exact composition of the mixture in a particular soot is dependent upon the type of material burned and the combustion conditions that existed when the soot was formed (Davidson, 1968; Wagner, 1981; Rivin & Smith, 1982; Gunn *et al.*, 1983; and this volume).

For historical reasons, carbon blacks (see IARC, 1984) are often regarded as another form of soot. However, carbon blacks are commercial products manufactured by the controlled pyrolysis of hydrocarbons, and they differ from the general class of soots in important chemical, physical and biological properties (European Committee for Biological Effects of Carbon Black, 1982; Rivin & Smith, 1982; Smith & Musch, 1982; Gunn *et al.*, 1983).

---

[1]This monograph deals only with soots from domestic and institutional heating fuels.

Soots are mixtures of various forms of particulate carbon, organic tars, resins and refractory inorganic materials produced from carbon-containing material during incomplete combustion. In contrast, carbon blacks are composed of carbon particles and have a low content of tars and other extraneous materials. Whereas soots collected from deposits on surfaces such as the walls of chimneys, fireboxes or exhaust pipes contain less than 50% by weight of particulate carbon and less than 1% by weight of aciniform carbon (defined below), carbon blacks have well-defined morphology and are almost entirely aciniform carbon. It is not unusual for the tarry component in soot to account for more than 25% by weight of the soot (European Committee For Biological Effects of Carbon Black, 1982; Rivin & Smith, 1982).

Within the family of soots, individual soots can be classified according to their major components. Four morphologically distinct forms of particulate carbons are found in commonly-occurring environmental soots: (1) aciniform carbon, (2) particulate carbonaceous xerogel, (3) carbon cenospheres and (4) coke and char fragments (European Committee for Biological Effects of Carbon Black, 1982).

Aciniform carbons are colloidal materials that occur as unique grape clusters called 'aciniform', which are typically formed by nucleation and deposition from the vapour phase. These aggregates are composed of spheroidal particles fused together in random configurations (European Committee for Biological Effects of Carbon Black, 1982; Medalia et al., 1983). Aciniform carbons are not present in significant quantities in deposited chimney soot; however, particulate emissions from fireplaces appear to consist mainly of aciniform carbon (Muhlbaier, 1981, quoted in Medalia et al., 1983).

Organic materials deposited on aciniform carbon can resinize and carbonize on further heating, causing aggregates to become cemented together and embedded in an amorphous carbonaceous matrix known as 'particulate carbonaceous xerogel'. Xerogel particles are prevalent in domestic chimney soots (European Committee for Biological Effects of Carbon Black, 1982).

Carbon cenospheres are hard, shiny, porous or hollow carbon spheres, typically of 10-100 μm in diameter. These are formed when liquid drops undergo carbonization without substantial change in shape. Carbon cenospheres have been recognized as a component of many soots, particularly those from the combustion of heavy fuel-oil sprays (European Committee for Biological Effects of Carbon Black, 1982).

Coke and char fragments are fragments of carbonized wood or coal, of dimensions ranging from μm to mm. These fragments appear to constitute the major form of carbon in domestic chimney soots from wood- and coal-burning fireplaces (European Committee for Biological Effects of Carbon Black, 1982).

Morphological methods (light and scanning electron microscopy) have been used to characterize different types of soot (see Fig. 1 for typical morphology of particulate soots from different fuels) (Fehrmann, 1982).

Major non-particulate components of soot that may be present are: (1) inorganic matter, (2) organic compounds that can be extracted with organic solvents, and (3) insoluble carbonaceous matter (European Committee for Biological Effects of Carbon Black, 1982).

Inorganic matter includes oxides and salts, metals, adsorbed liquids and gases (especially water), and sulphur and nitrogen compounds (European Committee for Biological Effects of Carbon Black, 1982).

The fraction of soots that can be extracted with organic solvents - the so-called soluble organic fraction (SOF) - includes several classes of compounds, including polynuclear aromatic compounds (PACs), their derivatives, and heterocyclic analogues of such compounds. PACs are present in large amounts in the high percentages of loosely-retained tarry matter in soots, and the bulk of the tarry matter is immediately soluble in organic solvents (Rivin & Smith, 1982; Smith & Musch, 1982).

Insoluble carbonaceous matter consists of: (1) resins that may be present as a coating, binder or separate entity, and (2) incompletely carbonized fuel fragments (European Committee for Biological Effects of Carbon Black, 1982).

1.3 Chemical composition and physical properties of soots

Analyses of some representative soots from various sources are summarized in Table 1. These data illustrate the wide range of chemical and physical properties of soots and their typically high content of inorganic material and/or soluble organic fraction (European Committee for Biological Effects of Carbon Black, 1982). The latter includes simple aliphatic hydrocarbons, naphthenes and polynuclear aromatic compounds (PAHs) and oxidized and heterocyclic derivatives. The total carbon content of chimney soots is <60% (Medalia et al., 1983).

The contents of certain individual PACs in the SOF of various types of chimney soots are presented in Table 2.

A large variation has been reported in the content of 11 PACs in different soot samples analysed within the same study (Fehrmann, 1982), as shown in Table 3. One of the major reasons for this variation was reported to be the different flame and smoke-gas temperatures of the different fuels; the PAC concentration was higher when the flame temperature was lower. The most frequently identified PAHs were phenanthrene and fluoranthene; benzo[a]pyrene was detected in only two of the 12 gas-oil soot samples. In the same study, 86 samples of soot swept out from chimneys were analysed for metals, sulphuric acid, sulphate and quartz. The elements, analysed by atomic absorption spectrophotometry, were magnesium, calcium, vanadium, chromium, manganese, iron, cobalt, nickel, copper, zinc, arsenic, selenium, molybdenum, cadmium, mercury and lead. Median values (% by weight) found in fuel-oil and gas-oil soots were, respectively, vanadium, 0.12 and <0.003; chromium, 0.14 and 0.011; iron, 15 and 4; nickel, 0.082 and 0.0062; copper, 0.054 and 0.0092; and lead, 0.072 and 0.0074. Coal-derived soot contained arsenic, 0.020; lead, 0.047; cobalt, 0.0053 and quartz, 2.4. The metal concentrations in hardwood soot were very low. In general, the concentrations of metals and PAHs were found to increase with decreasing particle size, except in fuel-oil soot. This was interpreted as a tendency for the chemicals to be adsorbed onto the particle surface of soots other than the fuel-oil soot (Fehrmann, 1982).

Emissions

A summary of the content of emissions from coal-fired residential stoves is given in IARC (1983).

The chemical composition of combustion emissions is extremely complex. The moderately polar fraction of wood-burning emissions contains about 23 identified PACs and includes such compounds as benzaldehyde, benzonitrile, phenols and methylated analogues, terpenes and resin acid derivatives (Ramdahl & Becher, 1982).

Fig. 1. Typical morphology of particulate soots from different fuels. a,b, fuel-oil soot particles; c,d, coal soot particles; e,f, hardwood soot particles; g,h, gas-oil (kerosene) soot particles[a]

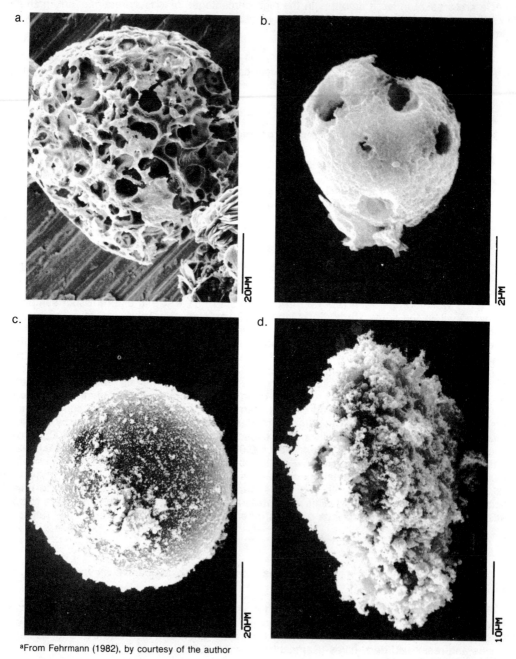

[a]From Fehrmann (1982), by courtesy of the author

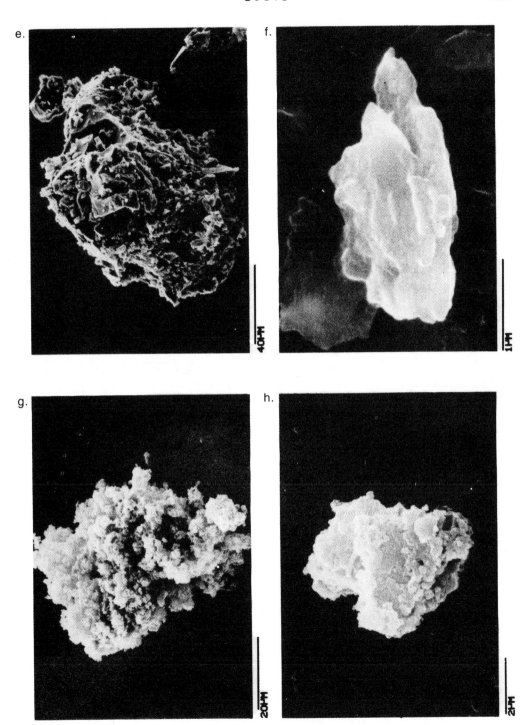

Table 1. Analytical properties of representative soot samples[a]

Sample number	Sample description	Ash (%) Original	Ash (%) Extracted[b]	SOF[c] (%)	MCD[d] (%)	Water extract (%)	TGA[e] weight loss (%)	Carbon (%) Aciniform	Carbon (%) Particulate[f]	Elemental composition[g] Carbon	Elemental composition[g] Hydrogen	Surface area[h] (m²/g)	Oil absorption[i] (cm³/g)	PCSA[j] (m/g)	Atomic ratio (H/C[k])
1.	Chimney soot from deposit near chimney top of hardwood-burning fireplace	21.8	22.9	15.8	15.0	14.2	48.0	0.024	47	51.0	4.6	3	0.52	1.4	1.08
2.	Chimney soot from deposit just above hardwood-burning fireplace	10.4	--	7.9	5.9	12.6	46.7	0.036	--	56.3	4.1	2	0.52	--	0.87
3.	Chimney soot from deposit just above hardwood-burning fireplace	24.0	--	17.0	16.0	13.0	48.0	--	--	--	--	3	0.61	--	--
4.	Blended chimney soot from London domestic coal fires	24.6	22.6	35.6	35.0	19.0	52.4	0.36	23	38.6	3.9	17	0.59	3.9	1.21
5.	Chimney soot from English coal-burning fireplace deposit	45.6	--	14.8	14.1	14.7	36.4	0.89	25	40.7	3.4	<1	0.84[l]	<0.3	1.00
6.	Chimney soot from deposit near chimney top of coal-burning fireplace with grate	73.6	--	1.6	1.4	9.3	16.5	1.4	--	16.2	1.6	21	0.42	--	1.18
7.	Chimney soot from deposit just above coal-burning fireplace with grate	73.2	--	1.4	1.2	2.8	19.0	--	--	--	--	13	0.44	--	--
8.	Chimney soot from coal-burning fireplace deposit	22.5	--	0.59	0.47	6.3	22.0	20.8	--	55.5	1.2	10	1.53	--	0.26
9.	Soot from 'soot box' of domestic oil furnace	53.8	40.7	0.64	0.50	50.7	43.7	0.83	8	--	--	32	0.25	2.6	--

[a]Data for sample numbers 1, 4, 5 and 9 are from European Committee for Biological Effects of Carbon Black (1982), unless otherwise indicated; all data for sample numbers 2, 3, 6, 7 and 8 are from Medalia et al. (1983)
[b]Ash content after extraction with dichloromethane, toluene and water; preferred to base weight of unextracted sample
[c]SOF, soluble organic fraction; the total extract obtained by successive Soxhlet extractions with dichloromethane (four hours) and toluene (48 hours)
[d]MCE, methylene chloride (dichloromethane) extract; this should contain most of the occluded and weakly-bound organic material from the successive Soxhlet extractions described in footnote c above
[e]TGA, thermogravimetric analysis; loss of weight upon heating to 910°C in nitrogen; data for all samples are from Medalia et al. (1983)
[f]Material remaining after correction for SOF, water extract and insoluble inorganic matter as estimated by ash (extracted) content
[g]Data for all samples are from Medalia et al. (1983)
[h]N₂ BET surface area (Brunauer, Emmett and Teller procedure for calculating surface area, determined by adsorption of nitrogen gas on sample) after extraction with dichloromethane, toluene and water, and acid de-ashing with HF, then HCl
[i]Data for all samples are from Medalia et al. (1983)
[j]PCSA, particulate carbon surface area; the surface area contribution in the original sample due to particulate carbon (PC); estimated as [PC fraction] x [surface area of de-ashed extracted sample]
[k]Data for sample numbers 1, 4, 5 and 9 are from Medalia et al. (1983)
[l]Oil absorption end-point not sharp

Table 2. Specific polynuclear aromatic compounds (PACs) found in the soluble organic fraction (SOF) of chimney soots

PAC	Type or source of soot	Concentration of PAC	Reference
Acephenanthrylene	Kerosene soot	Detected	Krishnan & Hites (1981)
Anthracene	Chimney soot from fuel-oil, kerosene, coal- and wood-burning chimneys	Variable amounts up to 16 µg/g of soot[a]	Fehrmann (1982)
Benz[a]anthracene	Chimney soot from hardwood-burning	Approximately 24 µg/g[b,c]	Rivin & Smith (1982)
	Chimney soot from hardwood-burning	2.3 µg/g of soot in one sample[a]	Fehrmann (1982)
Benzo[ghi]perylene	Chimney soot	Detected	Alexander et al. (1982)
	Chimney soot from hardwood-burning fireplace deposit	Approximately 14 µg/g[b,c]	Rivin & Smith (1982)
Benzo[c]phenanthrene	Soot from boiler	2.8 µg/g soot[a]	Fehrmann (1982)
Benzo[a]pyrene	Chimney soots: from deposit just above hardwood-burning fireplace	52 µg/g[d] in 48-h toluene extract	Medalia et al. (1983)
	from deposit near chimney top of hardwood-burning fireplace	25 µg/g[b,d]	
	from London coal fires	120 µg/g[b,d]	
	from coal-burning fireplace deposit	58 µg/g[b,d]	
	Chimney soot	Detected	Alexander et al. (1982)
	Chimney soot from hardwood-burning fireplace deposit	Approximately 22 µg/g[b,c]	Rivin & Smith (1982)
	Oil-shale soot	0.1 mg/g in benzene extract	Bogovski et al. (1970)
	Soot from boiler	Up to 1.6 µg/g of soot[a]	Fehrmann (1982)
Chrysene	Chimney soot	Detected	Alexander et al. (1982)
	Chimney soot from kerosene-, coal- and hardwood-burning chimneys	Variable amounts up to 9.0 µg/g of soot[a]	Fehrmann (1982)
Coronene	Chimney soot from hardwood-burning fireplace deposit	Approximately 3 µg/g[b,c]	Rivin & Smith (1982)
Dibenz[a,h]anthracene	Chimney soot	Detected	Alexander et al. (1982)
Fluoranthene	Chimney soot	Detected	Alexander et al. (1982)
	Chimney soot from hardwood-burning fireplace deposit	Approximately 120 µg/g[b,c]	Rivin & Smith (1982)
	Kerosene soot	Detected	Krishnan & Hites (1981)
	Soot from boiler and chimneys burning fuel oil, kerosene, coal and wood	Variable amounts up to 84 µg/g of soot[a]	Fehrmann (1982)

Table 2 (contd)

PAC	Type or source of soot	Concentration of PAC	Reference
Indeno[1,2,3-cd]pyrene	Chimney soot from hardwood-burning fireplace	Approximately 16 µg/g[b,c]	Rivin & Smith (1982)
Phenanthrene	Chimney soot	Detected	Alexander et al. (1982)
	Soot from chimneys and boilers burning fuel oil, kerosene, coal and wood	Variable amounts up to 64 µg/g of soot[a]	Fehrmann (1982)
	Chimney soot from hardwood-burning fireplace	Approximately 51 µg/g[b,c]	Rivin & Smith (1982)
Pyrene	Chimney soots: from deposit just above hardwood-burning fireplace	48 µg/g[d] in 48-h toluene extract	Medalia et al. (1983)
	from deposit near chimney top of hardwood-burning fireplace	60 µg/g[b,d]	
	from deposit near chimney top of coal-burning fireplace	11 µg/g[b,d]	
	from London coal fires	390 µg/g[b,d]	
	from coal-burning fireplace deposit	315 µg/g[b,d]	
	from coal-burning fireplace deposit	8 µg/g[d] in 48-h toluene extract	
	Chimney soots from burning of fuel oil, kerosene, coal and hardwood	Variable concentrations up to 59 µg/g of soot[a]	Fehrmann (1982)
	Chimney soot from hardwood-burning fireplace	57 µg/g[b,c]	Rivin & Smith (1982)
	Chimney soot	Detected	Alexander et al. (1982)

[a]Level in soot
[b]Found in four-hour dichloromethane extract
[c]Whether the level shown is for the soot or an extract is not clear from the original paper
[d]Level in extract

Table 3. Total polynuclear aromatic compounds (PACs) (µg/g of soot) in soots from different fuels[a]

Type of fuel	No. of samples	Mean	Range of results
Fuel oil	6	4.6	0.1-11
Gas oil (kerosene)	12	22	0.1-72
Coal	7	14	5.8-38
Wood	3	83	3.0-231

[a]The 11 PACs analysed were anthracene, benz[a]anthracene, benzo[c]phenanthrene, benzo[a]pyrene, biphenyl, chrysene, fluoranthene, fluorene, naphthalene, phenanthrene and pyrene.

An experimental study on wood-burning open fireplace emissions showed that average benzo[a]pyrene emissions are about 400 µg/kg of wood, with wide variations between different types of wood. A residential coal fireplace has been reported to emit benzo[a]pyrene at levels in the order of 25 000 µg/kg of coal (Dasch, 1982).

1.4 Technical products and impurities

Soots are generally formed as unwanted by-products of incomplete combustion or pyrolysis rather than as intentionally-manufactured products.

2. Production, Use, Occurrence and Analysis

2.1 Production and use

(a) Production

Chimney soots are not manufactured commercially.

(b) Use

The industrial uses of 'waste' soots are few and are mainly metallurgical (recovery of trace metals). Soot has been used for horticultural purposes. For this use, it must be 'weathered', presumably to allow acid to leach away or be neutralized by the ammonium compounds and other bases. Its undoubted, though small, fertilizing action is due to its content of nitrogenous matter; it may also supply trace metals essential for plant metabolism and for the formation of flower pigments. Its physical nature deters invasion by slugs, and its black colour absorbs solar radiation leading to warming of the soil (International Labour Office, 1983).

2.2 Occurrence and occupational exposures

Occupational exposure to soots can occur whenever any carbon-containing material is undergoing or has undergone combustion or pyrolysis and soots are formed. The scope of the present monograph is limited to chimney soots and the occupational exposures to such soots. Chimney-sweeps are the principal workers exposed to soots.

In Sweden (population, 8.2 million), there were 1400 chimney-sweeps in 1983. The number of chimney-sweeps in Denmark (population, 4.5 million) was 800 in 1984. In the USA, there were approximately 5000 chimney-sweeps in 1982, most of whom were part-time workers. In Japan, it is believed that there are 200-300 chimney-sweeps, most of whom work only two to three times a month.

Other occupations in which there is potential occasional exposure to chimney soots include heating-unit service personnel, brick masons and helpers, building demolition personnel, insulators and firemen.

The typical duties of chimney-sweeps include cleaning of: residential house chimneys (sometimes small boilers) without personal entry; industrial chimneys without personal entry; industrial boilers without personal entry; chimney ducts with personal entry; and chimneys with personal entry (Fehrmann, 1982). They are exposed to both gaseous and particulate combustion products *via* inhalation, ingestion of particles collected in the upper respiratory system, and skin contact. They may also be exposed to other types of soots, to various types of asbestos and mineral fibres, to organic solvents and to the materials specific to the industries in which they work (e.g., solvents used to remove fats from ducts over home and industrial kitchens).

Electron microscopic analysis shows that most of the aciniform particulate colloidal carbon is emitted from the chimney, whereas the coarse particles, typically in the form of carbonaceous microgel or xerogel, tend to deposit on the walls of the chimney (Rivin & Medalia, 1983).

The character of sweeping duties may vary considerably, and this significantly affects exposure. One hour's climb in chimney ducts may give an exposure to particulate dust 100 to 1000 times higher than that in other duties. Measurements of concentrations of particulate soot and respirable soot based on personal sampling during typical Danish working processes illustrated that the range of concentrations was very wide (Table 4). The highest peak concentration measured was 12 mg/l during cleaning of a chimney duct from coal-burning equipment. When the chimney systems were completely shut down during cleaning, the concentration of gases was found to be low. If the firing system was closed only partially, as in the case of coal-fired systems, the concentrations were very much higher: sulphur dioxide, up to 45 µg/l and carbon monoxide, up to 400 µg/l (Fehrmann, 1982).

Table 4. Median concentration (µg/l) of particulate soot in the breathing zone of chimney-sweeps during different duties (59 and 41 measurements for total soot and respirable soot, respectively; sampling times varied from 2 to 37 min)[a]

Type of chimney	Residential house	Industrial chimney	Chimney duct		Chimney with entry	
Type of fuel	Gas-oil (kerosene)	Gas-oil (kerosene)	Gas-oil and fuel-oil	Coal	Wood	Wood and other fuel
Total soot	5.3	4.1	63	388	14	190
Respirable[b] soot	2.2	1.1	8.1	18	5.0	25

[a]Data from Fehrmann (1982)
[b]Aerodynamic diameter <7 µm (obtained by cyclone preseparation)

In a Swedish study (Swedish National Board of Occupational Safety and Health, 1980), the airborne dust concentrations (mainly short-time samples) in the chimney-sweeps' breathing zone varied from 0-34 mg/m^3. The fibre contents of six samples were found to be very low in all but one sample (which contained 1.5 fibres/ml). The following airborne concentrations of elements were determined in one of the samples (mg/m^3): iron, 0.54; arsenic, 0.023; and zinc, 0.023; other samples had much lower concentrations.

No information was available on regulations covering soots as such.

2.3 Chemical analysis of chimney soots and emissions

Soots have been detected and analysed in air after collection by filtration, using gravimetric analysis (Fehrmann, 1982), thermal analysis (Novakov, 1980, 1982), photoacoustic spectroscopy (Killinger et al., 1980) and visible light absorption (Nolan, 1979; Weiss & Waggoner, 1982). Nolan (1979) specifically applied the latter technique to soots arising from the combustion of domestic heating fuels.

3. Biological Data Relevant to the Evaluation of Carcinogenic Risk to Humans

3.1 Carcinogenicity studies in animals

(a) Skin application

Mouse: Three fractions were prepared from a bituminous coal-derived household soot: fraction I was a diethyl ether extract; fraction II contained mainly ether-soluble bases and neutral substances; and fraction III contained mainly ether-soluble bases of water-soluble salts from fraction II. Three groups of 20, 40 and 20 'white' mice received fractions I, II and III (diluted in diethyl ether) by skin painting two or three times per week. At the end of six months, lesions described by the authors as 'warts' were observed in three mice and one mouse treated with fractions I and III, respectively, at which time these groups were killed. Fraction II induced 'warts' in 16/18 survivors after three months of treatment (37 applications). Treatment was continued, and nine mice subsequently developed carcinomas within one year of the start of treatment (Passey, 1922; Passey & Carter-Braine, 1925).

A benzene extract of a domestic chimney coal-derived soot was applied twice weekly for six months to the skin of 24 mice (male and female); 23 survived six months, 21 mice survived 10 months, five survived 18 months and none survived 24 months. Nine animals developed skin tumours, seven of which were carcinomas (Campbell, 1939).

An ethanol extract of wood (eucalyptus) soot from the smoking chamber of a sausage factory was applied daily on the neck skin of ten adult female mice for two years. No skin tumour was observed. Two mice developed para-urinary bladder sarcomas after five and 12 months, respectively, and one mouse developed a bladder carcinoma 21 months after the beginning of the experiment. No tumour was reported in 20 control mice after two years of observation (Sulman & Sulman, 1946). [The Working Group noted the small group size and the inadequate reporting concerning the control group.]

Oil-derived soots from the heating systems of nine commercial buildings were combined and extracted with benzene in soxhlet. A group of 36 (Ax Leaden) F_1 mice received twice-weekly applications of 0.025 ml of 4% oil-soot extract topically on the interscapular skin. A group of 24 mice received benzene alone. After 11 months (80 applications), no tumour was observed in the benzene-control group (22 survivors) or in the group given oil-soot extract (21 survivors) (Mittler & Nicholson, 1957). [The Working Group was unable to determine the precise concentration of the extract applied.]

A group of 100 white mice was painted twice weekly for 5.5 months (50 times) with one or two drops of a benzene extract of soot (obtained from burning solid oil shale) in vaseline. Skin tumours were observed in 58/74 survivors at five months; 36 had malignant tumours, and nine of these had metastases. A control group of 18 mice was painted with the solvent mixture alone; no skin tumour was observed, and no skin tumour occurred in a group of 100 mice that receive no treatment. Two additional groups of 93 and 95 mice were treated, the first only with a benzene extract of soot from shale-oil liquid-fuel combustion, and the second with the same extract mixed with benzene and vaseline. After the five months of treatment, small skin papillomas were noted in 15/71 and 11/70 survivors, respectively. Subsequent tumour regressions occurred after cessation of treatment in ten and seven mice.

Microscopically, four mice in the first group had papillomas and one other the initial stage of a squamous-cell carcinoma; in the second group, there were two papillomas and two carcinomas. No skin tumour occurred in 400 untreated controls. The authors noted that the strong carcinogenic action of benzene extracts of solid shale-oil soot could not be explained solely by the content of benzo[a]pyrene alone (Bogovski, 1961; Vösamäe, 1963, 1979).

(b) *Whole-body exposure*

Mouse: A group of 100 Buffalo strain mice, three months of age, was exposed to the dust of soot from a hospital furnace burning Kentucky bituminous coal. The soot was used as bedding material, and dust was raised by the movements of the mice and by shaking the cages two to three times daily. A group of 50 mice served as controls. Mortality, identical in both groups, at six, 12 and 18 months was 20%, 60% and 100%, respectively. Eight lung adenocarcinomas were reported in the exposed group and one in controls (Seelig & Benignus, 1936). [The Working Group noted the unusually high mortality in controls and that skin tumours were not reported.]

A group of 74 mice (male and female), three months of age, was exposed to soot from a domestic chimney ventilating an ordinary coal fire. A moderate cloud of soot was introduced into an inhalation chamber containing 8-10 mice once an hour for six hours, five days per week for one year, providing an estimated amount of soot of 0.150 g per day, or about 15 mg of soot or 2 mg of benzene-soluble fraction per mouse. At the end of two years, the incidence of lung tumours (eight benign and four malignant lung tumours) was not increased over that in six control groups surviving more than 10 months, in which the incidence of lung tumours varied from 8-20%. No skin tumour was found in treated mice (Campbell, 1939). [The Working Group noted the short duration of treatment.]

(c) *Intratracheal administration*

Rat: Two groups of 125 and 100 random-bred albino rats received ten intratracheal instillations at one-week intervals of 100 mg of tar extracted from soot obtained from the burning of oil-shale solid fuel (benzo[a]pyrene, 107 mg/kg) dissolved in Tween 40 (12.5% aqueous solution) or in peach oil. Control groups of 29 and 27 rats received, respectively, Tween 40 and peach oil. 'Some' animals from the first two groups received whole-body irradiation (400 rad single-dose X-ray) at the end of the intratracheal instillations. The irradiation had no effect on lung tumour incidence. In control groups, no lung tumour was reported. In 31/70 rats treated with the extract in Tween 40 and still alive at the time of the diagnosis of the first lung tumours, epidermoid lung neoplasms developed: eight had epidermoid carcinomas, two of which gave kidney metastases. Only three of 57 rats treated with the same tar extract dissolved in peach oil and surviving at the appearance of the first lung tumour showed benign lung epithelial tumours (Vösamäe, 1979). [The Working Group noted that the length of survival of the various groups was not given.]

(d) *Subcutaneous implantation*

Rat: Fragments of wood (eucalyptus) soot from the smoking chamber of a sausage factory, weighing from 5-20 mg, were implanted s.c. near the right axilla of 18 female rats (120 g) and in the scrotal sac of 18 male rats (150 g). No tumour was found in male rats after 2.5 years of observation. Three female rats developed sarcomas at the site of implantation with latent periods of 12, 17 and 24 months, respectively. In 36 male and female untreated controls observed during the same interval, no tumour was observed (Sulman & Sulman 1946). [The Working Group noted that survival data were not provided.]

3.2 Other relevant biological data

(a) *Experimental systems*

Toxic effects

Toxic effects were reported in several studies described in section 3.1. Skin painting of mice for six months with a diethyl ether extract of a coal-derived household soot caused irritation and ulceration of the skin (Passey & Carter-Braine, 1925). When an ethanolic extract of soot from eucalyptus wood was applied daily to the skin of 10 adult female mice, all animals developed cataracts within three months, attributed by the authors to naphthalenes (Sulman & Sulman, 1946). In a group of 100 mice bedded on soot dust taken from a hospital furnace, hyperplasia of the bronchial mucosa was found to be more prevalent in exposed than in unexposed animals; however, hyperplasia was more pronounced in those mice with pneumonia at autopsy (Seelig & Benignus, 1936).

No data were available to the Working Group on effects on reproduction and prenatal toxicity or on absorption, distribution, excretion and metabolism.

Mutagenicity and other short-term tests

Chimney-soot extracts

Only one study was available concerning the mutagenicity of soots from domestic sources. The content of polynuclear aromatic compounds of some of the samples tested is given in Table 1, section 1.2 (Medalia *et al.*, 1983).

Extracts (dichloromethane and/or toluene) of six different soot samples (wood and coal heating, from fireplace and chimney) were tested for mutagenicity in the *Salmonella*/microsome test with strain TA98 both with and without a rat-liver metabolic system (S9). At least one extract from each sample was mutagenic both with and without S9. Since the number and representativity of the samples tested was not sufficient, no conclusion could be drawn concerning possible differences in mutagenicity among the samples from different origins (Medalia *et al.*, 1983).

A chimney-soot extract was reported to transform Syrian hamster embryo cells [no detail given] (Hansen, 1984).

Smoke emissions

Mutagenic components of smoke emissions may be bound to the particulate phase but may also occur in the vapour phase. Test results on trapped samples depend largely on collection methods and on chemical and physical factors during the collection and treatment of samples (Yamasaki *et al.*, 1982; Lindskog, 1983). Furthermore, components of an emission sample may be changed by atmospheric reactions to chemically different compounds (Møller & Löfroth, 1982). In the studies described below, solvent extracts were prepared from emission samples for testing.

Most of the emission samples (total sample, particulates and/or vapours) analysed were tested in the *Salmonella*/microsome assay. Samples include emissions from small- and medium-size boilers (Löfroth, 1978; Møller & Alfheim, 1983), stationary stoves (Claxton & Huisingh, 1980; Lewtas, 1981; Ramdahl et al., 1982) and open fireplaces (Dasch, 1982), burning different types of fuels, (e.g., natural gas, fuel oil, diesel oils, coal, wood, wood charcoals, peat pellets and wood chips). All emission samples tested were mutagenic in the Salmonella/microsome assay, the response usually being highest in strains TA98 and TA100, in the presence of S9.

Incomplete combustion of wood has been shown to produce emissions which, when tested for mutagenicity in S. typhimurium TA98 (Ramdahl et al., 1982) and for induction of sister chromatid exchanges in Chinese hamster ovary cells (Hytönen et al., 1983), were more active than emissions from combustion at higher temperatures with normal or excess air. The activity in both test systems was higher in the absence of S9.

In a further study, extracts of the wood-combustion samples were fractionated by high-performance liquid chromatography; each fraction was tested for mutagenicity in the Salmonella/microsome assay (modified to microscale) and for induction of sister chromatid exchanges in the Chinese hamster ovary system both with and without S9. In both test systems, the most active fractions (in the presence of S9) were those presumably containing polynuclear aromatic hydrocarbons and the polar fractions presumably containing aza-arenes and aromatic amines (Alfheim et al., 1983; Salomaa et al., 1983). [The results from short-term tests on smoke emissions were not used by the Working Group in assessing the genetic activity of soots.]

Experimental soots

A soot extract was derived experimentally by burning kerosene fuel in a continuous flow combustor, water cooling the probe and extracting it with dichloromethane. The extract was mutagenic in the 8-azaguanine forward mutation system in S. typhimurium in the presence of a polychlorinated biphenyl-induced rat-liver metabolic system (Kaden et al., 1979). It was also mutagenic in human lymphoblasts (6-thioguanine resistance) in the presence of an induced rat-liver metabolic system (Skopek et al., 1979).

(b) *Humans*

Toxic effects

In a cohort mortality study of long-time employed chimney-sweeps (described in section 3.3), Hogstedt et al. (1982, 1983) found a statistically significant excess in the numbers of deaths from non-malignant respiratory-tract diseases (31 observed, 16.7 expected). In addition, significant increase in the observed number of deaths from chronic ischaemic heart disease (62 observed, 42.7 expected) was seen in chimney-sweeps of less than 65 years of age.

In a mortality study of 713 male chimney-sweeps in Denmark (described in section 3.3), ischaemic heart disease accounted for 12 deaths whereas 5.4 were expected; causes other than ischaemic heart disease and cancer were also statistically significantly increased (Hansen, 1983).

No data were available to the Working Group on effects on reproduction and prenatal toxicity, on absorption, distribution, excretion and metabolism, or on mutagenicity and chromosomal effects.

3.3 Case reports and epidemiological studies of carcinogenicity in humans

Soot was first noted as a cause of skin cancer in humans by Pott (1775), who wrote about chimney-sweeps' cancer as follows: 'It is a disease which always makes its first attack on, and its first appearance in the inferior part of the scrotum; where it produces a superficial, painful, ragged, ill-looking sore, with hard and rising edges. The trade call it the soot-wart.' Many case reports from several countries of scrotal and other skin cancers among chimney-sweeps appeared subsequently (e.g., Earle, 1808; Butlin, 1892a,b,c; Henry & Irvine, 1936). Henry wrote many reports on scrotal cancer (1937, 1946, 1947) as well as on cancer at other sites (Henry et al., 1931; Henry, 1947). A total of 1487 cases of scrotal cancer were reported to the Registrar General for England and Wales from 1911-1935 (Henry, 1937). Of these, 6.9% had occurred in chimney-sweeps; the estimated average number of chimney-sweeps in England and Wales in 1921 and 1931 was about 6000, representing 0.06% of all adult males.

Kennaway and Kennaway (1936, 1947) compared observed numbers of deaths from lung and laryngeal cancer among males in a large number of occupations, with expected numbers based upon national statistics in England and Wales during two time periods, 1921-1932 and 1933-1938. The occupations of the deceased were taken from death certificates, which gave the last occupation, but the total number of men in each occupation (the denominator) was approximated from the censuses of 1921 and 1931. For 1921-1932, the annual average during the period was calculated, but for 1933-1938, the 1931 census figures were used. The expected numbers were calculated with regard to seven age classes, and the ratios of observed to expected numbers are given. Chimney-sweeps had a non-significantly increased risk of lung cancer for 1921-1932 (11 observed, 6.5 expected) and a slightly decreased risk for 1933-1938 (10 observed, 11.1 expected). For laryngeal cancer, the chimney-sweeps had a significantly increased risk during the first period (15 observed, 7.8 expected) but not during the later period (6 observed, 4.6 expected).

Kupetz (1966) also reported an excess of respiratory-tract cancer among chimney-sweeps in the German Democratic Republic between 1954 and 1963.

A cohort study on mortality among 2071 members of the Swedish Chimney-Sweeps' Union who were active in 1950 or later and who had been members for at least 10 years was carried out by Hogstedt et al. (1982). Loss to follow-up was less than 1%. The observed numbers of deaths before the age of 80 years were compared with expected numbers specific for sex, single calendar year, and five-year age class obtained from the national statistics rates of 1951-1979. Underlying causes of deaths were recorded from death certificates. Overall, 230 deaths before the age of 80 were observed, whereas 197.6 were expected ($p < 0.05$). This excess mortality derived mainly from tumours at all sites (60 observed, 43.2 expected; $p < 0.05$), of the lung (16 observed, 6.9 expected; $p < 0.05$) and of the oesophagus (6 observed, 0.9 expected; $p < 0.05$). These excesses occurred only in persons with 30 or more years from first membership in the Union. There was also an overall excess mortality from non-malignant chronic respiratory diseases (15 observed, 7.6 expected; $p < 0.05$). Comparison of smoking habits among current Swedish chimney-sweeps in 1972 with the national average in 1963 indicated no large difference. The lack of excess mortality from liver cirrhosis was taken as an indication of no excess alcohol consumption. Exposure levels were not reported.

The Swedish study was expanded to include the 489 men who had been members of the Union for at least 10 years since its inception in 1918 but who had stopped working before 1950, to make a total of 2560 (Hogstedt et al., 1983). A total of 470 men had died, compared to 376 expected, without any upper limit on age at death. There were 126 deaths from all tumours, whereas 82.4 were expected ($p < 0.05$). Of these, 36 men had died from lung cancer, compared to 13 expected ($p < 0.05$), nine from oesophagal cancer with 1.7 expected ($p < 0.05$), eight from primary liver cancer with 1.5 expected ($p < 0.05$), and eight from leukaemia with 3.4 expected ($p < 0.05$). A non-statistically significant excess of liver cirrhosis (10 observed; 5.9 expected) was noted. Those who left the chimney-sweep profession before 1955, and had therefore been exposed mainly to wood, coal and coke soot, had the same excess cancer mortality (except for oesophageal cancer) as those exposed after that time, who were additionally exposed to oil soot. Two cases of pleural mesothelioma were registered among 470 deaths. [Like all people working in trades associated with heating, these workers are likely to have had some exposure to asbestos.]

In a cohort study, Hansen et al. (1982) reported on mortality among 106 members of the Pension Fund for Chimney-Sweeps in Copenhagen, Denmark, during 1958-1977. Eleven cancer deaths were noted, whereas 6.2 were expected, and the excess mortality from tumours was restricted to the age span 40-69 years (9 observed; 2.3 expected). Three deaths in people in this age span were due to lung cancer. One case each of myelomatosis, myeloid leukaemia and chronic lymphatic leukaemia were observed. No data on exposure or smoking were given.

In a larger and partly overlapping study (Hansen, 1983), 713 male chimney-sweeps identified from the 1970 census in Denmark were followed from 1970-1975 and their mortality compared with expected numbers calculated from age-, year- and cause-specific death rates among all employed Danish males in the same census. Twelve cancer deaths were found as against 5.3 expected ($p < 0.05$); lung cancer accounted for five of the cancer deaths observed, with 1.6 deaths expected ($p < 0.05$). No data on exposure or smoking were reported. An excess of deaths from ischaemic heart disease (12 observed, 5.4 expected) was also noted.

A population-based case-control interview study (McLaughlin et al., 1983) of cancer of the renal pelvis (74 cases and 697 controls) was conducted in the area around Minneapolis-St Paul, USA. After adjustment for age and cigarette smoking, elevated but non-significant odds ratios were found for occupational exposure to soot (odds ratio, 3.0; 95% confidence interval, 0.9-8.9). [The Working Group noted that the criteria for soot exposure were not given and that the cases were obtained from hospitals and the controls from the general population.]

4. Summary of Data Reported and Evaluation

4.1 Exposure data

Humans (primarily chimney-sweeps) are exposed to chimney soots in the course of chimney maintenance. Exposures occur to a limited extent in horticultural uses and in other occupations. The general public may be exposed to the particulates emitted from chimneys when domestic heating fuels are burned.

4.2 Experimental data

Coal soot was tested in two experiments in mice by whole-body exposure, but the studies were inadequate for evaluation.

Coal-soot extracts applied to the skin of mice produced skin tumours in two studies.

A wood-soot extract applied to the skin of mice was inadequately tested. In limited studies, subcutaneous implants of wood soot in female rats produced a few local sarcomas; similar implants in the scrotal sac of rats did not produce local tumours.

An extract of fuel-oil soot was inadequately tested by application to the skin of mice.

Extracts of soot from the combustion of oil shale produced skin tumours in mice after dermal application and lung tumours in rats after intratracheal instillation. Extracts of soot from the combustion of a heating oil produced from shale-oil produced skin tumours in mice in two experiments when applied to the skin.

In one study, extracts of soot samples from domestic sources were mutagenic in *Salmonella typhimurium*. Extracts of an experimentally-derived soot were mutagenic in forward mutation assays in *S. typhimurium* and in cultured human lymphoblasts.

Overall assessment of data from short-term tests on soot extracts[a]

	Genetic activity			Cell transformation
	DNA damage	Mutation	Chromosomal effects	
Prokaryotes		+[b,c]		
Fungi/green plants				
Insects				
Mammalian cells (*in vitro*)		+[c]		
Mammals (*in vivo*)				
Humans (*in vivo*)				
Degree of evidence in short-term tests for genetic activity: *limited*				Cell transformation: no data

[a]The groups into which the table is divided and the symbol + are defined on pp. 16-18 of the Preamble; the degrees of evidence are defined on p. 18.
[b]Soots from domestic sources
[c]Experimentally-derived soot

4.3 Human data

The carcinogenicity of soot is demonstrated by numerous case reports, dating back over 200 years, of skin cancer, particularly of the scrotum, among chimney-sweeps.

Cohort studies of mortality among chimney-sweeps in Sweden and Denmark have shown a significantly increased risk of lung cancer. Supporting evidence for an association with lung cancer was provided by two earlier epidemiological studies in the German Democratic Republic and the UK. The potentially confounding and interactive effects of smoking could not be evaluated; however, cigarette smoking is not believed to have seriously biased these estimates.

In addition to lung cancer, statistically significant excess mortality from oesophageal cancer, primary liver cancer and leukaemia was found among chimney-sweeps in one study.

4.4 Evaluation[1]

There is *sufficient evidence* for the carcinogenicity of coal-soot extract and of oil-shale soot extract in experimental animals.

There is *limited evidence* for the carcinogenicity of implanted wood soot and for that of an extract of a soot from heating oil produced from oil shale in experimental animals.

There is *inadequate evidence* for the carcinogenicity of coal soot, of an extract of wood soot and of an extract of a fuel-oil soot in experimental animals.

There is *sufficient evidence* that soot is carcinogenic to humans.

5. References

Alexander, R., Cumbers, K.M. & Kagi, R.I. (1982) Analysis of multiring aromatic hydrocarbons by HPLC using a reverse-phase radial compression column system. *Int. J. environ. anal. Chem.*, *12*, 161-167

Alfheim, I., Löfroth, G. & Møller, M. (1983) Bioassay of extracts of ambient particulate matter. *Environ. Health Perspect.*, *47*, 227-238

Bogovski, P.A. (1961) *The Cancerogenic Effect of Estonian Oil-Shale Processing Products*, Tallinn, Academy of Science Estonian SSR, pp. 245-250

Bogovski, P., Vösamäe, A. & Mirme, H. (1970) *Cocarcinogenicity studies on oil-shale processing products* (Abstract no. 118). In: *Proceedings of the Tenth International Cancer Congress, Houston, Texas, May 22-29, 1970*, Houston, TX, Medical Arts Publishing Co., p. 76

Butlin, H.T. (1892a) Three lectures on cancer of the scrotum in chimney sweeps and others. I. Secondary cancer without primary cancer. *Br. med. J.*, *ii*, 1341-1346

---

[1]For definitions of the italicized terms, see Preamble pp. 15-16 and 19.

Butlin, H.T. (1892b) Three lectures on cancer of the scrotum in chimney sweeps and others. II. Why foreign sweeps do not suffer from scrotal cancer. *Br. med. J., iii*, 1-6

Butlin, H.T. (1892c) Three lectures on cancer of the scrotum in chimney sweeps and others. III. Tar and paraffin cancer. *Br. med. J., iii*, 66-71

Campbell, J.A. (1939) Carcinogenic agents present in the atmosphere and incidence of primary lung tumours in mice. *Br. J. exp. Pathol.*, 20, 122-132

Claxton, L. & Huisingh, J.L. (1980) *Comparative mutagenic activity of organics from combustion sources*. In: Saunders, G.L., Gross, F.T., Dagle, G.E. & Mahaffey, J.A., eds, *Pulmonary Toxicology of Respirable Particles (DOE Symposium Series No. 53)*, Springfield, VA, US Department of Energy, Technical Information Center, pp. 453-465

Dasch, J.M. (1982) Particulate and gaseous emissions from wood-burning fireplaces. *Environ. Sci. Technol.*, 16, 639-645

Davidson, H.W. (1968) *The manufacture and properties of carbon blacks*. In: Davidson, H.W., Wiggs, P.K.C., Churchouse, A.H., Maggs, F.A.P. & Bradley, R.S., eds, *Manufactured Carbon*, London, Pergamon Press, pp. 2, 15-17

Earle, J. (1808) *The Chirurgical Works of Percivall Pott*, Vol. III, London, Wood & Innes, pp. 177-183

European Committee for Biological Effects of Carbon Black (1982) *A Comparative Study of Soot and Carbon Black (Bulletin No. 2)*, London

Fehrmann, R. (1982) *Survey of the Work Environment of Chimney Sweeps* (Dan.), Copenhagen, Danish Work Environment Fund

Gunn, E.F., Johnson, P.H., Nau, C.A. & Toeniskoetter, R.H. (1983) Distinguishing features of soots and carbon blacks. *Dangerous Prop. ind. Mater. Rep.*, 3, 11-13

Hansen, E.S. (1983) Mortality from cancer and ischemic heart disease in Danish chimney sweeps: A five-year follow-up. *Am. J. Epidemiol.*, 117, 160-164

Hansen, E.S., Olsen, J.H. & Tilt, B. (1982) Cancer and non-cancer mortality of chimney sweeps in Copenhagen. *Int. J. Epidemiol.*, 11, 356-361

Hansen, K. (1984) Chimney soot in the cell transformation assay. (Abstract no. II. 3D.21). *Mutat. Res.*, 130, 261

Henry, S.A. (1937) The study of fatal cases of cancer of the scrotum from 1911 to 1935 in relation to occupation, with special reference to chimney sweeping and cotton mule spinning. *Am. J. Cancer*, 31, 28-57

Henry, S.A. (1946) *Cancer of the Scrotum in Relation to Occupation*, London, Oxford University Press

Henry, S.A. (1947) Occupational cutaneous cancer attributable to certain chemicals in industry. *Br. med. Bull.*, 4, 389-401

Henry, S.A. & Irvine, E.D. (1936) Cancer of the scrotum in the Blackburn Reigstration District, 1837-1929. *J. Hyg.*, *36*, 310-337

Henry, S.A., Kennaway, N.M. & Kennaway, E.L. (1931) The incidence of cancer of the bladder and prostate in certain occupations. *J. Hyg.*, *31*, 125-137

Hogstedt, C., Andersson, K., Frenning, B. & Gustavsson A. (1982) A cohort study on mortality among long-time employed Swedish chimney sweeps. *Scand. J. Work Environ. Health*, *8*, Suppl. 1, 72-78

Hogstedt, C., Gustavsson, A. & Frenning, B. (1983) *Mortality and Exposure Among Swedish Chimney Sweeps* (Swed.) (*Sci. Series Arbete & Hälsa No. 16*), Solna, Arbetarskyddsstyrelsen

Hytönen, S., Alfheim, I. & Sorsa, M. (1983) Effect of emissions from residential wood stoves on SCE induction in CHO cells. *Mutat. Res.*, *118*, 69-75

IARC (1983) *IARC Monographs on the Evaluation of the Carcinogenic Risk of Chemicals to Humans*, Vol. 32, *Polynuclear Aromatic Compounds, Part 1, Chemical, Environmental and Experimental Data*, Lyon, p. 39

IARC (1984) *IARC Monographs on the Evaluation of the Carcinogenic Risk of Chemicals to Humans*, Vol. 33, *Polynuclear Aromatic Compounds, Part 2, Carbon Blacks, Mineral Oils and Some Nitroarenes*, Lyon, pp. 35-85

International Labour Office (1983) *Encyclopaedia of Occupational Health and Safety*, 3rd (rev.) ed., Vol. 2, Geneva, p. 2090

Kaden, D.A., Hites, R.A. & Thilly, W.G. (1979) Mutagenicity of soot and associated polycyclic aromatic hydrocarbons to *Salmonella typhimurium*. *Cancer Res.*, *39*, 4152-4159

Kennaway, E.L. & Kennaway, N.M. (1947) A further study of the incidence of cancer of the lung and larynx. *Br. J. Cancer*, *1*, 260-298

Kennaway, N.M. & Kennaway, E.L. (1936) A study of the incidence of cancer of the lung and larynx. *J. Hyg.*, *36*, 236-267

Killinger, D.K., Moore, J. & Japar, S.M. (1980) *The use of photoacoustic spectroscopy to characterize and monitor soot in combustion processes*. In: *ACS Symposium Series No. 134*, Washington DC, American Chemical Society, pp. 457-462

Krishnan, S. & Hites, R.A. (1981) Identification of acephenanthrylene in combustion effluents. *Anal. Chem.*, *53*, 342-343

Kupetz, G.-W. (1966) *Krebs des Atemtraktes und Schornsteinfegerberuf* (Tumour of the respiratory tract and chimney sweep occupation), Inaugural Dissertation, Berlin, Humboldt University

Lewtas, J. (1981) *Comparison of the mutagenic and potentially carcinogenic activity of particle bound organics from wood stoves, residential oil furnaces, and other combustion sources*. In: Cooper, J.A. & Malek, D., eds, *Residential Solid Fuels. Environmental Impacts and Solutions*, Beaverton, OR, Oregon Graduate Center, pp. 606-619

Lindskog, A. (1983) Transformation of polycyclic aromatic hydrocarbons during sampling. *Environ. Health Perspect.*, *47*, 81-84

Löfroth, G. (1978) Mutagenicity assay of combustion emissions. *Chemosphere*, *7*, 791-798

McLaughlin, J.K., Blot, W.J., Mandel, J.S., Schuman, L.M., Mehl, E.S. & Fraumeni, J.F., Jr (1983) Etiology of cancer of the renal pelvis. *J. natl Cancer Inst.*, *71*, 287-291

Medalia, A.I., Rivin, D. & Sanders, D.R. (1983) A comparison of carbon black with soot. *Sci. total Environ.*, *31*, 1-22

Mittler, S. & Nicholson, S. (1957) Carcinogenicity of atmospheric pollutants. *Ind. Med. Surg.*, *26*, 135-138

Møller, M. & Löfroth, G. (1982) *Genotoxic components in polluted air*. In: Sorsa, M. & Vainio, H., eds, *Mutagens in Our Environment*, New York, Alan R. Liss, Inc., pp. 221-234

Møller, M. & Alfheim, I. (1983) Mutagenicity of air samples from various combustion sources. *Mutat. Res.*, *116*, 35-46

Nolan, J.L. (1979) *Measurement of light absorbing aerosols from combustion sources*. In: *Lawrence Berkeley Laboratory Report No. LBL-9037, Carbonaceous Particulates in the Atmosphere*, Berkeley, CA, pp. 265-269

Novakov, T. (1980) *Soot in urban atmosphere*. In: *Lawrence Berkeley Laboratory Report No. LBL-10881, National Research Council Diesel Impacts Study Committee*, Berkeley, CA, pp. 1-6

Novakov, T. (1982) *Soot in the atmosphere*. In: Wolff, G.T. & Klimisch, R.L., eds, *Particulate Carbon: Atmospheric Life Cycle (Proceedings of an International Symposium)*, New York, Plenum Press, pp. 19-41

Passey, R.D. (1922) Experimental soot cancer. *Br. med. J.*, *ii*, 1112-1113

Passey, R.D. & Carter-Brain, J. (1925) Experimental soot cancer. *J. Pathol. Bacteriol.*, *28*, 133-144

Pott, P. (1775) *Chirurgical Observations Relative to the Cataract, the Polypus of the Nose, the Cancer of the Scrotum, the Different Kinds of Ruptures, and the Mortification of the Toes and Feet*, London, Hawes, Clarke & Collins, pp. 63-68

Ramdahl, T. & Becher, G. (1982) Characterization of polynuclear aromatic hydrocarbon derivatives in emissions from wood and cereal straw combustion. *Anal. chim. acta*, *144*, 83-91

Ramdahl, T., Alfheim, I., Rustad, S. & Olsen, T. (1982) Chemical and biological characterization of emissions from small residential stoves burning wood and charcoal. *Chemosphere*, *11*, 601-611

Rivin, D. & Smith, R.G. (1982) Environmental health aspects of carbon black. *Rubber Chem. Technol.*, *55*, 707-761

Rivin, D. & Medalia, A.I. (1983) *A comparative study of soot and carbon black*. In: Lahaye, J. & Prado, G., eds, *Soot in Combustion Systems and its Toxic Properties*, New York, Plenum Press, pp. 25-35

Salomaa, S., Sorsa, M. & Alfheim, I. (1983) *SCE induction by HPLC fractions from wood stove emission samples* (Abstract). In: *International Seminar on Methods of Monitoring Human Exposure to Carcinogenic and Mutagenic Agents, Espoo, Finland, 12-15 December 1983*, Helsinki, Institute of Occupational Health, p. 63

Seelig, M.G. & Benignus, E.L. (1936) Coal smoke soot and tumors of the lung in mice. *Am. J. Cancer, 28*, 96-111

Skopek, T.R., Liber, H.L., Kaden, D.A., Hites, R.A. & Thilly, W.G. (1979) Mutation of human cells by kerosene soot. *J. natl Cancer Inst., 63*, 309-312

Smith, R.G. & Musch, D.C. (1982) Occupational exposure to carbon black: A particulate sampling study. *Am. ind. Hyg. Assoc. J., 43*, 925-930

Sulman, E. & Sulman, F. (1946) The carcinogenicity of wood soot from the chimney of a smoked sausage factory. *Cancer Res., 6*, 366-367

Swedish National Board of Occupational Safety and Health (1980) *Chemical Health Risk during Chimney Sweeping* (Swed.) (*Uppdragsrapport D:nr 2603/79*), Solna, Sweden

Vösamäe, A.I. (1963) On the blastomogenic action of the Estonian shale oil soot and the soot of liquid fuel obtained from the processing of shale oil. *Acta unio int. contra cancrum, 19*, 739-741

Vösamäe, A.I. (1979) Carcinogenicity studies of Estonian oil shale soots. *Environ. Health Perspect., 30*, 173-176

Wagner, H.G. (1981) *Soot formation - An overview*. In: Siegla, D.C. & Smith, G.W., eds, *Particulate Carbon: Formation and Combustion* (*Proceedings of an International Symposium*), New York, Plenum Press, pp. 1-29

Weiss, R.E. & Waggoner, A.P. (1982) *Optical measurements of airborne soot in urban, rural and remote locations*. In: Wolff, G.T. & Klimisch, R.L., eds, *Particulate Carbon: Atmospheric Life Cycle* (*Proceedings of an International Symposium*), New York, Plenum Press, pp. 317-325

Yamasaki, H., Kuwata, K. & Miyamoto, H. (1982) Effects of ambient temperature on aspects of airborne polycyclic aromatic hydrocarbons. *Environ. Sci. Technol., 16*, 189-194

GLOSSARY

ACINIFORM CARBON. Carbon particles in the form of grape clusters, which are major components of commercial CARBON BLACKS but are not present in significant quantities in deposited chimney soots

CARBON BLACKS. Finely divided carbon particles produced commercially by the controlled pyrolysis of hydrocarbons. They have a low content of tars compared to chimney soots.

CARBON CENOSPHERES. Hard carbon spheres formed when liquid drops undergo carbonization without substantial change in shape. They are components of many soots.

PARTICULATE CARBONACEOUS XEROGEL. Aggregates of carbon particles formed when organic materials on ACINIFORM CARBON are resinized and carbonized on heating. They are prevalent in domestic chimney soots.

SOLUBLE ORGANIC FRACTION (SOF). The fraction of soots that can be extracted by organic solvents. Used to characterize different soots and CARBON BLACKS

APPENDIX:

Some case reports of skin cancer in workers exposed to soots, coal-tars, pitch, creosote, bitumens and shale-oils[a]

Reference	Date of case ascertainment	No. of skin cancer cases	Location of skin cancer	Occupation and/or exposure	Duration of work or exposure	Country
Bell (1876)	1874 and 1875	2	Scrotum	Paraffin in shale works	-	UK
Manouvriez (1876)	1874	6	3 scrotum 1 scrotum and nose 2 other	Tar distillery	5-13 years	France
Tillmans (1880)	1877	1	Scrotum and arm	Brown-coal tar and paraffin	21 years	FRG
Ball (1885)	1881	2	1 scrotum 1 head	Tar works (naphtha, creosote oil, pitch) Tar distillery	9 years 15 years	UK
Butlin (1892)	Unknown	39	39 scrotum	29 chimney-sweeps, 1 tar worker, 1 pitch and asphalt worker, 1 gasfitter, 7 other	Unknown-18 years	UK
Liebe (1892)	1890	2	Scrotum	Tar	11-16 years	FRG
Oliver (1908)	1905-1908	12-13	3 hands and arms 3 scrotum 6 or 7 (site not given)	3 coal grease 2 asphalt, 1 chimney-sweep Tar	Unknown-40 years 13, 20 years, >26 years Unknown	UK
Green (1910)	1885-1910	7	4 scrotum 1 penis and scrotum 2 unconfirmed	1 labourer, 2 spinners, 1 carrier 1 labourer, 2 spinners, 1 carrier	- -	USA
Schamberg (1910)	1910	5	4 hand and arm 1 scrotum and hand	All employed in tar-paper saturation	13-55 years	USA
O'Donovan (1920)	1903-1919	16	5 scrotum 11 other	3 creosote, 1 tar, 1 gas worker; 3 tar, 2 gas workers, 3 pitch, 2 others, 1 creosote	Unknown-40 years	UK
Scott (1922)	1900-1921	65	31 scrotum 34 others	19 shale paraffin (3 scrotum, 16 other), 46 other oil shale (retortmen, etc.)	-	UK
Southam & Wilson (1922)	1902-1922	141	Scrotum	69 mule-spinners, 22 tar and paraffin, 1 sweep, 49 others	-	UK

Reference	Date of case ascertainment	No. of skin cancer cases	Location of skin cancer	Occupation and/or exposure	Duration of work or exposure	Country
Kennaway (1925)	1900-1921 (shale-oil) 1910-1912 (soot) 1910-1923 (pitch)	274 cancers	90 scrotum only 184 other	16 soot, 32 pitch, 11 tar, 31 shale-oil 78 pitch, 23 tar, 49 soot, 34 shale-oil	-	UK
de Vries (1928)	1913-1927	4 cancers	2 scrotum 1 face and hand 1 hand	All employed in a briquette factory	11-20 years	Netherlands
Teutschlaender (1929)	1922 and 1927	2	1 scrotum 1 nose	Pitch	12-16 years	FRG
Heller (1930)	1920-1928	40	4 scrotum 36 other	1 gas works, 2 gas-works pitch, 1 coke-oven pitch 11 gas works, 4 roofers, 2 tar distillation, 1 road worker, 13 gas-works pitch, 3 coke-oven pitch, 2 lampblack	-	USA
Shambaugh (1935)	1932-1933	8	Lip, all epidermoid	Tar needle held in lips, sunlight	6-60 years	USA
Henry & Irvine (1936)	1837-1929	132	Scrotum	89 cotton mule-spinners, 16 ex-mule-spinners, 2 gas retortmen, 2 tar distillers, 2 chimney-sweeps, 21 other	10-50 years	UK
Henry (1937)	1911-1935	1487	Scrotum	345 cotton mule-spinners, 103 chimney-sweeps, 1039 other	-	UK
Henry[b] (1947)	1920-1945	3921	1421 scrotum 2500 other	141 pitch, 48 tar, 299 pitch and tar, 12 creosote oil, 899 mineral oil, 15 shale-oil, 7 other 606 pitch, 66 tar, 1147 pitch and tar, 27 creosote oil, 5 anthracene, 594 mineral oil, 38 shale-oil, 17 other	-	UK
Dean (1948)	1926-1947	27	Scrotum	5 petroleum refiners, 6 machinists, 1 roofer, 1 boiler and chimney cleaner, 14 other (1 pitch exposure)	10-40 years	USA
Ross (1948)	1936-1947	23 (16 persons)	4 scrotum 4 arm and hand 15 head and face	All tar and pitch	6 months-43 years	UK
Cruickshank & Gourevitch (1952)	1941-1950	44	Hands and forearms	18 oils, 6 pitch, 13 other	-	UK
Smith (1952)	1950	1	Forearm	Paraffin process (shale-oil)	-	UK

APPENDIX: CASE REPORTS

Reference	Year	Cases	Site	Agent	Duration	Country
Alexander & Macrosson (1954)	1952	1	Leg	20% coal-tar cream for psoriasis	9 years	UK
Patch (1954)	1952	1	Neck	Pitch, bitumen, slate powder and asbestos	27 years	UK
Lenson (1956)	1955	1	Face	Creosote Paints	3 years 41 years	USA
Fife (1962)	1944-1959	3259	Skin (not further specified)	2672 tar and pitch, 491 mineral oil (cotton industries), 96 mineral oil (other industries)	-	UK
Spink et al. (1964)	1941-1959	3	2 scrotum 1 other	Pipe jointers using a 20% tar mixture	17-35 years	UK
Doig (1970)	1944-1968 (Scotland only)	85	48 scrotum	4 jute industry, 11 engineering, 5 soot, 1 arsenic, 9 tar, 14 pitch, 4 shale industry	24-40 years	UK
			37 other	10 tar, 13 pitch, 1 bitumen, 5 shale-oil, 2 jute, 4 engineering, 2 other	24-40 years	
Milne (1970)	1944-1968 (UK)	4471	All skin	3704 tar and pitch, 767 oil	-	UK
	1966-1967	5	Scrotum	2 shale-oil plus others, 1 oil and grease, 1 gas works, 1 clerk	50-70 years	Australia
Lee et al. (1972)	1962-1968	103 (89 with useful information)	Scrotum	40 mule-spinning only 11 mule-spinning and others 6 pitch, tar or its products, 11 oil, 21 others	1-9 years 3-47 years 6-55 years	UK
Durkin et al. (1978)	1974	1	Back	Coal-tar, ultraviolet light, steroids for psoriasis	20 years	USA
Jarvis (1980)	Not given	1	Scrotum	Pitch	14 years	UK

[a]This table is not a comprehensive listing of all case reports of skin cancer

[b]Numbers of 'cases' or 'sites' are taken from Henry (1947); these may not tally with the numbers of 'patients' or 'cases' cited in the text.

References

Alexander, J.O'D. & Macrosson, K.I. (1954) Squamous epithelioma probably due to tar ointment in a case with psoriasis. *Br. med. J.*, *iv*, 1089

Ball, C.B. (1885) Tar cancer. *Trans. Acad. Med. (Irel.)*, *3*, 318-321

Bell, J. (1876) Paraffin epithelioma of the scrotum. *Edinb. med. J.*, *22*, 135-137

Butlin, H.T. (1892) Cancer of the scrotum in chimney-sweeps and others. I. Secondary cancer without primary cancer; II. Why foreign sweeps do not suffer from scrotal cancer; III. Tar and paraffin cancer. *Br. med. J.*, *i*, 1341-1346; *ii*, 1-6, 66-71

Cruickshank, C.N.D. & Gourevitch, A. (1952) Skin cancer of the hand and forearm. *Br. J. ind. Med.*, *9*, 74-79

Dean, A.L. (1948) Epithelioma of scrotum. *J. Urol.*, *60*, 508-518

Doig, A.T. (1970) Epithelioma of the scrotum in Scotland in 1967. *Health Bull.*, *28*, 45-51

Durkin, W., Sun, N., Link, J., Paroly, W. & Schweitzer, R. (1978) Melanoma in a patient treated for psoriasis. *South. med. J.*, *71*, 732-733

Fife, J.G. (1962) Carcinoma of the skin in machine tool setters. *Br. J. ind. Med.*, *19*, 123-125

Green, R.M. (1910) Cancer of the scrotum. *Boston med. surg. J.*, *163*, 792-797

Heller, I. (1930) Occupational cancers. Cancer caused by coal tar and coal tar products. *J. ind. Hyg.*, *12*, 169-196

Henry, S.A. (1937) The study of fatal cases of cancer of the scrotum from 1911 to 1935 in relation to occupation, with special reference to chimney sweeping and cotton mule spinning. *Am. J. Cancer*, *31*, 28-57

Henry, S.A. (1947) Occupational cutaneous cancer attributable to certain chemicals in industry. *Br. med. Bull.*, *4*, 389-401

Henry, S.A. & Irvine, E.D. (1936) Cancer of the scrotum in the Blackburn Registration District, 1837-1929. *J. Hyg.*, *36*, 310-337

Jarvis, H. (1980) Scrotal cancer in a pitch worker. *J. soc. occup. Med.*, *30*, 61-62

Kennaway, E.L. (1925) The anatomical distribution of the occupational cancers. *J. ind. Hyg.*, *7*, 69-93

Lee, W.R., Alderson, M.R. & Downes, J.E. (1972) Scrotal cancer in the north-west of England, 1962-68. *Br. J. ind. Med.*, *29*, 188-195

Lenson, N. (1956) Multiple cutaneous carcinoma after creosote exposure. *New Engl. J. Med.*, *254*, 520-523

Liebe, G. (1892) Tar and paraffin cancer (Ger.). *Schmidt's Jahrbuch*, *236*, 71-80

Manouvriez, A. (1876) Diseases and hygiene of workers making tar and pitch briquettes (Fr.). *Ann. Hyg. publ.*, 45, 459-482

Milne, J.E.H. (1970) Carcinoma of the scrotum. *Med. J. Austr.*, 2, 13-16

O'Donovan, W.J. (1920) Epitheliomatous ulceration among tar workers. *Br. J. Dermatol. Syphilis*, 32, 215-228

Oliver, T. (1908) Tar and asphalt workers' epithelioma and chimney-sweeps' cancer. *Br. med. J.*, ii, 493-494

Patch, I.L. (1954) Pitch and pulmonary carcinoma. *Br. J. Tubercul. Dis. Chest*, 48, 145-150

Ross, P. (1948) Occupational skin lesions due to pitch and tar. *Br. med. J.*, ii, 369-374

Schamberg, J.F. (1910) Cancer in tar workers. *J. cutan. Dis.*, 28, 644-662

Scott, A. (1922) On the occupation cancer of the paraffin and oil workers of the Scottish shale oil industry. *Br. med. J.*, iv, 1108-1109

Shambaugh, P. (1935) Tar cancer of the lip in fishermen. *J. Am. med. Assoc.*, 104, 2326-2329

Smith, W.E. (1952) Survey of some current British and European studies of occupational tumor problems. *Arch. ind. Hyg. occup. Med.*, 5, 242-263

Southam, A.H. & Wilson, S.R. (1922) Cancer of the scrotum: The etiology, clinical features, and treatment of the disease. *Br. med. J.*, iv, 971-973

Spink, M.S., Baynes, A.H. & Tombleson, J.B.L. (1964) Skin carcinoma in the process of 'Stanford jointing'. *Br. J. ind. Med.*, 21, 154-157

Teutschlaender, A. (1929) The 'Fohr-Kleinschmidt' pitch pulverization process as a process for prevention of pitch tumours in briquette factories. (Results of a visit to the Engelsburg mine near Bochum) (Ger.). *Z. Krebsforsch.*, 30, 231-240

Tillmanns, H. (1880) On tumours induced by tar, soot and tobacco (Ger.). *Dtsch. Z. Chir.*, 13, 519-535

de Vries, W.M. (1928) *Pitch cancer in the Netherlands*. In: *Report of the International Conference on Cancer*, London, Fowler Wright Ltd, pp. 290-292

SUPPLEMENTARY CORRIGENDA TO VOLUMES 1-34

Corrigenda covering Volumes 1-6 appeared in Volume 7; others appeared in Volumes 8, 10-13, 15-34.

Volume 26

p. 137   Structural formula   *replace* ⎔ *by* ⬡

Supplement 4

p. 56   B. Evidence of carcinogenicity to animals   *replace* (limited) *by* (inadequate)** *and add footnote :* ** More recent data would provide an evaluation of *limited evidence* (*IARC Monographs, 29*, 93-148, 1982)

CUMULATIVE INDEX TO IARC MONOGRAPHS ON THE EVALUATION OF THE CARCINOGENIC RISK OF CHEMICALS TO HUMANS

Numbers in italics indicate volume, and other numbers indicate page. References to corrigenda are given in parentheses. Compounds marked with an asterisk(*) were considered by the working groups in the year indicated, but monographs were not prepared because adequate data on carcinogenicity were not available.

A

Acetaldehyde formylmethylhydrazone	*31*, 163
Acetamide	*7*, 197
Acetylsalicyclic acid (1976)*	
Acridine orange	*16*, 145
Acriflavinium chloride	*13*, 31
Acrolein	*19*, 479
Acrylic acid	*19*, 47
Acrylic fibres	*19*, 86
Acrylonitrile	*19*, 73
	Suppl. *4*, 25
Acrylonitrile-butadiene-styrene copolymers	*19*, 91
Actinomycins	*10*, 29 (corr. *29*, 399; *34*, 197)
	Suppl. *4*, 27
Adipic acid (1978)*	
Adriamycin	*10*, 43
	Suppl. *4*, 29
AF-2	*31*, 47
Aflatoxins	*1*, 145 (corr. *7*, 319)
	(corr. *8*, 349)
	10, 51
	Suppl. *4*, 31
Agaritine	*31*, 63
Aldrin	*5*, 25
	Suppl. *4*, 35
Aluminium production	*34*, 37
Amaranth	*8*, 41
5-Aminoacenaphthene	*16*, 243
2-Aminoanthraquinone	*27*, 191
para-Aminoazobenzene	*8*, 53
ortho-Aminoazotoluene	*8*, 61 (corr. *11*, 295)
para-Aminobenzoic acid	*16*, 249
4-Aminobiphenyl	*1*, 74 (corr. *10*, 343)
	Suppl. *4*, 37
3-Amino-1,4-dimethyl-5*H*-pyrido-[4,3-*b*] — indole and its acetate	*31*, 247
1-Amino-2-methylanthraquinone	*27*, 199
3-Amino-1-methyl-5*H*-pyrido-[4,3-*b*] — indole and its acetate	*31*, 255
2-Amino-5-(5-nitro-2-furyl)-1,3,4-thiadiazole	*7*, 143

4-Amino-2-nitrophenol	*16*, 43
2-Amino-4-nitrophenol (1977)*	
2-Amino-5-nitrophenol (1977)*	
2-Amino-5-nitrothiazole	*31*, 71
6-Aminopenicillanic acid (1975)*	
Amitrole	*7*, 31
	Suppl. *4*, 38
Amobarbital sodium (1976)*	
Anaesthetics, volatile	*11*, 285
	Suppl. *4*, 41
Aniline	*4*, 27 (corr. *7*, 320)
	27, 39
	Suppl. *4*, 49
Aniline hydrochloride	*27*, 40
ortho-Anisidine and its hydrochloride	*27*, 63
para-Anisidine and its hydrochloride	*27*, 65
Anthanthrene	*32*, 95
Anthracene	*32*, 105
Anthranilic acid	*16*, 265
Apholate	*9*, 31
Aramite®	*5*, 39
Arsenic and arsenic compounds	*1*, 41
	2, 48
	23, 39
	Suppl. *4*, 50
Arsanilic acid	
Arsenic pentoxide	
Arsenic sulphide	
Arsenic trioxide	
Arsine	
Calcium arsenate	
Dimethylarsinic acid	
Lead arsenate	
Methanearsonic acid, disodium salt	
Methanearsonic acid, monosodium salt	
Potassium arsenate	
Potassium arsenite	
Sodium arsenate	
Sodium arsenite	
Sodium cacodylate	
Asbestos	*2*, 17 (corr. *7*, 319)
	14 (corr. *15*, 341)
	(corr. *17*, 351)
	Suppl. *4*, 52
Actinolite	
Amosite	
Anthophyllite	
Chrysotile	
Crocidolite	
Tremolite	
Asiaticoside (1975)*	
Auramine	*1*, 69 (corr. *7*, 319)
	Suppl. *4*, 53 (corr. *33*, 223)

CUMULATIVE INDEX

Aurothioglucose	*13*, 39
5-Azacytidine	*26*, 37
Azaserine	*10*, 73 (corr. *12*, 271)
Azathioprine	*26*, 47
	Suppl. 4, 55
Aziridine	*9*, 37
2-(1-Aziridinyl)ethanol	*9*, 47
Aziridyl benzoquinone	*9*, 51
Azobenzene	*8*, 75

B

Benz[*a*] acridine	*32*, 123
Benz[*c*]acridine	*3*, 241
	32, 129
Benzal chloride	*29*, 65
	Suppl. 4, 84
Benz[*a*]anthracene	*3*, 45
	32, 135
Benzene	*7*, 203 (corr. *11*, 295)
	29, 93, 391
	Suppl. 4, 56 (corr. *35*, 249)
Benzidine and its salts	*1*, 80
	29, 149, 391
	Suppl. 4, 57
Benzo[*b*]fluoranthene	*3*, 69
	32, 147
Benzo[*j*]fluoranthene	*3*, 82
	32, 155
Benzo [*k*]fluoranthene	*32*, 163
Benzo[*ghi*]fluoranthene	*32*, 171
Benzo[*a*]fluorene	*32*, 177
Benzo[*b*]fluorono	*32*, 183
Benzo[*c*]fluorene	*32*, 189
Benzo[*ghi*]perylene	*32*,195
Benzo[*c*]phenanthrene	*32*, 205
Benzo[*a*]pyrene	*3*, 91
	Suppl. 4, 227
	32, 211
Benzo[*e*]pyrene	*3*, 137
	32, 225
para-Benzoquinone dioxime	*29*, 185
Benzotrichloride	*29*, 73
	Suppl. 4, 84
Benzoyl chloride	*29*, 83
	Suppl. 4, 84
Benzyl chloride	*11*, 217 (corr. *13*, 243)
	29, 49 (corr. *30*, 407)
	Suppl. 4, 84
Benzyl violet 4B	*16*, 153

Beryllium and beryllium compounds   1, 17
                                      23, 143 (corr. 25, 392)
   Bertrandite   Suppl. 4, 60
   Beryllium acetate
   Beryllium acetate, basic
   Beryllium-aluminium alloy
   Beryllium carbonate
   Beryllium chloride
   Beryllium-copper alloy
   Beryllium-copper-cobalt alloy
   Beryllium fluoride
   Beryllium hydroxide
   Beryllium-nickel alloy
   Beryllium oxide
   Beryllium phosphate
   Beryllium silicate
   Beryllium sulphate and its tetrahydrate
   Beryl ore
   Zinc beryllium silicate

Bis(1-aziridinyl)morpholinophosphine sulphide   9, 55
Bis(2-chloroethyl)ether   9, 117
N,N-Bis(2-chloroethyl)-2-naphthylamine (chlornaphazine)   4, 119 (corr. 30, 407)
                                                 Suppl. 4, 62
Bischloroethyl nitrosourea (BCNU)   26, 79
                                                 Suppl. 4, 63
Bis-(2-chloroisopropyl)ether (1976)*
1,2-Bis(chloromethoxy)ethane   15, 31
1,4-Bis(chloromethoxymethyl)benzene   15, 37
Bis(chloromethyl)ether   4, 231 (corr. 13, 243)
                                                 Suppl. 4, 64
Bitumens   35, 39
Bleomycins   26, 97
                                                 Suppl. 4, 66
Blue VRS   16, 163
Boot and shoe manufacture and repair   25, 249
                                                 Suppl. 4, 138
Brilliant blue FCF diammonium and disodium salts   16, 171 (corr. 30, 407)
1,4-Butanediol dimethanesulphonate (Myleran)   4, 247
                                                 Suppl. 4, 68
Butyl benzyl phthalate   29, 194 (corr. 32, 455)
Butyl-cis-9,10-epoxystearate (1976)*
β-Butyrolactone   11, 225
γ-Butyrolactone   11, 231

C

Cadmium and cadmium compounds   2, 74
                                                 11, 39 (corr. 27, 320)
   Cadmium acetate   Suppl. 4, 71
   Cadmium chloride
   Cadmium oxide
   Cadmium sulphate
   Cadmium sulphide

Calcium cyclamate	22, 58 (corr. *25*, 391)
	Suppl. 4, 97
Calcium saccharin	22, 120 (corr. *25*, 391)
	Suppl. 4, 225
Cantharidin	*10*, 79
Caprolactam	*19*, 115 (corr. *31*, 293)
Captan	*30*, 295
Carbaryl	*12*, 37
Carbazole	*32*, 239
Carbon blacks	*3*, 22
	33, 35
Carbon tetrachloride	*1*, 53
	20, 371
	Suppl. 4, 74
Carmoisine	*8*, 83
Carpentry and joinery	*25*, 139
	Suppl. 4, 139
Carrageenans (native)	*10*, 181 (corr. *11*, 295)
	31, 79
Catechol	*15*, 155
Chloramben (1982)*	
Chlorambucil	*9*, 125
	26, 115
	Suppl. 4, 77
Chloramphenicol	*10*, 85
	Suppl. 4, 79
Chlordane	*20*, 45 (corr. *25*, 391)
	Suppl. 4, 80
Chlordecone (Kepone)	*20*, 67
Chlordimeform	*30*, 61
Chlorinated dibenzodioxins	*15*, 41
	Suppl. 4, 211, 238
Chlormadinone acetate	*6*, 149
	21, 365
	Suppl. 4, 192
Chlorobenzilate	*5*, 75
	30, 73
1-(2-Chloroethyl)-3-cyclohexyl-1-nitrosourea (CCNU)	*26*, 137 (corr. *35*, 249)
	Suppl. 4, 83
Chloroform	*1*, 61
	20, 401
	Suppl. 4, 87
Chloromethyl methyl ether	*4*, 239
	Suppl. 4, 64
4-Chloro-*ortho*-phenylenediamine	*27*, 81
4-Chloro-*meta*-phenylenediamine	*27*, 82
Chloroprene	*19*, 131
	Suppl. 4, 89
Chloropropham	*12*, 55
Chloroquine	*13*, 47
Chlorothalonil	*30*, 319
para-Chloro-*ortho*-toluidine and its hydrochloride	*16*, 277
	30, 61

5-Chloro-*ortho*-toluidine (1977)*
Chlorotrianisene   *21*, 139
Chlorpromazine (1976)*
Cholesterol   *10*, 99
                                                    31, 95
Chromium and chromium compounds   *2*, 100
                                                    23, 205
                                                    Suppl. 4, 91
   Barium chromate
   Basic chromic sulphate
   Calcium chromate
   Chromic acetate
   Chromic chloride
   Chromic oxide
   Chromic phosphate
   Chromite ore
   Chromium carbonyl
   Chromium potassium sulphate
   Chromium sulphate
   Chromium trioxide
   Cobalt-chromium alloy
   Ferrochromium
   Lead chromate
   Lead chromate oxide
   Potassium chromate
   Potassium dichromate
   Sodium chromate
   Sodium dichromate
   Strontium chromate
   Zinc chromate
   Zinc chromate hydroxide
   Zinc potassium chromate
   Zinc yellow
Chrysene   *3*, 159
                                                    32, 247
Chrysoidine   *8*, 91
C.I. Disperse Yellow 3   *8*, 97
Cinnamyl anthranilate   *16*, 287
                                                    31, 133
Cisplatin   *26*, 151
                                                    Suppl. 4, 93
Citrus Red No. 2   *8*, 101 (corr. *19*, 495)
Clofibrate   *24*, 39
                                                    Suppl. 4, 95
Clomiphene and its citrate   *21*, 551
                                                    Suppl. 4, 96
Coal gasification   *34*, 65
Coal-tars and derived products   *35*, 83
Coke production   *34*, 101
Conjugated œstrogens   *21*, 147
                                                    Suppl. 4, 179
Copper 8-hydroxyquinoline   *15*, 103
Coronene   *32*, 263

CUMULATIVE INDEX

Coumarin	*10*, 113
meta-Cresidine	*27*, 91
para-Cresidine	*27*, 92
Cycasin	*1*, 157 (corr. *7*, 319)
	10, 121
Cyclamic acid	*22*, 55 (corr. *25*, 391)
Cyclochlorotine	*10*, 139
Cyclohexylamine	*22*, 59 (corr. *25*, 391)
	Suppl. 4, 97
Cyclopenta[*cd*]pyrene	*32*, 269
Cyclophosphamide	*9*, 135
	26, 165
	Suppl. 4, 99

D
2,4-D and esters	*15*, 111
	Suppl. 4, 101, 211
Dacarbazine	*26*, 203
	Suppl. 4, 103
D and C Red No. 9	*8*, 107
Dapsone	*24*, 59
	Suppl. 4, 104
Daunomycin	*10*, 145
DDT and associated substances	*5*, 83 (corr. *7*, 320)
	Suppl. 4, 105
DDD (TDE)	
DDE	
Diacetylaminoazotoluene	*8*, 113
N,N'-Diacetylbenzidine	*16*, 293
Diallate	*12*, 69
	30, 235
2,4-Diaminoanisole and its sulphate	*16*, 51
	27, 103
2,5-Diaminoanisole (1977)*	
4,4'-Diaminodiphenyl ether	*16*, 301
	29, 203
1,2-Diamino-4-nitrobenzene	*16*, 63
1,4-Diamino-2-nitrobenzene	*16*, 73
2,4-Diaminotoluene	*16*, 83
2,5-Diaminotoluene and its sulphate	*16*, 97
Diazepam	*13*, 57
Diazomethane	*7*, 223
Dibenz[*a,h*]acridine	*3*, 247
	32, 277
Dibenz[*a,j*]acridine	*3*, 254
	32, 283
Dibenz[*a,c*]anthracene	*32*, 289 (corr. *34*, 197)
Dibenz[*a,h*]anthracene	*3*, 178
	32, 299
Dibenz[*a, j*]anthracene	*32*, 309
7H-Dibenzo[*c,g*]carbazole	*3*, 260
	32, 315
Dibenzo[*a,e*]fluoranthene	*32*, 321

Dibenzo[h,rst]pentaphene	3, 197
Dibenzo[a,e]pyrene	3, 201
	32, 327
Dibenzo[a,h]pyrene	3, 207
	32, 331
Dibenzo[a,i]pyrene	3, 215
	32, 337
Dibenzo[a,l]pyrene	3, 224
	32, 343
1,2-Dibromo-3-chloropropane	15, 139
	20, 83
ortho-Dichlorobenzene	7, 231
	29, 213
	Suppl. 4, 108
para-Dichlorobenzene	7, 231
	29, 215
	Suppl. 4, 108
3,3'-Dichlorobenzidine and its dihydrochloride	4, 49
	29, 239
	Suppl. 4, 110
trans-1,4-Dichlorobutene	15, 149
3,3'-Dichloro-4,4'-diaminodiphenyl ether	16, 309
1,2-Dichloroethane	20, 429
Dichloromethane	20, 449
	Suppl. 4, 111
Dichlorvos	20, 97
Dicofol	30, 87
Dicyclohexylamine	22, 60 (corr. 25, 391)
Dieldrin	5, 125
	Suppl. 4, 112
Dienoestrol	21, 161
	Suppl. 4, 183
Diepoxybutane	11, 115 (corr. 12, 271)
Di-(2-ethylhexyl) adipate	29, 257
Di-(2-ethylhexyl) phthalate	29, 269 (corr. 32, 455)
1,2-Diethylhydrazine	4, 153
Diethylstilboestrol	6, 55
	21, 173 (corr. 23, 417)
	Suppl. 4, 184
Diethylstilboestrol dipropionate	21, 175
Diethyl sulphate	4, 277
	Suppl. 4, 115
Diglycidyl resorcinol ether	11, 125
Dihydrosafrole	1, 170
	10, 233
Dihydroxybenzenes	15, 155
Dihydroxymethylfuratrizine	24, 77
Dimethisterone	6, 167
	21, 377
	Suppl. 4, 193
Dimethoate (1977)*	
Dimethoxane	15, 177

3,3'-Dimethoxybenzidine (*ortho*-Dianisidine)	*4*, 41
	Suppl. 4, 116
para-Dimethylaminoazobenzene	*8*, 125 (corr. *31*, 293)
para-Dimethylaminobenzenediazo sodium sulphonate	*8*, 147
trans-2[(Dimethylamino)methylimino]-5-[2-(5-nitro-2-furyl)vinyl]-1,3,4-oxadiazole	*7*, 147 (corr. *30*, 407)
3,3'-Dimethylbenzidine (*ortho*-Tolidine)	*1*, 87
Dimethylcarbamoyl chloride	*12*, 77
	Suppl. 4, 118
1,1-Dimethylhydrazine	*4*, 137
1,2-Dimethylhydrazine	*4*, 145 (corr. *7*, 320)
1,4-Dimethylphenanthrene	*32*, 349
Dimethyl sulphate	*4*, 271
	Suppl. 4, 119
Dimethylterephthalate (1978)*	
1,8-Dinitropyrene	*33*, 171
Dinitrosopentamethylenetetramine	*11*, 241
1,4-Dioxane	*11*, 247
	Suppl. 4, 121
2,4'-Diphenyldiamine	*16*, 313
Diphenylthiohydantoin (1976)*	
Direct Black 38	*29*, 295 (corr. *32*, 455)
	Suppl. 4, 59
Direct Blue 6	*29*, 311
	Suppl. 4, 59
Direct Brown 95	*29*, 321
	Suppl. 4, 59
Disulfiram	*12*, 85
Dithranol	*13*, 75
Dulcin	*12*, 97

E
Endrin	*5*, 157
Enflurane (1976)*	
Eosin and its disodium salt	*15*, 183
Epichlorohydrin	*11*, 131 (corr. *18*, 125)
	(corr. *26*, 387)
	Suppl. 4, 122
	(corr. *33*, 223)
1-Epoxyethyl-3,4-epoxycyclohexane	*11*, 141
3,4-Epoxy-6-methylcyclohexylmethyl-3,4-epoxy-6-methyl-cyclohexane carboxylate	*11*, 147
cis-9,10-Epoxystearic acid	*11*, 153
Ethinyloestradiol	*6*, 77
	21, 233
	Suppl. 4, 186
Ethionamide	*13*, 83
Ethyl acrylate	*19*, 57
Ethylene	*19*, 157
Ethylene dibromide	*15*, 195
	Suppl. 4, 124

Ethylene oxide	*11*, 157
	Suppl. 4, 126
Ethylene sulphide	*11*, 257
Ethylenethiourea	*7*, 45
	Suppl. 4, 128
Ethyl methanesulphonate	*7*, 245
Ethyl selenac	*12*, 107
Ethyl tellurac	*12*, 115
Ethynodiol diacetate	*6*, 173
	21, 387
	Suppl. 4, 194
Evans blue	*8*, 151

F

Fast green FCF	*16*, 187
Ferbam	*12*, 121 (corr. *13*, 243)
Fluometuron	*30*, 245
Fluoranthene	*32*, 355
Fluorene	*32*, 365
Fluorescein and its disodium salt (1977)*	
Fluorides (inorganic, used in drinking-water and dental preparations)	*27*, 237
Fluorspar	
Fluosilicic acid	
Sodium fluoride	
Sodium monofluorophosphate	
Sodium silicofluoride	
Stannous fluoride	
5-Fluorouracil	*26*, 217
	Suppl. 4, 130
Formaldehyde	*29*, 345
	Suppl. 4, 131
2-(2-Formylhydrazino)-4-(5-nitro-2-furyl)thiazole	*7*, 151 (corr. *11*, 295)
Furazolidone	*31*, 141
The furniture and cabinet-making industry	*25*, 99
	Suppl. 4, 140
2-(2-Furyl)-3-(5-nitro-2-furyl)acrylamide	*31*, 47
Fusarenon-X	*11*, 169
	31, 153

G

L-Glutamic acid-5-[2-(4-Hydroxymethyl) phenylhydrazide)	*31*, 63
Glycidaldehyde	*11*, 175
Glycidyl oleate	*11*, 183
Glycidyl stearate	*11*, 187
Griseofulvin	*10*, 153
Guinea green B	*16*, 199
Gyromitrin	*31*, 163

H

Haematite	*1*, 29
	Suppl. *4*, 254
Haematoxylin (1977)*	
Hair dyes, epidemiology of	*16*, 29
	27, 307
Halothane (1976)*	
Heptachlor and its epoxide	*5*, 173
	20, 129
	Suppl. *4*, 80
Hexachlorobenzene	*20*, 155
Hexachlorobutadiene	*20*, 179
Hexachlorocyclohexane (α-,β-,δ-,ε-,technical HCH and lindane)	*5*, 47
	20, 195 (corr. *32*, 455)
	Suppl. *4*, 133
Hexachloroethane	*20*, 467
Hexachlorophene	*20*, 241
Hexamethylenediamine (1978)*	
Hexamethylphosphoramide	*15*, 211
Hycanthone and its mesylate	*13*, 91
Hydralazine and its hydrochloride	*24*, 85
	Suppl. *4*, 135
Hydrazine	*4*, 127
	Suppl. *4*, 136
Hydroquinone	*15*, 155
4-Hydroxyazobenzene	*8*, 157
17α-Hydroxyprogesterone caproate	*21*, 399 (corr. *31*, 293)
	Suppl. *4*, 195
8-Hydroxyquinoline	*13*, 101
Hydroxysenkirkine	*10*, 265

I

Indeno[1,2,3-*cd*]pyrene	*3*, 229
	32, 373
Iron and steel founding	*34*, 133
Iron-dextran complex	*2*, 161
	Suppl. *4*, 145
Iron-dextrin complex	*2*, 161 (corr. *7*, 319)
Iron oxide	*1*, 29
Iron sorbitol-citric acid complex	*2*, 161
Isatidine	*10*, 269
Isoflurane (1976)*	
Isonicotinic acid hydrazide	*4*, 159
	Suppl. *4*, 146
Isophosphamide	*26*, 237
Isoprene (1978)*	
Isopropyl alcohol	*15*, 223
	Suppl. *4*, 151
Isopropyl oils	*15*, 223
	Suppl. *4*, 151
Isosafrole	*1*, 169
	10, 232

J
Jacobine — *10*, 275

K
Kaempferol — *31*, 171

L
Lasiocarpine — *10*, 281
Lead and lead compounds — *1*, 40 (corr. *7*, 319)
2, 52 (corr. *8*, 349)
2, 150
23, 39, 205, 325
Suppl. 4, 149

    Lead acetate and its trihydrate
    Lead carbonate
    Lead chloride
    Lead naphthenate
    Lead nitrate
    Lead oxide
    Lead phosphate
    Lead subacetate
    Lead tetroxide
    Tetraethyllead
    Tetramethyllead

The leather goods manufacturing industry (other than boot and shoe manufacture and tanning) — *25*, 279
Suppl. 4, 142
The leather tanning and processing industries — *25*, 201
Suppl. 4, 142
Ledate — *12*, 131
Light green SF — *16*, 209
Lindane — *5*, 47
20, 196
The lumber and sawmill industries (including logging) — *25*, 49
Suppl. 4, 143
Luteoskyrin — *10*, 163
Lynoestrenol — *21*, 407
Suppl. 4, 195
Lysergide (1976)*

M
Magenta — *4*, 57 (corr. *7*, 320)
Suppl. 4, 152
Malathion — *30*, 103
Maleic hydrazide — *4*, 173 (corr. *18*, 125)
Maneb — *12*, 137
Mannomustine and its dihydrochloride — *9*, 157
MCPA — *Suppl. 4*, 211
30, 255
Medphalan — *9*, 168

Medroxyprogesterone acetate	6, 157
	21, 417 (corr. 25, 391)
	Suppl. 4, 196
Megestrol acetate	21, 431
	Suppl. 4, 198
Melphalan	9, 167
	Suppl. 4, 154
6-Mercaptopurine	26, 249
	Suppl. 4, 155
Merphalan	9, 169
Mestranol	6, 87
	21, 257 (corr. 25, 391)
	Suppl. 4, 188
Methacrylic acid (1978)*	
Methallenoestril (1978)*	
Methotrexate	26, 267
	Suppl. 4, 157
Methoxsalen	24, 101
	Suppl. 4, 158
Methoxychlor	5, 193
	20, 259
Methoxyflurane (1976)*	
Methylacrylate	19, 52
2-Methylaziridine	9, 61
Methylazoxymethanol	10, 121
Methylazoxymethanol acetate	1, 164
	10, 131
Methyl bromide (1978)*	
Methyl carbamate	12, 151
1-,2-,3-,4-,5-and 6-Methylchrysenes	32, 379
N-Methyl-N,4-dinitrosoaniline	1, 141
4,4'-Methylene bis(2-chloroaniline)	4, 65 (corr. 7, 320)
4,4'-Methylene bis(N,N-dimethyl)benzenamine	27, 119
4,4'-Methylene bis(2-methylaniline)	4, 73
4,4'-Methylenedianiline	4, 79 (corr. 7, 320)
4,4'-Methylenediphenyl diisocyanate	19, 314
2-and 3-Methylfluoranthenes	32, 399
Methyl iodide	15, 245
Methyl methacrylate	19, 187
Methyl methanesulphonate	7, 253
2-Methyl-1-nitroanthraquinone	27, 205
N-Methyl-N'-nitro-N-nitrosoguanidine	4, 183
Methyl parathion	30, 131
1-Methylphenanthrene	32, 405
Methyl protoanemonin (1975)*	
Methyl red	8, 161
Methyl selenac	12, 161
Methylthiouracil	7, 53
Metronidazole	13, 113
	Suppl. 4, 160
Mineral oils	3, 30
	Suppl. 4, 227
	33, 87

Mirex	5, 203
	20, 283 (corr. 30, 407)
Miristicin (1982)*	
Mitomycin C	10, 171
Modacrylic fibres	19, 86
Monocrotaline	10, 291
Monuron	12, 167
5-(Morpholinomethyl)-3-[(5-nitrofurfurylidene)amino]-2-oxazolidinone	7, 161
Mustard gas	9, 181 (corr. 13, 243)
	Suppl. 4, 163

N

Nafenopin	24, 125
1,5-Naphthalenediamine	27, 127
1,5-Naphthalene diisocyanate	19, 311
1-Naphthylamine	4, 87 (corr. 8, 349)
	(corr. 22, 187)
	Suppl. 4, 164
2-Naphthylamine	4, 97
	Suppl. 4, 166
1-Naphthylthiourea (ANTU)	30, 347
Nickel and nickel compounds	2, 126 (corr. 7, 319)
	11, 75
	Suppl. 4, 167
Nickel acetate and its tetrahydrate	
Nickel ammonium sulphate	
Nickel carbonate	
Nickel carbonyl	
Nickel chloride	
Nickel-gallium alloy	
Nickel hydroxide	
Nickelocene	
Nickel oxide	
Nickel subsulphide	
Nickel sulphate	
Nihydrazone (1982)*	
Niridazole	13, 123
Nithiazide	31, 179
5-Nitroacenaphthene	16, 319
5-Nitro-*ortho*-anisidine	27, 133
9-Nitroanthracene	33, 179
6-Nitrobenzo[a]pyrene	33, 187
4-Nitrobiphenyl	4, 113
6-Nitrochrysene	33, 195
Nitrofen	30, 271
3-Nitrofluoranthene	33, 201
5-Nitro-2-furaldehyde semicarbazone	7, 171
1[(5-Nitrofurfurylidene)amino]-2-imidazolidinone	7, 181
N-[4-(5-Nitro-2-furyl)-2-thiazolyl]acetamide	1, 181
	7, 185

Nitrogen mustard and its hydrochloride	9, 193
	Suppl. 4, 170
Nitrogen mustard N-oxide and its hydrochloride	9, 209
2-Nitropropane	29, 331
1-Nitropyrene	33, 209
N-Nitrosatable drugs	24, 297 (corr. 30, 407)
N-Nitrosatable pesticides	30, 359
N-Nitrosodi-n-butylamine	4, 197
	17, 51
N-Nitrosodiethanolamine	17, 77
N-Nitrosodiethylamine	1, 107 (corr. 11, 295)
	17, 83 (corr. 23, 417)
N-Nitrosodimethylamine	1, 95
	17, 125 (corr. 25, 391)
N-Nitrosodiphenylamine	27, 213
para-Nitrosodiphenylamine	27, 227 (corr. 31, 293)
N-Nitrosodi-n-propylamine	17, 177
N-Nitroso-N-ethylurea	1, 135
	17, 191
N-Nitrosofolic acid	17, 217
N-Nitrosohydroxyproline	17, 304
N-Nitrosomethylethylamine	17, 221
N-Nitroso-N-methylurea	1, 125
	17, 227
N-Nitroso-N-methylurethane	4, 211
N-Nitrosomethylvinylamine	17, 257
N-Nitrosomorpholine	17, 263
N'-Nitrosonornicotine	17, 281
N-Nitrosopiperidine	17, 287
N-Nitrosoproline	17, 303
N-Nitrosopyrrolidine	17, 313
N-Nitrososarcosine	17, 327
N-Nitrososarcosine ethyl ester (1977)*	
Nitrovin	31, 185
Nitroxolne (1976)*	
Nivalenol (1976)*	
Norethisterone and its acetate	6, 179
	21, 441
	Suppl. 4, 199
Norethynodrel	6, 191
	21, 461 (corr. 25, 391)
	Suppl. 4, 201
Norgestrel	6, 201
	21, 479
	Suppl. 4, 202
Nylon 6	19, 120
Nylon 6/6 (1978)*	

O
Ochratoxin A	10, 191
	31, 191 (corr. 34, 197)

Oestradiol-17β	6, 99
	21, 279
	Suppl. 4, 190
Oestradiol 3-benzoate	21, 281
Oestradiol dipropionate	21, 283
Oestradiol mustard	9, 217
Oestradiol-17β-valerate	21, 284
Oestriol	6, 117
	21, 327
Oestrone	6, 123
	21, 343 (corr. 25, 391)
	Suppl. 4, 191
Oestrone benzoate	21, 345
	Suppl. 4, 191
Oil Orange SS	8, 165
Orange I	8, 173
Orange G	8, 181
Oxazepam	13, 58
Oxymetholone	13, 131
	Suppl. 4, 203
Oxyphenbutazone	13, 185

P

Panfuran S (Dihydroxymethylfuratrizine)	24, 77
Parasorbic acid	10, 199 (corr. 12, 271)
Parathion	30, 153
Patulin	10, 205
Penicillic acid	10, 211
Pentachlorophenol	20, 303
	Suppl. 4, 88, 205
Pentobarbital sodium (1976)*	
Perylene	32, 411
Petasitenine	31, 207
Phenacetin	13, 141
	24, 135
	Suppl. 4, 47
Phenanthrene	32, 419
Phenazopyridine [2,6-Diamino-3-(phenylazo)pyridine] and its hydrochloride	8, 117
	24, 163 (corr. 29, 399)
	Suppl. 4, 207
Phenelzine and its sulphate	24, 175
	Suppl. 4, 207
Phenicarbazide	12, 177
Phenobarbital and its sodium salt	13, 157
	Suppl. 4, 208
Phenoxybenzamine and its hydrochloride	9, 223
	24, 185
Phenylbutazone	13, 183
	Suppl. 4, 212
ortho-Phenylenediamine (1977)*	

meta-Phenylenediamine and its hydrochloride	*16*, 111
para-Phenylenediamine and its hydrochloride	*16*, 125
N-Phenyl-2-naphthylamine	*16*, 325 (corr. *25*, 391)
	Suppl. *4*, 213
ortho-Phenylphenol and its sodium salt	*30*, 329
N-Phenyl-para-phenylenediamine (1977)*	
Phenytoin and its sodium salt	*13*, 201
	Suppl. *4*, 215
Piperazine oestrone sulphate	*21*, 148
Piperonyl butoxide	*30*, 183
Polyacrylic acid	*19*, 62
Polybrominated biphenyls	*18*, 107
Polychlorinated biphenyls	*7*, 261
	18, 43
	Suppl. *4*, 217
Polychloroprene	*19*, 141
Polyethylene (low-density and high-density)	*19*, 164
Polyethylene terephthalate (1978)*	
Polyisoprene (1978)*	
Polymethylene polyphenyl isocyanate	*19*, 314
Polymethyl methacrylate	*19*, 195
Polyoestradiol phosphate	*21*, 286
Polypropylene	*19*, 218
Polystyrene	*19*, 245
Polytetrafluoroethylene	*19*, 288
Polyurethane foams (flexible and rigid)	*19*, 320
Polyvinyl acetate	*19*, 346
Polyvinyl alcohol	*19*, 351
Polyvinyl chloride	*7*, 306
	19, 402
Polyvinylidene fluoride (1978)*	
Polyvinyl pyrrolidone	*19*, 463
Ponceau MX	*8*, 189
Ponceau 3R	*8*, 199
Ponceau SX	*8*, 207
Potassium bis (2-hydroxyethyl)dithiocarbamate	*12*, 183
Prednisone	*26*, 293
	Suppl. *4*, 219
Procarbazine hydrochloride	*26*, 311
	Suppl. *4*, 220
Proflavine and its salts	*24*, 195
Progesterone	*6*, 135
	21, 491 (corr. *25*, 391)
	Suppl. *4*, 202
Pronetalol hydrochloride	*13*, 227 (corr. *16*, 387)
1,3-Propane sultone	*4*, 253 (corr. *13*, 243)
	(corr. *20*, 591)
Propham	*12*, 189
β-Propiolactone	*4*, 259 (corr. *15*, 341)
n-Propyl carbamate	*12*, 201
Propylene	*19*, 213
Propylene oxide	*11*, 191

Propylthiouracil	7, 67
	Suppl. 4, 222
The pulp and paper industry	25, 157
	Suppl. 4, 144
Pyrazinamide (1976)*	
Pyrene	32, 431
Pyrimethamine	13, 233
Pyrrolizidine alkaloids	10, 333

Q

Quercitin	31, 213
Quinoestradol (1978)*	
Quinoestrol (1978)*	
para-Quinone	15, 255
Quintozene (Pentachloronitrobenzene)	5, 211

R

Reserpine	10, 217
	24, 211 (corr. 26, 387)
	(corr. 30, 407)
	Suppl. 4, 222
Resorcinol	15, 155
Retrorsine	10, 303
Rhodamine B	16, 221
Rhodamine 6G	16, 233
Riddelliine	10, 313
Rifampicin	24, 243
Rotenone (1982)*	
The rubber industry	28 (corr. 30, 407)
Rugulosin (1975)*	Suppl. 4, 144

S

Saccharated iron oxide	2, 161
Saccharin	22, 111 (corr. 25, 391)
	Suppl. 4, 224
Safrole	1, 169
	10, 231
Scarlet red	8, 217
Selenium and selenium compounds	9, 245 (corr. 12, 271)
	(corr. 30, 407)
Semicarbazide hydrochloride	12, 209 (corr. 16, 387)
Seneciphylline	10, 319
Senkirkine	10, 327
	31, 231
Shale-oils	35, 161
Simazine (1982)*	
Sodium cyclamate	22, 56 (corr. 25, 391)
	Suppl. 4, 97
Sodium diethyldithiocarbamate	12, 217
Sodium equilin sulphate	21, 148

Sodium oestrone sulphate	*21*, 147
Sodium saccharin	*22*, 113 (corr. *25*, 391)
	Suppl. 4, 224
Soot and tars	*3*, 22
	Suppl. 4, 227
Soots	*35*, 219
Spironolactone	*24*, 259
	Suppl. 4, 229
Sterigmatocystin	*1*, 175
	10, 245
Streptozotocin	*4*, 221
	17, 337
Styrene	*19*, 231
	Suppl. 4, 229
Styrene-acrylonitrile copolymers	*19*, 97
Styrene-butadiene copolymers	*19*, 252
Styrene oxide	*11*, 201
	19, 275
	Suppl. 4, 229
Succinic anhydride	*15*, 265
Sudan I	*8*, 225
Sudan II	*8*, 233
Sudan III	*8*, 241
Sudan brown RR	*8*, 249
Sudan red 7B	*8*, 253
Sulfafurazole (Sulphisoxazole)	*24*, 275
	Suppl. 4, 233
Sulfallate	*30*, 283
Sulfamethoxazole	*24*, 285
	Suppl. 4, 234
Sulphamethazine (1982)*	
Sunset yellow FCF	*8*, 257
Symphytine	*31*, 239

T
2,4,5-T and esters	*15*, 273
	Suppl. 4, 211, 235
Tannic acid	*10*, 253 (corr. *16*, 387)
Tannins	*10*, 254
Terephthalic acid (1978)*	
Terpene polychlorinates (Strobane^R)	*5*, 219
Testosterone	*6*, 209
	21, 519
Testosterone oenanthate	*21*, 521
Testosterone propionate	*21*, 522
2,2',5,5'-Tetrachlorobenzidine	*27*, 141
Tetrachlorodibenzo-*para*-dioxin (TCDD)	*15*, 41
	Suppl. 4, 211, 238
1,1,2,2-Tetrachloroethane	*20*, 477
Tetrachloroethylene	*20*, 491
	Suppl. 4, 243
Tetrachlorvinphos	*30*, 197

Tetrafluoroethylene	*19*, 285
Thioacetamide	*7*, 77
4,4'-Thiodianiline	*16*, 343
	27, 147
Thiouracil	*7*, 85
Thiourea	*7*, 95
Thiram	*12*, 225
2,4-Toluene diisocyanate	*19*, 303
2,6-Toluene diisocyanate	*19*, 303
ortho-Toluenesulphonamide	*22*, 121
	Suppl. 4, 224
ortho-Toluidine and its hydrochloride	*16*, 349
	27, 155
	Suppl. 4, 245
Toxaphene (Polychlorinated camphenes)	*20*, 327
Treosulphan	*26*, 341
	Suppl. 4, 246
Trichlorphon	*30*, 207
1,1,1-Trichloroethane	*20*, 515
1,1,2-Trichloroethane	*20*, 533
Trichloroethylene	*11*, 263
	20, 545
	Suppl. 4, 247
2,4,5- and 2,4,6-Trichlorophenols	*20*, 349
	Suppl. 4, 88, 249
Trichlorotriethylamine hydrochloride	*9*, 229
Trichlorphon	*30*, 207
T_2-Trichothecene	*31*, 265
Triethylene glycol diglycidyl ether	*11*, 209
Trifluralin (1982)*	
2,4,5-Trimethylaniline and its hydrochloride	*27*, 177
2,4,6-Trimethylaniline and its hydrochloride	*27*, 178
Triphenylene	*32*, 447
Tris(aziridinyl)-*para*-benzoquinone (Triaziquone)	*9*, 67
	Suppl. 4, 251
Tris(1-aziridinyl)phosphine oxide	*9*, 75
Tris(1-aziridinyl)phosphine sulphide (Thiotepa)	*9*, 85
	Suppl. 4, 252
2,4,6-Tris(1-aziridinyl)-*s*-triazine	*9*, 95
1,2,3-Tris(chloromethoxy)propane	*15*, 301
Tris(2,3-dibromopropyl)phosphate	*20*, 575
Tris(2-methyl-1-aziridinyl)phosphine oxide	*9*, 107
Trp-P-1	*31*, 247
Trp-P-2	*31*, 255
Trypan blue	*8*, 267

U

Uracil mustard	*9*, 235
	Suppl. 4, 256
Urethane	*7*, 111

V

Vinblastine sulphate	*26*, 349 (corr. *34*, 197)
	Suppl. 4, 257
Vincristine sulphate	*26*, 365
	Suppl. 4, 259
Vinyl acetate	*19*, 341
Vinyl bromide	*19*, 367
Vinyl chloride	*7*, 291
	19, 377
	Suppl. 4, 260
Vinyl chloride-vinyl acetate copolymers	*7*, 311
	19, 412
4-Vinylcyclohexene	*11*, 277
Vinylidene chloride	*19*, 439
	Suppl. 4, 262 (corr. *31*, 293)
Vinylidene chloride-vinylchloride copolymers	*19*, 448
Vinylidene fluoride (1978)*	
N-Vinyl-2-pyrrolidone	*19*, 461

X

2,4-Xylidine and its hydrochloride	*16*, 367
2,5-Xylidine and its hydrochloride	*16*, 377
2,6-Xylidine (1977)*	

Y

Yellow AB	*8*, 279
Yellow OB	*8*, 287

Z

Zearalenone	*31*, 279
Zectran	*12*, 237
Zineb	*12*, 245
Ziram	*12*, 259

IARC SCIENTIFIC PUBLICATIONS

Available from Oxford University Press, Walton Street, Oxford OX2 6DP, UK and in London, New York, Toronto, Delhi, Bombay, Calcutta, Madras, Karachi, Kuala Lumpur, Singapore, Hong Kong, Tokyo, Nairobi, Dar es Salaam, Cape Town, Melbourne, Auckland and associated companies in Beirut, Berlin, Ibadan, Mexico City, Nicosia.

Liver Cancer	No. 1, 1971; 176 pages $10.—
Oncogenesis and Herpesviruses	No. 2, 1972; 515 pages £30.—
N-Nitroso Compounds, Analysis and Formation	No. 3, 1972; 140 pages £8.50
Transplacental Carcinogenesis	No. 4, 1973; 181 pages £11.95
Pathology of Tumours in Laboratory Animals—Volume I—Tumours of the Rat, Part 1	No. 5, 1973; 214 pages £17.50
Pathology of Tumours in Laboratory Animals—Volume I—Tumours of the Rat, Part 2	No. 6, 1976; 319 pages £17.50
Host Environment Interactions in the Etiology of Cancer in Man	No. 7, 1973; 464 pages £30.—
Biological Effects of Asbestos	No. 8, 1973; 346 pages £25.—
N-Nitroso Compounds in the Environment	No. 9, 1974; 243 pages £15.—
Chemical Carcinogenesis Essays	No. 10, 1974; 230 pages £15.—
Oncogenesis and Herpesviruses II	No. 11, 1975; Part 1, 511 pages £30.— Part 2, 403 pages £30.—
Screening Tests in Chemical Carcinogenesis	No. 12, 1976; 666 pages £30.—
Environmental Pollution and Carcinogenic Risks	No. 13, 1976; 454 pages £17.50
Environmental N-Nitroso Compounds—Analysis and Formation	No. 14, 1976; 512 pages £35.—
Cancer Incidence in Five Continents—Volume III	No. 15, 1976; 584 pages £35.—
Air Pollution and Cancer in Man	No. 16, 1977; 331 pages £30.—
Directory of On-Going Research in Cancer Epidemiology 1977	No. 17, 1977; 599 pages (OUT OF PRINT)
Environmental Carcinogens—Selected Methods of Analysis, Vol. 1: Analysis of Volatile Nitrosamines in Food	No. 18, 1978; 212 pages £30.—
Environmental Aspects of N-Nitroso Compounds	No. 19, 1978; 566 pages £35.—
Nasopharyngeal Carcinoma: Etiology and Control	No. 20, 1978; 610 pages £35.—
Cancer Registration and Its Techniques	No. 21, 1978; 235 pages £11.95
Environmental Carcinogens—Selected Methods of Analysis, Vol. 2: Methods for the Measurement of Vinyl Chloride in Poly(vinyl chloride), Air, Water and Foodstuffs	No. 22, 1978; 142 pages £35.—
Pathology of Tumours in Laboratory Animals—Volume II—Tumours of the Mouse	No. 23, 1979; 669 pages £35.—
Oncogenesis and Herpesviruses III	No. 24, 1978; Part 1, 580 pages £20.— Part 2, 522 pages $20.—
Carcinogenic Risks—Strategies for Intervention	No. 25, 1979; 283 pages £20.—
Directory of On-Going Research in Cancer Epidemiology 1978	No. 26, 1978; 550 pages (OUT OF PRINT)
Molecular and Cellular Aspects of Carcinogen Screening Tests	No. 27, 1980; 371 pages £20.—
Directory of On-Going Research in Cancer Epidemiology 1979	No. 28, 1979; 672 pages (OUT OF PRINT)
Environmental Carcinogens—Selected Methods of Analysis, Vol. 3: Analysis of Polycyclic Aromatic Hydrocarbons in Environmental Samples	No. 29, 1979; 240 pages £17.50
Biological Effects of Mineral Fibres	No. 30, 1980; Volume 1, 494 pages £25.— Volume 2, 513 pages £25.—
N-Nitroso Compounds: Analysis, Formation and Occurrence	No. 31, 1980; 841 pages £30.—
Statistical Methods in Cancer Research, Vol. 1: The Analysis of Case-Control Studies	No. 32, 1980; 338 pages £17.50
Handling Chemical Carcinogens in the Laboratory—Problems of Safety	No. 33, 1979; 32 pages £3.95
Pathology of Tumours in Laboratory Animals—Volume III—Tumours of the Hamster	No. 34, 1982; 461 pages £30.—
Directory of On-Going Research in Cancer Epidemiology 1980	No. 35, 1980; 660 pages (OUT OF PRINT)
Cancer Mortality by Occupation and Social Class 1851-1971	No. 36, 1982; 253 pages £20.—
Laboratory Decontamination and Destruction of Aflatoxins B_1, B_2, G_1, G_2 in Laboratory Wastes	No. 37, 1980; 59 pages £5.95
Directory of On-Going Research in Cancer Epidemiology 1981	No. 38, 1981; 696 pages (OUT OF PRINT)
Host Factors in Human Carcinogenesis	No. 39, 1982; 583 pages £35.—
Environmental Carcinogens—Selected Methods of Analysis, Vol. 4: Some Aromatic Amines and Azo Dyes in the General and Industrial Environment	No. 40, 1981; 347 pages £20.—
N-Nitroso Compounds: Occurrence and Biological Effects	No. 41, 1982; 755 pages £35.—
Cancer Incidence in Five Continents—Volume IV	No. 42, 1982; 811 pages £35.—
Laboratory Decontamination and Destruction of Carcinogens in Laboratory Wastes: Some N-Nitrosamines	No. 43, 1982; 73 pages £6.50

Environmental Carcinogens—Selected Methods of Analysis, Vol. 5: Mycotoxins	No. 44, 1983; 455 pages £ 20.—	
Environmental Carcinogens—Selected Methods of Analysis, Vol. 6: N-Nitroso Compounds	No. 45, 1983; 508 pages £ 20.—	
Directory of On-Going Research in Cancer Epidemiology 1982	No. 46, 1982; 722 pages (OUT OF PRINT)	
Cancer Incidence in Singapore	No. 47, 1982; 174 pages £ 10.—	
Cancer Incidence in the USSR Second Revised Edition	No. 48, 1982; 75 pages £ 10.—	
Laboratory Decontamination and Destruction of Carcinogens in Laboratory Wastes: Some Polycyclic Aromatic Hydrocarbons	No. 49, 1983; 81 pages £ 7.95	
Directory of On-Going Research in Cancer Epidemiology 1983	No. 50, 1983; 740 pages (OUT OF PRINT)	
Modulators of Experimental Carcinogenesis	No. 51, 1983; 307 pages £ 25.—	
Second Cancers Following Radiation Treatment for Cancer of the Uterine Cervix: The Results of a Cancer Registry Collaborative Study	No. 52, 1984; 207 pages £ 17.50	
Nickel in the Human Environment	No. 53, 1984; 529 pages £ 30.—	
Laboratory Decontamination and Destruction of Carcinogens in Laboratory Wastes: Some Hydrazines	No. 54; 1983; 87 pages £ 6.95	
Laboratory Decontamination and Destruction of Carcinogens in Laboratory Wastes: Some N-Nitrosamides	No. 55, 1984; 65 pages £ 6.95	
Models, Mechanisms and Etiology of Tumour Promotion	No. 56, 1985 (in press)	
N-Nitroso Compounds: Occurrence, Biological Effects and Relevance to Human Cancer	No. 57, 1984; 1013 pages £ 75.—	
Directory of On-Going Research in Cancer Epidemiology 1984	No. 62, 1984; 728 pages £ 18.—	

NON-SERIAL PUBLICATIONS

Available from IARC

Alcool et Cancer	1978; 42 pages Fr. fr. 35-; Sw. fr. 14.-
Cancer Morbidity and Causes of Death Among Danish Brewery Workers	1980, 145 pages US$ 25.00; Sw.fr. 45.-

IARC MONOGRAPHS ON THE EVALUATION OF THE CARCINOGENIC RISK OF CHEMICALS TO HUMANS

Available from WHO Sales Agents. See addresses on back cover

Some Inorganic Substances, Chlorinated Hydrocarbons, Aromatic Amines, N-Nitroso Compounds, and Natural Products	Volume 1, 1972; 184 pages (out of print)
Some Inorganic and Organometallic Compounds	Volume 2, 1973; 181 pages US$ 3.60; Sw. fr. 12.-- (out of print)
Certain Polycyclic Aromatic Hydrocarbons and Heterocyclic Compounds	Volume 3, 1973; 271 pages (out of print)
Some Aromatic Amines, Hydrazine and Related Substances, N-Nitroso Compounds and Miscellaneous Alkylating Agents	Volume 4, 1974; 286 pages US$ 7.20; Sw. fr. 18.--
Some Organochlorine Pesticides	Volume 5, 1974; 241 pages US$ 7.20; Sw. fr. 18.-- (out of print)
Sex Hormones	Volume 6, 1974; 243 pages US$ 7.20; Sw. fr. 18.--
Some Anti-thyroid and Related Substances, Nitrofurans and Industrial Chemicals	Volume 7, 1974; 326 pages US$ 12.80; Sw. fr. 32.--
Some Aromatic Azo Compounds	Volume 8, 1975; 357 pages US$ 14.40; Sw. fr. 36.--
Some Aziridines, N-, S- and O-Mustards and Selenium	Volume 9, 1975; 268 pages US$ 10.80; Sw. fr. 27.--
Some Naturally Occurring Substances	Volume 10, 1976; 353 pages US$ 15.00; Sw. fr. 38.--
Cadmium, Nickel, Some Epoxides, Miscellaneous Industrial Chemicals and General Considerations on Volatile Anaesthetics	Volume 11, 1976; 306 pages US$ 14.00; Sw. fr. 34.--
Some Carbamates, Thiocarbamates and Carbazides	Volume 12, 1976; 282 pages US$ 14.00; Sw. fr. 34.--
Some Miscellaneous Pharmaceutical Substances	Volume 13, 1977; 255 pages US$ 12.00; Sw. fr. 30.--
Asbestos	Volume 14, 1977; 106 pages US$ 6.00; Sw. fr. 14.--
Some Fumigants, the Herbicides 2,4-D and 2,4,5-T, Chlorinated Dibenzodioxins and Miscellaneous Industrial Chemicals	Volume 15, 1977; 354 pages US$ 20.00; Sw. fr. 50.--
Some Aromatic Amines and Related Nitro Compounds - Hair Dyes, Colouring Agents and Miscellaneous Industrial Chemicals	Volume 16, 1978; 400 pages US$ 20.00; Sw. fr. 50.--
Some N-Nitroso Compounds	Volume 17, 1978; 365 pages US$ 25.00; Sw. fr. 50.--
Polychlorinated Biphenyls and Polybrominated Biphenyls	Volume 18, 1978; 140 pages US$ 13.00; Sw. fr. 20.--
Some Monomers, Plastics and Synthetic Elastomers, and Acrolein	Volume 19, 1979; 513 pages US$ 35.00; Sw. fr. 60.--
Some Halogenated Hydrocarbons	Volume 20, 1979; 609 pages US$ 35.00; Sw. fr. 60.--
Sex Hormones (II)	Volume 21, 1979; 583 pages US$ 35.00; Sw. fr. 60.--
Some Non-nutritive Sweetening Agents	Volume 22, 1980; 208 pages US$ 15.00; Sw. fr. 25.--
Some Metals and Metallic Compounds	Volume 23, 1980; 438 pages US$ 30.00; Sw. fr. 50.--
Some Pharmaceutical Drugs	Volume 24, 1980; 337 pages US$ 25.00; Sw. fr. 40.--
Wood, Leather and Some Associated Industries	Volume 25, 1980; 412 pages US$ 30.00; Sw. fr. 60.--
Some Anticancer and Immunosuppressive Drugs	Volume 26, 1981; 411 pages US$ 30.00; Sw. fr. 62.--
Some Aromatic Amines, Anthraquinones and Nitroso Compounds and Inorganic Fluorides Used in Drinking-Water and Dental Preparations	Volume 27, 1982; 341 pages US$ 25.00; Sw. fr. 40.--
The Rubber Industry	Volume 28, 1982; 486 pages US$ 35.00; Sw. fr. 70.--
Some Industrial Chemicals and Dyestuffs	Volume 29, 1982; 416 pages US$ 30.00; Sw. fr. 60.--
Miscellaneous Pesticides	Volume 30, 1983; 424 pages US$ 30.00; Sw. fr. 60.--
Some Feed Additives, Food Additives and Naturally Occurring Substances	Volume 31, 1983; 314 pages US$ 30.00; Sw. fr. 60. --
Chemicals and Industrial Processes Associated with Cancer in Humans (IARC Monographs 1-20)	Supplement 1, 1979; 71 pages (out of print)
Long-term and Short-term Screening Assays for Carcinogens: A Critical Appraisal	Supplement 2, 1980; 426 pages US$ 25.00; Sw. fr. 40.--
Cross Index of Synonyms and Trade Names in Volumes 1 to 26	Supplement 3, 1982; 199 pages US$ 30.00; Sw. fr. 60.--
Chemicals, Industrial Processes and Industries Associated with Cancer in Humans (IARC Monographs Volumes 1 to 29)	Supplement 4, 1982; 292 pages US$ 30.00; Sw. fr. 60.--
Polynuclear Aromatic Compounds, Part 1, Chemical, Environmental and Experimental Data	Volume 32, 1983 ; 477 pages US $ 35.00 ; Sw.fr. 70
Polynuclear Aromatic Compounds, Part 2, Carbon Blacks, Mineral Oils and Some Nitroarenes	Volume 33, 1984; 245 pages US$ 25.00; Sw. fr. 50.--
Polynuclear Aromatic Compounds, Part 3, Industrial Exposures in Aluminium Production, Coal Gasification, Coke Production, and Iron and Steel Founding	**Volume 34, 1984 ; 219 pages** US$ 20.00 ; Sw. fr. 48.—
Polynuclear Aromatic Compounds, Part 4, Bitumens, Coal-tars and Derived Products, Shale-oils and Soots	Volume 35, 1985; 271 pages US$ 25.00; Sw. fr. 70.—

Composition, impression et façonnage
Groupe MCP-Mame
Dépôt légal : Avril 1985